RELATED KAPLAN BOOKS

College Admissions and Financial Aid
Conquer the Cost of College
Parent's Guide to College Admissions
The Unofficial, Unbiased, Insider's Guide to the 320 Most Interesting Colleges
Yale Daily News Guide to Succeeding in College

Test Preparation
ACT
AP Biology
AP Calculus AB
AP Chemistry
AP English Literature & Composition
AP Macroeconomics/Microeconomics
AP Physics
AP Statistics
AP U.S. Government & Politics
AP U.S. History
PSAT
SAT & PSAT
SAT Vocabulary Words Flip-O-Matic
SAT Math Mania
SAT Math Workbook
SAT Verbal Velocity
SAT Verbal Workbook
SAT II: Biology E/M
SAT II: Literature
SAT II: Mathematics
SAT II: Physics
SAT II: Spanish
SAT II: World History
SAT II: Writing

SAT* II
CHEMISTRY
2003–2004 Edition

By Claire Aldridge, Ph.D.,
Karl Lee, M.A.
and the
Staff of Kaplan, Inc.

Simon & Schuster

NEW YORK · LONDON · SINGAPORE · SYDNEY · TORONTO

*SAT is a registered trademark of the College Entrance Examination Board, which is not affiliated with this book.

Kaplan Publishing
Published by Simon & Schuster
1230 Avenue of the Americas
New York, New York 10020

For bulk sales to schools, colleges, and universities, please contact: Order Department, Simon and Schuster, 100 Front Street, Riverside, NJ 08075. Phone: (800) 223-2336. Fax: (800) 943-9831.

The material in this book is up-to-date at the time of publication. However, the College Entrance Examination Board and Educational Testing Service (ETS) may have instituted changes in the test after this book was published. Be sure to carefully read the materials you receive when you register for the test.

Contributing Editor: Seppy Basili
Project Editor: Ruth Baygell
Cover Design: Cheung Tai
Interior Page Design: Dave Chipps, Laurel Douglas
Production Editor: Maude Spekes
Desktop Publishing Manager: Michael Shevlin
Editorial Coordinator: Déa Alessandro
Executive Editor: Del Franz

Special thanks to Rebecca Engle, Rodolfo Robles, and Sara Pearl.

Manufactured in the United States of America.
Published simultaneously in Canada.

March 2003
10 9 8 7 6 5 4 3 2 1

ISBN: 0-7432-4122-3

CONTENTS

About the Authors

Claire Aldridge received her bachelor of science degree in biomedical science from Texas A&M University. She earned her Ph.D. from the Department of Immunology and the University Program in Genetics at Duke University in 1996. The author of two Kaplan books, *SAT II Biology* and *Microbiology and Immunology*, as well as of numerous scientific articles, she is adjunct faculty in the biology department at the University of North Texas in Denton, Texas. She lives in Dallas, Texas.

Karl Lee received a bachelor's degree in chemistry from Amherst College, and earned an M.A. in physical chemistry from Harvard University in 1996. Until recently, he was Kaplan's director of pre-health curriculum.

HOW TO USE THIS BOOK

For more than 50 years, Kaplan has prepared students to take SATs. Our team of researchers and editors knows more about SAT preparation than anyone else, and you'll find their accumulated experience and knowledge in this book. As you work your way through the chapters, we'll show you precisely what knowledge and skills you'll need in order to do your very best on the SAT II: Chemistry Subject Test. You'll discover the most effective way to tackle each type of question, and you'll reinforce your studies with lots of practice questions. At the beginning of the chemistry review section, you'll find a lengthy diagnostic practice test, and each chapter in this section ends with a short quiz. At the back of the book you'll find two full-length, formatted tests with answer keys, scoring instructions, and detailed explanations. In addition, the Ready, Set, Go! section contains helpful tips on beating stress while you prepare for the test and on pulling off a victory on Test Day.

Get Ready to Prep

If possible, work your way through this book bit by bit over the course of a few weeks. Cramming the week before the test is not a good idea; you probably won't absorb much information, and it's sure to make you more anxious.

If you find that your anxiety about the test is interfering with your ability to study, start off by reading the Stress Management chapter in this book. It provides many practical tips to help you stay calm and centered. Use these tips before, during, and after the test.

Learn the Basics

The first thing you need to do is find out what's on the SAT II: Chemistry Test. In the first section of this book, "The Basics," we'll provide you with background information about the SAT II: Subject Test and what it's used for. We'll also give you the lowdown on the kinds of questions that are typically seen on the test and how best to tackle them.

The Best Prep

Kaplan's three full-length practice tests give you a great prep experience for the SAT II: Chemistry Subject Test.

Chemistry Review

Once you have the big picture, it's time to focus on the chemistry. The "Chemistry Review" gives you a succinct review of the chemistry you'll need to know on Test Day. Each chapter within this section deals with a major subdivision of chemistry and focuses on concepts crucial to a full understanding of the subject.

The chemistry review section begins with a full–length diagnostic test. If your time is limited, you can use the diagnostic test to bypass the material you already know well enough and to zero in on what you need to work on. Each chapter in this section also ends with a follow-up quiz, complete with answers and explanations. When you feel you have mastered the material in a chapter, take the follow-up quiz to make sure. Don't forget to use the glossary at the end of the book to brush up on vital definitions.

Finding Your Way

How you use this book depends on how much time you have. Let's take a look at three typical students who are planning to take the SAT II: Chemistry Subject Test. Note that our hypothetical students use this book in three different ways. Which study plan best matches your situation?

 "I'm taking the Chemistry Subject Test a month from today."

Angela has plenty of time to prep for the test. If you're like Angela, and you have at least two weeks to prepare, then we recommend that you work through the entire book.

 "I'm taking the Chemistry Subject Test in a week."

If you're like Bill, you'll need a shortcut. If you have fewer than two weeks but more than two days to prepare, then we recommend that you use the diagnostic test to determine which chapters you can safely skim, or even skip.

 "Help! It's two days before Test Day!"

Eric is in a panic. But you don't need to freak, even if you're in Eric's situation. There's still a lot you can do to improve your potential score. Use our Panic Plan to get through this book. First and foremost, you should become familiar with the test. Read the introductory section in this book. And if you do nothing else, you should at least sit down and work through one of the full-length practice tests at the back of this book under reasonably testlike conditions.

When you finish the practice test, check your answers and look at the explanations to the questions you didn't get right. When you come across a topic that you only half remember, turn to the appropriate chapter for a quick review. When you come across a topic you don't remember or understand at all, skip it. You don't have time to learn and assimilate completely new material. At least you'll know to skip any similar question you might encounter on the actual SAT II: Chemistry Subject Test.

The Icons

As you work your way through this book, you'll see the following helpful icons used repeatedly in the sidebars in the margins of the text. Here's what they mean.

Basic Concept. This icon highlights a basic concept that is explained in more detail in the accompanying text. Use these sidebars to quickly determine what is being discussed in the text, and to get a wider perspective on the topic at hand.

A Closer Look. This sidebar displays additional facts and/or examples associated with a particular topic. Often, you'll be surprised by what you read here.

Don't Mix These Up on Test Day. This sidebar points out easily confused concepts that you'll likely see on the test, as well as mnemonics that can help you to memorize important facts. You'll be provided with information that can help you grasp these concepts more easily.

Test Strategy. This icon highlights a Kaplan test-taking strategy that can help you boost your score.

Quick Quiz. You'll find this icon features a "quick quiz"—questions that will get your brain working on tough concepts.

Take the Practice Tests

At the back of the book are two full-length chemistry practice tests. The best way to use these tests is to take them under testlike conditions. Don't just drop in and do a random question here and there. Use these tests to gain experience with the complete testing experience, including pacing and endurance. You can do these tests at any time. You don't have to save them all until after you've read this whole book. Just be sure to save at least one test for your dress rehearsal some time in the last week before Test Day.

Take a Break Before Test Day

A day or two before the test, be sure to review chapter 18: The Final Countdown. It includes essential "how-to" tips that will help you get the upper hand on test day. Then, relax! Read a book or watch a moivie. And most important, get a good night's sleep. How you approach the days leading up to the test really does matter!

On the morning of the test, eat a light breakfast (nothing too heavy to make you sleepy!) and quickly review a few questions if you feel like it (just enough to get you focused). Walk into the test center with confidnce—you're ready for the challenge!

A Special Note for
International Students

If you are an international student considering attending an American university, you are not alone. Over 582,000 international students pursued academic degrees at the undergraduate, graduate, or professional school level at U.S. universities during the 2001-2002 academic year, according to the Institute of International Education's Open Doors 2001 report. Almost 50 percent of these students were studying for a bachelor's or first university degree. This number of international students pursuing higher education in the United States is expected to continue to grow. Business, management, engineering, and the physical and life sciences are particularly popular majors for students coming to the United States from other countries.

If you are not a U.S. citizen and you are interested in attending college or university in the United States, here is what you'll need to get started.

- If English is not your first language, you'll probably need to take the TOEFL (Test of English as a Foreign Language) or provide some other evidence that you are proficient in English. Colleges and universities in the United States will differ on what they consider to be an acceptable TOEFL score. A minimum TOEFL score of 213 (550 on the paper-based TOEFL) or better is often required by more prestigious and competitive institutions. Because American undergraduate programs require all students to take a certain number of general education courses, all students-even math and computer science students-need to be able to communicate well in spoken and written English.

- You may also need to take the SAT or the ACT. Many undergraduate institutions in the United States require both the SAT and TOEFL of international students.

- There are over 3,400 accredited colleges and universities in the United States, so selecting the correct undergraduate school can be a confusing task for anyone. You will need to get help from a good advisor or at least a good college guide that gives you detailed information on the different schools available. Since admission to many undergraduate programs is quite competitive, you may want to select three or four colleges and complete applications for each school.

- You should begin the application process at least a year in advance. An increasing number of schools accept applications year round. In any case, find out the application deadlines and plan accordingly. Although September (the fall semester) is the traditional time to begin university study in the United States, you can begin your studies at many schools in January (the spring semester).

In addition, you will need to obtain an I-20 Certificate of Eligibility from the school you plan to attend if you intend to apply for an F-1 Student Visa to study in the United States.

Kaplan International Programs

If you need more help with the complex process of university admissions, assistance preparing for the SAT, ACT, or TOEFL, or help building your English language skills in general, you may be interested in Kaplan's programs for international students.

Kaplan International Programs were designed to help students and professionals from outside the United States meet their educational and career goals. At locations throughout the United States, international students take advantage of Kaplan's programs to help them improve their academic and conversational English skills, raise their scores on the TOEFL, SAT, ACT, and other standardized exams, and gain admission to the schools of their choice. Our staff and instructors give international students the individualized attention they need to succeed. Here is a brief description of some of Kaplan's programs for international students:

General Intensive English

Kaplan's General Intensive English classes are designed to help you improve your skills in all areas of English and to increase your fluency in spoken and written English. Classes are available for beginning to advanced students, and the average class size is 12 students.

English for TOEFL and Academic English

This course provides you with the skills you need to improve your TOEFL score and succeed in an American university or graduate program. It includes advanced reading, writing, listening, grammar and conversational English. You will also receive training for the TOEFL using Kaplan's exclusive computer-based practice materials.

SAT Test Preparation Course

The SAT is an important admission criterion for American colleges and universities. A high score can help you stand out from other applicants. This course includes the skills you need to succeed on each section of the SAT, as well as access to Kaplan's exclusive practice materials.

Other Kaplan Programs

Since 1938, more than 3 million students have come to Kaplan to advance their studies, prepare for entry to American universities, and further their careers. In addition to the above programs, Kaplan offers courses to prepare for the ACT, GMAT, GRE, MCAT, DAT, USMLE, NCLEX, and other standardized exams at locations throughout the United States.

Applying to Kaplan International Programs

To get more information, or to apply for admission to any of Kaplan's programs for international students and professionals, contact us at:

Kaplan International Programs
700 South Flower, Suite 2900
Los Angeles, CA 90017 USA
Phone (if calling from within the United States: (800) 818-9128
Phone (if calling from outside the United States: (213) 452-5800
Fax: (213) 892-1364
Website: www.kaplaninternational.com
Email: world@kaplan.com

Kaplan is authorized under federal law to enroll nonimmigrant alien students. Kaplan is accredited by ACCET (Accrediting Council for Continuing Education and Training).

Test names are registered trademarks of their respective owners.

THE BASICS

CHAPTER 1

ABOUT THE SAT II: SUBJECT TESTS

You're serious about going to the college of your choice. You wouldn't have opened this book otherwise. You've made a wise choice, because this book can help you to achieve your goal. It'll show you how to score your best on the SAT II: Chemistry Subject Test. But before turning to the chemistry review, let's look at the SAT II as a whole.

Frequently Asked Questions

The following background information about the SAT II is important to keep in mind as you get ready to prep for the SAT II: Chemistry Subject Test. Remember, though, that sometimes the test makers change the test policies after a book has gone to press. The information here is accurate at the time of publication but it's a good idea to check the test information on the College Board Web site at www.collegeboard.com.

What Is the SAT II?

The SAT II is actually a set of more than 20 different Subject Tests. These tests are designed to measure what you have learned in such subjects as Literature, American History and Social Studies, Biology, and Spanish. Each test lasts one hour and consists entirely of multiple-choice questions, except for the Writing Test, which has a 20-minute essay section and a 40-minute multiple-choice section. On any one test date, you can take up to three Subject Tests.

How Does the SAT II Differ from the SAT I?

SAT I is largely a test of verbal and math skills. True, you need to know some vocabulary and some formulas for the SAT I; but it's designed to measure how well you read and think rather than how much you remember. The SAT II tests are very different. They're designed to measure what you know about specific disciplines. Sure, critical reading and thinking skills play a part on these tests, but their main purpose is to determine exactly what you know about writing, math, history, chemistry, and so on.

"What Does That Spell?"

Originally, *SAT* stood for *Scholastic Aptitude Test*. Then, when the test changed in the mid-1990s, the official name was changed to *Scholastic Assessment Test*. Finally, in 1997, the test makers announced that *SAT* no longer stands for anything, officially.

Dual Role

Colleges use your SAT II scores in both admissions and placement decisions.

How Do Colleges Use the SAT II?

Many people will tell you that the SATs (I and II alike) measure only your ability to perform on standardized exams—that they measure neither your reading and thinking skills nor your level of knowledge. Maybe they're right. But these people don't work for colleges. Those schools that require SATs feel that they are an important indicator of your ability to succeed in college. Specifically, they use your scores in one or both of two ways: to help them make admissions and/or placement decisions.

Like the SAT I, the SAT II tests provide schools with a standard measure of academic performance, which they use to compare you with applicants from different high schools and different educational backgrounds. This information helps them to decide whether you're ready to handle their curriculum.

SAT II scores may also be used to decide what course of study is appropriate for you once you've been admitted. A low score on the Writing Test, for example, might mean that you have to take a remedial English course. Conversely, a high score on an SAT II: Mathematics Test might mean that you'll be exempted from an introductory math course.

Call Your Colleges

Many colleges require you to take certain SAT II tests. Check with all of the schools you're interested in applying to before deciding which tests to take.

Which SAT II Tests Should I Take?

The simple answer is: those that you'll do well on. High scores, after all, can only help your chances for admission. Unfortunately, many colleges demand that you take particular tests, usually the Writing Test and/or one of the Mathematics Tests. Some schools will give you a degree of choice in the matter, especially if they want you to take a total of three tests. Before you register to take any tests, therefore, check with the colleges you're interested in to find out exactly which tests they require. Don't rely on high school guidance counselors or admissions handbooks for this information. They might not give you accurate or current information.

When Are the SAT II Tests Administered?

Count to Three

You can take up to three SAT II tests in one day. The Writing Test must be taken first.

Most of the SAT II Tests are administered six times a year: in October, November, December, January, May, and June. A few of the tests are offered less frequently. Due to admissions deadlines, many colleges insist that you take the SAT II no later than December or January of your senior year in high school. You may even have to take it sooner if you're interested in applying for "early admission" to a school. Those schools that use scores for placement decisions only may allow you to take the SAT II as late as May or June of your senior year. You should check with colleges to find out which test dates are most appropriate for you.

How Do I Register for the SAT II?

The College Board administers the SAT II tests, so you must sign up for the tests with them. The easiest way to register is to obtain copies of the *SAT Registration Bulletin* and *Taking the SAT II: Subject Tests*. These publications contain all of the necessary information, including current test dates and fees. They can be obtained at any high school guidance office or directly from the College Board at www.collegeboard.com.

You can also reregister by telephone if you have previously registered for an SAT I or SAT II test. If you choose this option, you should still read the College Board publications carefully before you make any decisions.

How Are the SAT II Tests Scored?

Like the SAT I, the SAT II tests are scored on a 200–800 scale.

What's a "Good" Score?

That's tricky. The obvious answer is: the score that the colleges of your choice demand. Keep in mind, though, that SAT II scores are just one piece of information that colleges will use to evaluate you. The decision to accept or reject you will be based on many criteria, including your high school transcript, your SAT I scores, your recommendations, your personal statement, your interview (where applicable), your extracurricular activities, and the like. So, failure to achieve the necessary score doesn't automatically mean that your chances of getting in have been damaged. For those who really want a numerical benchmark, a score of 600 is considered very solid. (Note that as of 2002, the Score Choice option to withhold a score you're unhappy with has been eliminated for the SAT II subject tests.)

What Should I Bring to the SAT II?

It's a good idea to get your test materials together the day before the tests. You'll need an admission ticket; a form of identification (check the *Registration Bulletin* to find out what is permissible); a few sharpened No. 2 pencils; a good eraser; and a scientific calculator (for Math Level IC or IIC). If you'll be registering as a standby, collect the appropriate forms beforehand. Also, make sure that you know how to get to the test center.

Do the Legwork

Ask your school counselor's office for the SAT Registration Bulletin, which contains a Registration Form, test dates, fees, and instructions.

By mail: Mail in the Registration Form to the College Board. The address will be on the Registration Form.

Online: With a credit card, you can register online at www.collegeboard.com/sat/html/satform.html

By phone: You can register by phone *only if* you have registered for an SAT test in the past.
(800) 728–7267 (automated)
(609) 771–7600 (customer service)

Real World

The mean score of the 52,200 students who took the SAT II: Chemistry test in 2002 was 610.

Pack Your Bag

Gather your test materials the day before the test. You'll need:

- Your admission ticket

- A proper form of I.D.

- Some sharpened No. 2 pencils

- A good eraser

Don't Get Lost

Learn SAT II directions as you prepare for the tests. That way, you'll have more time to spend answering the questions on Test Day.

SAT II Mastery

Now that you know a little about the SAT II tests, it's time to let you in on a few basic test taking skills and strategies that can improve your performance on them. You should practice these skills and strategies as you prepare for the SAT II.

Use the Structure of the Test to Your Advantage

The SAT II tests are different from the tests that you're used to taking. On your high school tests, you probably go through the questions in order. You probably spend more time on hard questions than on easy ones, since hard questions are generally worth more points. And you often show your work, since your teachers tell you that how you approach questions is as important as getting the right answers.

None of this applies to the SAT II tests. You can benefit from moving around within the tests, hard questions are worth the same as easy ones, and it doesn't matter how you answer the questions—only what your answers are.

The SAT II tests are highly predictable. Because the format and directions of the SAT II tests remain unchanged from test to test, you can learn the setup of each test in advance. On Test Day, the various question types on each test shouldn't be new to you.

One of the easiest things you can do to help your performance on the SAT II tests is to understand the directions before taking the test. Since the instructions are always the same, there's no reason to waste a lot of time on Test Day reading them. Learn them beforehand as you work through this book and the College Board publications.

Not all of the questions on the SAT II tests are equally difficult. The questions often get harder as you work through different parts of a test. This pattern can work to your benefit. Try to be aware of where you are in a test.

When working on more basic problems, you can generally trust your first impulse—the obvious answer is likely to be correct. As you get to the end of a test section, you need to be a bit more suspicious. Now the answers probably won't come as quickly and easily—if they do, look again because the obvious answers may be wrong. Watch out for answers that just "look right." They may be distractors—wrong answer choices deliberately meant to entice you.

There's no mandatory order to the questions on the SAT II. You're allowed to skip around on the SAT II tests. High scorers know this fact. They move through the tests efficiently. They don't dwell on any one question, even a hard one, until they've tried every question at least once.

When you run into questions that look tough, circle them in your test booklet and skip them for the time being. Go back and try again after you've answered the easier ones if you've got time. After a second look, trouble-some questions can turn out to be remarkably simple.

If you've started to answer a question but get confused, quit and go on to the next question. Persistence might pay off in high school, but it usually hurts your SAT II scores. Don't spend so much time answering one hard question that you use up three or four questions' worth of time. That'll cost you points, especially if you don't even get the hard question right.

You can use the so-called guessing penalty to your advantage. You might have heard it said that the SAT II has a "guessing penalty." That's a mis-nomer. It's really a *wrong-answer penalty*. If you guess wrong, you get a small penalty. If you guess right, you get full credit.

The fact is, if you can eliminate one or more answer choices as definitely wrong, you'll turn the odds in your favor and actually come out ahead by guessing. The fractional points that you lose are meant to offset the points you might get "accidentally" by guessing the correct answer. With practice, however, you'll see that it's often easy to eliminate *several* answer choices on some of the questions.

The answer grid has no heart. It sounds simple, but it's extremely impor-tant: Don't make mistakes filling out your answer grid. When time is short, it's easy to get confused going back and forth between your test booklet and your grid. If you know the answers, but misgrid, you won't get the points. Here's how to avoid mistakes.

Always circle the questions you skip. Put a big circle in your test booklet around any question numbers that you skip. When you go back, these ques-tions will be easy to relocate. Also, if you accidentally skip a box on the grid, you'll be able to check your grid against your booklet to see where you went wrong.

Always circle the answers you choose. Circling your answers in the test book-let makes it easier to check your grid against your booklet.

Grid five or more answers at once. Don't transfer your answers to the grid after every question. Transfer them after every five questions. That way, you won't keep breaking your concentration to mark the grid. You'll save time and gain accuracy.

Approaching SAT II Questions

Apart from knowing the setup of the SAT II tests that you'll be taking, you've got to have a system for attacking the questions. You wouldn't trav-el around an unfamiliar city without a map, and you shouldn't approach the SAT II without a plan. What follows is the best method for approaching SAT II questions systematically.

Think about the questions before you look at the answers. The test mak-ers love to put distractors among the answer choices. Distractors are

Leap Ahead

Do the questions in the order that's best for you. Skip hard questions until you've gone through every question once. Don't pass up the opportunity to score easy points by wast-ing time on hard questions. Come back to them later.

Guessing Rule

Don't guess, unless you can eliminate at least one answer choice. Don't leave a question blank unless you have absolutely no idea how to answer it.

Hit the Spot

A common mistake is filling in all of the questions with the right answers—in the wrong spots. Whenever you skip a question, circle it in your test booklet and make doubly sure that you skip it on the answer grid as well.

Think First

Try to think of the answer to a question before you shop among the answer choices. If you've got some idea of what you're looking for, you'll be less likely to be fooled by "trap" choices.

Speed Limit

Work quickly on easier questions to leave more time for harder questions. But not so quickly that you make careless errors. And it's okay to leave a few questions blank if you have to—you can still get a high score.

answers that look like they're correct, but aren't. If you jump right into the answer choices without thinking first about what you're looking for, you're much more likely to fall for one of these traps.

Guess—when you can eliminate at least one answer choice. You already know that the "guessing penalty" can work in your favor. Don't simply skip questions that you can't answer. Spend some time with them in order to see whether you can eliminate any of the answer choices. If you can, it pays for you to guess.

Pace yourself. The SAT II tests give you a lot of questions in a short period of time. To get through the tests, you can't spend too much time on any single question. Keep moving through the tests at a good speed. If you run into a hard question, circle it in your test booklet, skip it, and come back to it later if you have time.

You don't have to spend the same amount of time on every question. Ideally, you should be able to work through the easier questions at a brisk, steady clip, and use a little more time on the harder questions. One caution: Don't rush through basic questions just to save time for the harder ones. The basic questions are points in your pocket, and you're better off not getting to some harder questions if it means losing easy points because of careless mistakes. Remember, you don't earn any extra credit for answering hard questions.

Locate quick points if you're running out of time. Some questions can be done more quickly than others because they require less work or because choices can be eliminated more easily. If you start to run out of time, look for these quicker questions.

When you take the SAT II: Subject Tests, you have one clear objective in mind: to score as many points as you can. It's that simple. The rest of this book is dedicated to helping you to do that on the SAT II: Chemistry Subject Test.

CHAPTER 2

GETTING READY FOR THE SAT II: CHEMISTRY TEST

Now that you know the basics about the SAT II: Subject Tests, it's time to focus on the Chemistry test. What's on it? How is it scored? After reading this chapter, you'll know what to expect on test day.

Content

The SAT II: Chemistry Subject test expects you to have a mastery of the concepts and principles covered in a one-year, college-prep chemistry class. This one-hour exam consists of 85 multiple-choice questions covering the topics in the table on the next page.

Every edition of the chemistry subject test contains approximately five questions on equation balancing and/or predicting products of chemical reactions. These are distributed among the various content categories.

Three basic skills are tested on this exam. First, the ability to recall knowledge forms the basis for approximately 20 percent of the questions. This means that your ability to remember specific facts, your mastery of terminology, and your comfort with straightforward knowledge will be examined on approximately 17 of the questions.

The second skill tested is your ability to apply your chemistry knowledge to unfamiliar situations. This is the basis for approximately 38 (45 percent) of the questions on the exam. These questions test how well you understand concepts and can reformulate information into equivalent forms. They also test how well you can solve problems, particularly those dealing with mathematical relationships.

The third question type explores your proficiency at synthesizing information. These questions, comprising about 35 percent of the exam (30 questions), require you to infer and deduce from qualitative and quantitative data, such as that which you might accumulate doing an experiment in the laboratory, and then integrate that data to form conclusions. The data may be in paragraph form, like those word problems you hated in fourth grade, or it may be in graph or chart form. These questions also test your ability to

Topics	Percentage of the test
Atomic and Molecular Structure	25%
Atomic Theory and Structure	
Nuclear Reactions	
Chemical Bonding and Molecular Structure	
States of Matter	15%
Kinetic Molecular Theory of Gases	
Solution Chemistry	
Concentration Units	
Solubility	
Conductivity	
Colligative Properties	
Reaction Types	14%
Acids and Bases	
Oxidation-Reduction	
Stoichiometry	12%
Mole concept and Avogadro's Number	
Empirical and Molecular Formulas	
Percentage Composition	
Stoichiometric Calculations	
Limiting Reagents	
Equilibrium and Reaction Rates	7%
Equilibrium	
Mass Action Expressions	
Ionic Equilibria	
Le Châtelier's Principle	
Reaction Rates	
Thermodynamics	6%
Energy Changes in Chemical Reactions	
Randomness	
Spontaneity	
Descriptive Chemistry	13%
Physical and Chemical Properties of the Elements and Familiar Compounds	
Periodic Properties	
Laboratory	8%
Equipment	
Procedures	
Observations	
Calculations and Interpretation of Results	
Safety	

recognize unstated assumptions, so you need to be prepared to think about what is implied in the setup of the experiment or the question stem.

As you can see from the table on the previous page, the SAT II: Chemistry Subject test covers a broad range of topics. It requires you to think about those topics in ways you may not have done before. As a result, it's likely that some test questions will be topics that you did not cover in your chemistry class. Don't be alarmed; there is so much to chemistry that you cannot possibly cover everything in a year. If, while you are taking the diagnostic, you discover that there are areas you haven't covered in school, plan to spend a little extra time on those chapters that cover your areas of weakness.

While preparing for the exam, also make sure that you understand common algebraic concepts such as ratios and proportions and, more important, can apply them to word problems and data interpretations questions. You will not be allowed to use a calculator on this exam, but don't worry; the math should be nothing more than simple calculations—nothing more complicated than multiplication or division.

Basic Skills

Three basic skills are tested on the SAT II: Chemistry Test:

- Recalling information
- Applying knowledge
- Synthesizing information

Scoring Information

This exam is scored in a range from 200–800, just like a section of the SAT I. Your raw score is calculated by subtracting $\frac{1}{4}$ of the number of questions you got wrong from the number of questions you got right. If you answered 70 questions correctly and 15 incorrectly, your raw score would be:

Number correct	70
$\frac{1}{4} \times$ Number incorrect	-3.75
Raw Score	66.25 (66)

This raw score is then compared to all the other test takers to come up with a scaled score. This scaling takes into account any slight variations between test administrations. On a recent administration, you could miss ten questions and still receive a scaled score of 800. A raw score of 65 translated into a 730. So, you can miss a few questions and still receive a competitive score.

Test Tip

Our test strategies won't make up for a weakness in a given area, but they will help you to manage your time effectively and maximize points.

Question Types

On your test, there will be three main question types: classification questions, relationship analysis questions, and five-choice completion questions. Make sure you feel comfortable with all of them and their directions before test day. Don't waste time reading directions when you are being timed!

Test Strategy

On Test Day, do classification questions first; they require less reading and will give you the most points for your time invested.

Test Strategy

On classification questions, don't eliminate an answer choice just because you've used it. Answer choices can be used more than once (or not at all).

Classification Questions

Classification questions consist of five lettered choices—typically, ideas, chemical laws, graphs, or some other type of data presentation. Following the five choices will be three to five statements that can be functions of the choices, definitions, descriptive characteristics, or conditions that would favor that data set. The five choices may be used more than once, so do not eliminate an answer just because you have used it.

Read through the directions and attempt to answer questions 1–4 below. Check your answers against the explanations that follow the question set.

Directions: Each set of lettered choices below refers to the numbered statements immediately following it. Select the one lettered choice that best fits each statement and then fill in the corresponding oval on the answer sheet. In each set, a choice may be used once, more than once, or not at all.

Questions 1–4

(A) Electronegativity
(B) Electron affinity
(C) Ionization potential
(D) Standard electrode potential
(E) Atomic radius

1. Measures an energy consuming process

2. Measures an energy releasing process

3. Measured against the hydrogen half cell reaction

4. Decreases as you travel right and up the periodic chart

Ionization potential is a measure of the energy required to form a cation from the ground state. Electron affinity is the measure of the energy released when the atom accepts an electron to form an anion. The standard electrode potential is defined to be zero volts for the standard hydrogen electrode in which H^+ is reduced to H_2 gas. The atomic radius is determined by the principal quantum number of the valence shell and the interaction of the positive nucleus and the negative electrons. Larger atoms have electrons in higher level shells (larger principal quantum number) and atoms with more positive charge in the nucleus attract valence electrons more strongly and

KAPLAN

thus draw them in tighter, decreasing radius. In light of this, the correct answers for this question set are 1.(C); 2.(B); 3.(D); 4.(E).

Relationship Analysis Questions

These questions consist of a specific statement (I) followed by an explanation (II). You must determine if both are true and whether or not the explanation properly explains the statement.

Directions: Each question below consists of two statements, statement I in the left-hand column and statement II in the right-hand column. For each question, determine whether statement I is true or false and whether statement II is true or false. Then, fill in the corresponding T or F ovals on your answer sheet. Fill in oval CE only if statement II is a correct explanation of statement I.

Questions 5–6

	I		II
	I		**II**
5.	A 2M solution of HBr has a low pH	BECAUSE	Bromine is very acidic.
6.	Two electrons in the same orbital will have opposite spins	BECAUSE	they have the same n, l, and m_l values.

For question 5, HBr is a strong acid. Low pH indicates a strong acid because of a high hydrogen ion concentration. Bromine (Br^-) is the conjugate base of HBr. So statement I is true, but statement II is false.

For question 6, you know from the Pauli Exclusion Principle that no two electrons in the same atom can have identical quantum numbers. So, with n, l, and m_l the same, they may have different m_s numbers only. The n value indicates that these electrons are in the same shell. The value of n also limits the value of l, defining the possible subshells the electrons can occupy. If they have the same l value, they must be in the same subshell if they are in the same atom. Well, the l value limits the m_l value, which defines the specific spatial orientation of the orbital the electrons are in. If the m_l value is the same for two electrons, they are in the same orbital. So, since the identical m_l value puts these two electrons in the same orbital, the Pauli principle states that we can distinguish them by their m_s value. One will have an m_s value of $+\frac{1}{2}$ and the other will be $-\frac{1}{2}$. So both statements are true, and statement II is a correct explanation of statement I.

Five-Choice Completion Questions

These questions are your common multiple-choice questions; there are three types of them.

Directions: Each of the incomplete statements or questions below is followed by five suggested completions or answers. In each case, select the one that is best and fill in the corresponding oval on the answer sheet.

Type 1 Questions

These have a unique solution, often the only correct answer or the best answer. Sometimes, though, the most inappropriate answer will be correct; these question types will have NOT, EXCEPT, LEAST somewhere in the stimulus.

Questions 7–8

7. A 330 cc sample of ideal gas weighing 72 grams at 18°C and 748 torr is placed in an evacuated vessel of volume 1320 cc. If the gas is to fill the whole vessel at 748 torr, to what temperature must the assembly be heated?

   (A) 891° C

   (B) 1437 K

   (C) 18° C

   (D) 1164° C

   (E) None of the above

8. Which of the following is NOT a state function?

   (A) Temperature

   (B) Density

   (C) Work

   (D) Viscosity

   (E) All of the above

Question 7 tests your ability to use the equation

$$\frac{P_1 V_1}{T_1} = \frac{P_2 V_2}{T_2}$$

Substitute the values and convert temperatures to Kelvin from the problem and you have

$$\frac{(748)(330)}{(18+273)} = \frac{(748)(1320)}{(T_2)}$$

$$T_2 = \frac{(1320)(291)}{(330)}$$

$$T_2 = 1164 \text{ K}$$

$$T_2 = 891° \text{ C}$$

So choice A is correct.

In question 8, a state function is one whose value depends only on the initial and final states of the system; it is independent of the path taken in going from the initial to the final states. Examples of state functions are pressure, volume, temperature, density, viscosity, enthalpy, internal energy, entropy, and free energy. Two important functions that are not state functions are work and q, the amount of heat transferred. So choice C is correct.

Type 2 Questions

These typically have three to five Roman numeral statements following each question. One or more of these statements may be the correct answer. Following each statement, there are five lettered choices with various combinations of the Roman numerals. You must select the combination that includes all of the correct answers and excludes all of the incorrect answers.

Question 9

9.   Which of the following phenomena are considered to illustrate colligative properties?

    I.   The lowering of the vapor pressure of a solvent by a solute

    II.   The raising of the boiling point of a solvent by a solute

    III. The lowering of the freezing point of a solvent by a solute

    IV.   Osmotic pressure

    (A) I only

    (B) IV only

    (C) I and IV only

    (D) II and III only

    (E) I, II, III, and, IV

All of the statements are examples of colligative properties. Colligative properties are properties of solutions that depend only on the number of solute particles present but not on the nature of those particles. Examples of colligative properties are boiling point elevation, freezing point depression, osmotic pressure, and vapor-pressure lowering. They are usually associated with dilute solutions. Choice E is correct.

Type 3 Questions

These questions are also organized in sets, but center on an experiment, chart, graph, or other experimental data presentation. They assess how well you apply science to unfamiliar situations. Each question is independent of the others; in addition, these questions are typically found in the latter part of the test. Most students find them to be the most difficult questions on the test.

Type 3 questions test your ability to identify a problem; evaluate experimental situations; suggest hypotheses; interpret data such as graphs or mathematical expressions; make inferences and draw conclusions; check the logical consistency of hypotheses based on your observations; convert information to graphical form; apply mathematical relationships; and select the appropriate procedure for further study.

Questions 10–13 are based on the following equation:

$$N_2(g) + 3H_2\,(g) \rightleftharpoons 2NH_3\,(g)\quad \Delta H = -30kJ/mole$$

10. How will the equilibrium of the following reaction be affected if the temperature is increased?

    (A) It will be shifted to the right.

    (B) It will be shifted to the left.

    (C) It will be unaffected.

    (D) The effect on the equilibrium cannot be determined without more information.

    (E) The equilibrium will be squared.

11. The equilibrium constant is given by the expression

    (A) $\dfrac{[NH_3]}{[H_2]^2[N_2]}$

    (B) $\dfrac{[NH_3]^2}{[H_2]^3[N_2]}$

    (C) $\dfrac{[NH_3]^2}{[H_2][N_2]}$

    (D) $\dfrac{[NH_3]}{[H_2][N_2]^2}$

    (E) None of the above

12. Suppose that 1.5 moles of N_2 is converted to NH_3. The amount of heat produced would be

(A) 45 kJ

(B) 30 kJ

(C) 15 kJ

(D) 60 kJ

(E) None of the above

13. Which compound(s) has basic qualities?

(A) H_2

(B) N_2

(C) N_2 and NH_3

(D) NH_3

(E) All of them have basic qualities.

In question 10, if the temperature of an exothermic reaction is increased, the reaction shifts to the left. If the temperature of an endothermic reaction is increased, it shifts to the right. The reaction has a negative enthalpy of reaction, meaning that it is an exothermic reaction. If the temperature of this reaction is increased it will be shifted to the left, making choice B correct.

For question 11, the equilibrium constant is the ratio of the concentration of the products to the concentration of the reactants for a certain reaction at equilibrium, all raised to their stoichiometric coefficients. Properties of the equilibrium constant include: Pure solids and liquids do not appear in the equilibrium constant expression; K_{eq} is characteristic of a given system at a given temperature; if the value of K_{eq} is very large compared to 1, an equilibrium mixture of reactant and products will contain very little of the reactants compared to the products; if the value of K_{eq} is very small compared to 1, an equilibrium mixture of reactants and products will contain very little of the products compared to the reactants. So, choice B is correct.

In the exothermic equation (negative ΔH) in question 12, 30kJ/mole are given off. If one mole of N_2 produces 30kJ/mole, then 1.5 moles of N_2 will produce 1.5(30 kJ/mole), which equals 45 kJ, choice A.

To answer question 13, you need to investigate the Brønsted-Lowry definition of acids and bases. The Brønsted-Lowry acid is a species that donates protons, while a Brønsted-Lowry base accepts protons. For example, NH_3 and Cl^- are both Brønsted-Lowry bases because they accept protons. They are not Arrhenius bases because they do not produce OH^- in aqueous solutions. The advantage of the Brønsted-Lowry concept of acids and bases is that it is not limited to aqueous solutions. Choice D is correct.

Strategies

There are ways to approach this test that will allow you to maximize your score. Read through these strategies before you begin your diagnostic practice test. Try to internalize each of them so that on test day they will be second nature to you. If you accomplish this, you will be rewarded with a higher score on your SAT II: Chemistry Test.

1. Do classification questions first; they require less reading and you'll get the most points for your time invested.

2. Next, do the Type 1 and Type 2 five-choice completion questions. Again, you will get more points with less time invested.

3. In the relationship analysis questions, look at each statement individually. If either statement is false, then you will never bubble in CE. You are required to determine if it is the correct explanation when only both statements are true.

4. This test emphasizes general trends and basic chemistry concepts, so you probably won't see a question/graph that would take a rocket scientist 30 minutes to figure out. Look for trends and outliers in graphs. If a value or a plot is vastly different from the others, it is likely that there will be a question about it.

5. Look for opposing answers in the answer selections. If two answers are close in wording or if they contain opposite ideas, there is a strong possibility that one of them is the right answer.

6. By the same token, if two answers mean basically the same thing, then they both cannot be correct and you can eliminate both answer choices.

7. Use the structure of a Roman numeral question to your advantage. Eliminate choices as soon as you find them to be inconsistent with the truth or falsehood of a statement in the stimulus. Similarly, consider only those choices that include a statement that you've already determined to be true.

8. Predict your answer before you go to the answer choices so you don't get persuaded by the wrong answers. This helps protect you from persuasive or tricky incorrect choices. Most wrong answer choices are logical twists on the correct choice.

9. Eliminate answers and guess.

10. Think, don't compute!

Now you're ready to tackle our practice diagnostic test. This test will probe your mastery of the various chemistry topics covered on the SAT II: Chemistry exam. Use it to identify areas in which you need to refresh your knowledge, and plan to review the chapters that deal with these topics carefully. Good luck!

CHEMISTRY
REVIEW

PRACTICE TEST 1:
DIAGNOSTIC

Periodic Table of the Elements

1 H 1.0																	2 He 4.0
3 Li 6.9	4 Be 9.0											5 B 10.8	6 C 12.0	7 N 14.0	8 O 16.0	9 F 19.0	10 Ne 20.2
11 Na 23.0	12 Mg 24.3											13 Al 27.0	14 Si 28.1	15 P 31.0	16 S 32.1	17 Cl 35.5	18 Ar 39.9
19 K 39.1	20 Ca 40.1	21 Sc 45.0	22 Ti 47.9	23 V 50.9	24 Cr 52.0	25 Mn 54.9	26 Fe 55.8	27 Co 58.9	28 Ni 58.7	29 Cu 63.5	30 Zn 65.4	31 Ga 69.7	32 Ge 72.6	33 As 74.9	34 Se 79.0	35 Br 79.9	36 Kr 83.8
37 Rb 85.5	38 Sr 87.6	39 Y 88.9	40 Zr 91.2	41 Nb 92.9	42 Mo 95.9	43 Tc (98)	44 Ru 101.1	45 Rh 102.9	46 Pd 106.4	47 Ag 107.9	48 Cd 112.4	49 In 114.8	50 Sn 118.7	51 Sb 121.8	52 Te 127.6	53 I 126.9	54 Xe 131.3
55 Cs 132.9	56 Ba 137.3	57 La* 138.9	72 Hf 178.5	73 Ta 180.9	74 W 183.9	75 Re 186.2	76 Os 190.2	77 Ir 192.2	78 Pt 195.1	79 Au 197.0	80 Hg 200.6	81 Tl 204.4	82 Pb 207.2	83 Bi 209.0	84 Po (209)	85 At (210)	86 Rn (222)
87 Fr (223)	88 Ra 226.0	89 Act 227.0	104 Unq (261)	105 Unp (262)	106 Unh (263)	107 Uns (262)	108 Uno (265)	109 Une (267)									

	58 Ce 140.1	59 Pr 140.9	60 Nd 144.2	61 Pm (145)	62 Sm 150.4	63 Eu 152.0	64 Gd 157.3	65 Tb 158.9	66 Dy 162.5	67 Ho 164.9	68 Er 167.3	69 Tm 168.9	70 Yb 173.0	71 Lu 175.0
†	90 Th 232.0	91 Pa (231)	92 U 238.0	93 Np (237)	94 Pu (244)	95 Am (243)	96 Cm (247)	97 Bk (247)	98 Cf (251)	99 Es (252)	100 Fm (257)	101 Md (258)	102 No (259)	103 Lr (260)

ANSWER SHEET

1 Ⓐ Ⓑ Ⓒ Ⓓ Ⓔ	19 Ⓐ Ⓑ Ⓒ Ⓓ Ⓔ	37 Ⓐ Ⓑ Ⓒ Ⓓ Ⓔ	55 Ⓐ Ⓑ Ⓒ Ⓓ Ⓔ
2 Ⓐ Ⓑ Ⓒ Ⓓ Ⓔ	20 Ⓐ Ⓑ Ⓒ Ⓓ Ⓔ	38 Ⓐ Ⓑ Ⓒ Ⓓ Ⓔ	56 Ⓐ Ⓑ Ⓒ Ⓓ Ⓔ
3 Ⓐ Ⓑ Ⓒ Ⓓ Ⓔ	21 Ⓐ Ⓑ Ⓒ Ⓓ Ⓔ	39 Ⓐ Ⓑ Ⓒ Ⓓ Ⓔ	57 Ⓐ Ⓑ Ⓒ Ⓓ Ⓔ
4 Ⓐ Ⓑ Ⓒ Ⓓ Ⓔ	22 Ⓐ Ⓑ Ⓒ Ⓓ Ⓔ	40 Ⓐ Ⓑ Ⓒ Ⓓ Ⓔ	58 Ⓐ Ⓑ Ⓒ Ⓓ Ⓔ
5 Ⓐ Ⓑ Ⓒ Ⓓ Ⓔ	23 Ⓐ Ⓑ Ⓒ Ⓓ Ⓔ	41 Ⓐ Ⓑ Ⓒ Ⓓ Ⓔ	59 Ⓐ Ⓑ Ⓒ Ⓓ Ⓔ
6 Ⓐ Ⓑ Ⓒ Ⓓ Ⓔ	24 Ⓐ Ⓑ Ⓒ Ⓓ Ⓔ	42 Ⓐ Ⓑ Ⓒ Ⓓ Ⓔ	60 Ⓐ Ⓑ Ⓒ Ⓓ Ⓔ
7 Ⓐ Ⓑ Ⓒ Ⓓ Ⓔ	25 Ⓐ Ⓑ Ⓒ Ⓓ Ⓔ	43 Ⓐ Ⓑ Ⓒ Ⓓ Ⓔ	61 Ⓐ Ⓑ Ⓒ Ⓓ Ⓔ
8 Ⓐ Ⓑ Ⓒ Ⓓ Ⓔ	26 Ⓐ Ⓑ Ⓒ Ⓓ Ⓔ	44 Ⓐ Ⓑ Ⓒ Ⓓ Ⓔ	62 Ⓐ Ⓑ Ⓒ Ⓓ Ⓔ
9 Ⓐ Ⓑ Ⓒ Ⓓ Ⓔ	27 Ⓐ Ⓑ Ⓒ Ⓓ Ⓔ	45 Ⓐ Ⓑ Ⓒ Ⓓ Ⓔ	63 Ⓐ Ⓑ Ⓒ Ⓓ Ⓔ
10 Ⓐ Ⓑ Ⓒ Ⓓ Ⓔ	28 Ⓐ Ⓑ Ⓒ Ⓓ Ⓔ	46 Ⓐ Ⓑ Ⓒ Ⓓ Ⓔ	64 Ⓐ Ⓑ Ⓒ Ⓓ Ⓔ
11 Ⓐ Ⓑ Ⓒ Ⓓ Ⓔ	29 Ⓐ Ⓑ Ⓒ Ⓓ Ⓔ	47 Ⓐ Ⓑ Ⓒ Ⓓ Ⓔ	65 Ⓐ Ⓑ Ⓒ Ⓓ Ⓔ
12 Ⓐ Ⓑ Ⓒ Ⓓ Ⓔ	30 Ⓐ Ⓑ Ⓒ Ⓓ Ⓔ	48 Ⓐ Ⓑ Ⓒ Ⓓ Ⓔ	66 Ⓐ Ⓑ Ⓒ Ⓓ Ⓔ
13 Ⓐ Ⓑ Ⓒ Ⓓ Ⓔ	31 Ⓐ Ⓑ Ⓒ Ⓓ Ⓔ	49 Ⓐ Ⓑ Ⓒ Ⓓ Ⓔ	67 Ⓐ Ⓑ Ⓒ Ⓓ Ⓔ
14 Ⓐ Ⓑ Ⓒ Ⓓ Ⓔ	32 Ⓐ Ⓑ Ⓒ Ⓓ Ⓔ	50 Ⓐ Ⓑ Ⓒ Ⓓ Ⓔ	68 Ⓐ Ⓑ Ⓒ Ⓓ Ⓔ
15 Ⓐ Ⓑ Ⓒ Ⓓ Ⓔ	33 Ⓐ Ⓑ Ⓒ Ⓓ Ⓔ	51 Ⓐ Ⓑ Ⓒ Ⓓ Ⓔ	69 Ⓐ Ⓑ Ⓒ Ⓓ Ⓔ
16 Ⓐ Ⓑ Ⓒ Ⓓ Ⓔ	34 Ⓐ Ⓑ Ⓒ Ⓓ Ⓔ	52 Ⓐ Ⓑ Ⓒ Ⓓ Ⓔ	
17 Ⓐ Ⓑ Ⓒ Ⓓ Ⓔ	35 Ⓐ Ⓑ Ⓒ Ⓓ Ⓔ	53 Ⓐ Ⓑ Ⓒ Ⓓ Ⓔ	
18 Ⓐ Ⓑ Ⓒ Ⓓ Ⓔ	36 Ⓐ Ⓑ Ⓒ Ⓓ Ⓔ	54 Ⓐ Ⓑ Ⓒ Ⓓ Ⓔ	

	I	II CE*		I	II CE*
101	Ⓣ Ⓕ	Ⓣ Ⓕ ◯	109	Ⓣ Ⓕ	Ⓣ Ⓕ ◯
102	Ⓣ Ⓕ	Ⓣ Ⓕ ◯	110	Ⓣ Ⓕ	Ⓣ Ⓕ ◯
103	Ⓣ Ⓕ	Ⓣ Ⓕ ◯	111	Ⓣ Ⓕ	Ⓣ Ⓕ ◯
104	Ⓣ Ⓕ	Ⓣ Ⓕ ◯	112	Ⓣ Ⓕ	Ⓣ Ⓕ ◯
105	Ⓣ Ⓕ	Ⓣ Ⓕ ◯	113	Ⓣ Ⓕ	Ⓣ Ⓕ ◯
106	Ⓣ Ⓕ	Ⓣ Ⓕ ◯	114	Ⓣ Ⓕ	Ⓣ Ⓕ ◯
107	Ⓣ Ⓕ	Ⓣ Ⓕ ◯	115	Ⓣ Ⓕ	Ⓣ Ⓕ ◯
108	Ⓣ Ⓕ	Ⓣ Ⓕ ◯	116	Ⓣ Ⓕ	Ⓣ Ⓕ ◯

Use the answer key following the test to count up the number of questions you got right and the number you got wrong. (Remember not to count omitted questions as wrong.)

right

wrong

PART A

Directions: Each set of lettered choices below refers to the numbered formulas or statements immediately following it. Select the one lettered choice that best fits each formula or statement; then fill in the corresponding oval on the answer sheet. In each set, a choice may be used once, more than once, or not at all.

Note: For all questions involving solutions and/or chemical equations, you can assume that the system is in water unless otherwise stated.

Questions 1–5

(A) Atomic mass
(B) Atomic number
(C) Atomic radius
(D) Electronegativity
(E) Ionization potential

1. Takes into account various isotopes of an atom

2. Determines how electron density is shared when an atom forms a bond

3. Average distance between the nucleus and the outermost electron

4. Number of protons in an element

5. Energy required to remove an electron

Questions 6–8

(A) Na
(B) Fe
(C) Cl^-
(D) Rb
(E) Ca^+

6. Has a filled $4p$ orbital

7. All of its electrons are paired in this ion

8. Electron configuration is

$$1s^2 2s^2 2p^6 3s^2 3p^6 3d^6 4s^2$$

Questions 9–12

(A) Cation
(B) Inert Gas
(C) Crystal
(D) Anion
(E) Element

9. Cannot be further broken down by chemical means

10. Ionic species with a positive charge

11. Has an octet of valence electrons

12. Defined by the number of protons

GO ON TO THE NEXT PAGE

Questions 13–16

 (A) Centrifuge
 (B) Barometer
 (C) Balance
 (D) Calorimeter
 (E) Battery

13. Apparatus used to measure the heat absorbed or released by a reaction

14. Apparatus used to measure atmospheric pressure

15. Apparatus used to measure weight

16. Apparatus used to sediment particles in suspension

Questions 17–19

 (A) Gibbs Free Energy
 (B) Heat of Formation
 (C) Specific Heat
 (D) Heinsenberg Uncertainty Principle
 (E) Heat of Vaporization

17. The energy of a system available to do work

18. Heat required to raise one unit mass of a substance by 1 degree Celsius

19. Heat absorbed or released during production of a substance from elements in their standard states

Questions 20–23

What happens when the following are dissolved in water?

 (A) $NaOH$
 (B) $CsCl$
 (C) HBr
 (D) CH_3COOH
 (E) O_2

20. Forms a strong base

21. Forms a weak acid

22. Forms a strong acid

23. Forms an ionic solution with a neutral pH

GO ON TO THE NEXT PAGE

PART B

Directions: Each question below consists of two statements, statement I in the left-hand column and statement II in the right-hand column. For each question, determine whether statement I is true or false and whether statement II is true or false. Then, fill in the corresponding T or F ovals on your answer sheet. Fill in oval CE only if statement II is a correct explanation of statement I.

Examples:

I		II
EX 1. HCl is a strong acid	**BECAUSE**	HCl contains sulfur.
EX 2. An atom of nitrogen is electrically neutral	**BECAUSE**	a nitrogen atom contains the same number of protons and electrons.

	I	II	CE
SAMPLE ANSWERS			
EX 1	● Ⓕ	Ⓣ ●	○
EX 1	● Ⓕ	● Ⓕ	●

	I		II
101.	Diamond and graphite are both substances made up of carbon but have different properties	BECAUSE	they are composed of different isotopes.
102.	Na and Cl form an ionic bond	BECAUSE	Cl donates an electron to Na.
103.	AgCl will not dissolve in a concentrated NaCl solution	BECAUSE	the chloride ions from NaCl suppress the solubility of AgCl.
104.	Hydrogen and deuterium are different elements	BECAUSE	they have a different number of protons.
105.	On the periodic chart, atomic radius increases from left to right	BECAUSE	the number of protons is increasing.
106.	An element with an atomic number of X and mass number of N has N-X neutrons	BECAUSE	elements have more neutrons than protons.
107.	Radioactive decay has a characteristic half life	BECAUSE	first order kinetics are found in radioactive decay.
108.	Sulfur chemically resembles oxygen	BECAUSE	they are in the same period.

GO ON TO THE NEXT PAGE

109. A nonelectrolyte does not ionize in water   BECAUSE   the solution does not conduct electricity.

110. SO_3 diffuses more slowly than CO_2 BECAUSE   it has smaller bond angles.

111. Hydrogen bonding is stronger between noble gases than small electronegative atoms such as fluorine   BECAUSE   noble gases have a large dipole moment.

112. An electron in a $3s$ subshell may be excited to jump into the $3p$ BECAUSE   the Heisenberg uncertainty principle states that one cannot know what orbital an electron is in.

113. A positive ΔG tells you that the reaction is spontaneous   BECAUSE   entropy always decreases in an isolated system.

114. An increase in pressure in a closed container with an ideal gas leads to a decrease in volume   BECAUSE   pressure and volume are proportional.

115. An amphoteric species acts as either an acid or a base   BECAUSE   it contains both hydrophobic and hydrophilic regions.

116. An indicator will allow you to determine whether a solution is acidic or basic   BECAUSE   it will change colors in solutions with different pHs.

RETURN TO THE SECTION OF YOUR ANSWER SHEET YOU STARTED FOR CHEMISTRY AND ANSWER QUESTIONS 24–69.

PART C

Directions: Each of the incomplete statements or questions below is followed by five suggested completions or answers. Select the one that is best for each case and fill in the corresponding oval on the answer sheet.

24. What is the formal charge on the nitrogen atom in HNO_3?

(A)   –1
(B)   +1
(C)   0
(D)   +2
(E)   +3

25. What volume of a 1M solution of hydrochloric acid is required to neutralize 80mL of a 0.5M NaOH solution?

(A) 320mL
(B) 160mL
(C) 80mL
(D) 40mL
(E) 20mL

GO ON TO THE NEXT PAGE

26. Which of the following would be classified as a strong electrolyte?

    (A) Benzoic acid

    (B) Water

    (C) Hydrofluoric acid

    (D) Potassium Chloride

    (E) Glucose

27. Which atom has an ionic radius that is larger than its atomic radius?

    (A) Na

    (B) K

    (C) Mg

    (D) Ca

    (E) Cl

28. The basic structure of crystalline substances is called

    (A) unit cell

    (B) molecule

    (C) lattice

    (D) geode

    (E) matrix

29. The oxidation state of nitrogen is most negative in which of the following compounds?

    (A) N_2

    (B) N_2O

    (C) NH_3

    (D) NO_2

    (E) NO_3

30. An insulated tube with a movable piston at one end had 500J of heat added to it. If, during the experiment, the piston moves and does 75J of work on the atmosphere, what is the change in energy of the tube system?

    (A) 575J

    (B) 500J

    (C) 425J

    (D) –75J

    (E) –425J

31. The reaction of NaOH and H_2SO_4 in water goes to completion because

    (A) it is a neutralization reaction

    (B) water is a strong electrolyte

    (C) sodium sulfate quickly precipitates

    (D) a volatile product is formed

    (E) sulfuric acid is a very strong acid

32. Which of the following would be different in a ground state and an excited state neon atom?

    (A) The number of neutrons

    (B) The number of electrons

    (C) The atomic weight

    (D) The electronic configuration

    (E) Everything would remain the same

GO ON TO THE NEXT PAGE

33. The K_{sp} of $Mg(OH)_2$ in water is 1.2×10^{-11}. If the Mg^{2+} concentration in an acid solution is 1.2×10^{-5} mol/L, what is the pH at which $Mg(OH)_2$ just begins to precipitate?

    (A) 3

    (B) 4

    (C) 5

    (D) 11

    (E) 12

34. Which of the following states has the highest average translational kinetic energy?

    (A) solid

    (B) liquid

    (C) gas

    (D) colloid

    (E) None of the above

35. Which of the following will favor the melting of ice in a closed container if all other parameters are kept constant?

    (A) Adding water with a temperature of 0°C

    (B) Lowering the temperature below 0°C

    (C) Lowering the pressure

    (D) Raising the pressure

    (E) Decreasing the amount of ice

36. Which of the following has the most polar bond?

    (A) N–O

    (B) C–H

    (C) C–C

    (D) H–F

    (E) None of the bonds are polar

37. The rate law expression for the reaction

    $$N_2 + 3H_2 \rightarrow 2NH_3$$

    (A) can be represented by rate $= \dfrac{[NH_3]^2}{[N_2][H_2]^3}$

    (B) can be represented by rate $= k[NH_3]$

    (C) can be represented by rate $= \dfrac{k[NH_3]^2}{[N_2][H_2]^3}$

    (D) can be represented by rate $= \dfrac{k[NH_3]^2}{[N_2]}$

    (E) cannot be determined from the information given

38. 100mL of 10N H_2SO_4 are diluted to 800 ml. What is the molarity of the dilute acid solution?

    (A) 16/10M

    (B) 8/10M

    (C) 10/8M

    (D) 10/12M

    (E) 5/8M

GO ON TO THE NEXT PAGE

39. Which of the following molecules contains both ionic and covalent bonds?

(A) C_6H_{14}

(B) $MgCl_2$

(C) $(NH_4)_2SO_4$

(D) H_2O

(E) C_2H_4

40. Gas A is at 30°C and gas B is at 20°C. Both gases are at 1 atmosphere. What is the ratio of the volume of 1 mole of gas A to 1 mole of gas B?

(A) 1:1

(B) 2:3

(C) 3:2

(D) 303:293

(E) 606:293

41. How will the equilibrium of the following reaction be affected if more chlorine is added?

$$PCl_5 (g) \leftrightarrows PCl_3 (g) + Cl_2 (g)$$

(A) It will be shifted to the right.

(B) It will be shifted to the left.

(C) It will be unaffected.

(D) The effect on the equilibrium cannot be determined without more information.

(E) More PCl_3 will be produced.

42. After balancing the equation

$$...BrO_3^- (aq) + ...Br^- (aq) + ...H^+(aq) \rightarrow$$
$$...Br_2 (l) + ...H_2O$$

the ratio of BrO_3^- to Br^- is

(A) 1:5

(B) 1:3

(C) 1:2

(D) 1:1

(E) 2:3

43. What mass of sodium carbonate, Na_2CO_3, (formula weight = 106 amu), is needed to make 120mL of a 1.5 M solution?

(A) 295g

(B) 9.5g

(C) 19g

(D) 589g

(E) 19,000g

44. Equimolar amounts of hydrogen and oxygen gas, at the same temperature, are released into a large container. The ratio of the rate of diffusion of the hydrogen molecules to that of the molecules of oxygen would be

(A) 256:1

(B) 16:1

(C) 1:16

(D) 4:1

(E) 1:4

GO ON TO THE NEXT PAGE

45. What does X represent in the following nuclear reaction?

$$^{27}_{13}Al + {}^{4}_{2}He \rightarrow {}^{30}_{15}P + X$$

(A) β particle

(B) positron

(C) α particle

(D) neutron

(E) γ ray

46. When chromium metal is used to form $K_2Cr_2O_7$, the oxidation state of chromium changes from

(A) 0 to 4

(B) 3 to 6

(C) 2 to 6

(D) 0 to 6

(E) 2 to 4

47. Electron density studies have revealed that X and Y have an equal number of electrons. Which of the following could X and Y be?

(A) Ca^+ and K

(B) H^+ and He

(C) Cl and F

(D) O^- and S^+

(E) None of the above

48. All halogens have similar reactivity because

(A) they have the same number of protons

(B) they have the same number of electrons

(C) they have similar outer shell electron configurations

(D) they have valence electrons with the same quantum numbers

(E) they have the same number of neutrons

49. K^+ and Cl^- have the same

(A) atomic weight

(B) electronic configuration

(C) ionization potential

(D) number of protons and neutrons

(E) atomic radius

50. Which of the following has the largest ionic radius?

(A) Na^+

(B) K^+

(C) Mg^{++}

(D) Al^{3+}

(E) Cl^-

GO ON TO THE NEXT PAGE

51. When the following reaction is balanced, what is the net ionic charge on the right side of the equation?

$$...H^+ + ...MnO_4^- + ...Fe^{2+} \rightarrow ...Mn^{2+} + ...Fe^{3+} + ...H_2O$$

(A) +5

(B) +7

(C) +10

(D) +17

(E) The net ionic charge on either side must be zero.

Questions 52–54 refer to the following equation.

$$...Ag(NH_3)_2^+ \rightarrow ...Ag^+ + ...NH_3$$

52. What is the sum of the coefficients once the equation is balanced?

(A) 1

(B) 2

(C) 3

(D) 4

(E) 5

53. How many moles of $Ag(NH_3)_2^+$ are required to produce 11 moles of ammonia?

(A) 1

(B) 2

(C) 5.5

(D) 11

(E) 22

54. What is the percent composition by weight of Ag in $Ag(NH_3)_2^+$?

(A) 4.2

(B) 19.7

(C) 76.1

(D) 80.3

(E) 95.8

55. If 88 g of C_3H_8 and 160 g of O_2 are allowed to react maximally to form CO_2 and H_2O, how many grams of CO_2 will be formed?

(A) 33

(B) 66

(C) 132

(D) 264

(E) None of the above

56. A 200 ml flask contains oxygen at 200mm Hg, and a 300 ml flask contains neon at 100 mmHg. The two flasks are connected so that each gas fills their combined volumes. Assuming no change in temperature, what is the partial pressure of neon in the final mixture?

(A) 60mm Hg

(B) 80mm Hg

(C) 100mm Hg

(D) 150mm Hg

(E) 200mm Hg

GO ON TO THE NEXT PAGE

57. What is the value of Z in the beta decay reaction below:

$$^{60}_{27}Co \rightarrow {}^{60}_{Z}X + e^-$$

(A) 25

(B) 26

(C) 27

(D) 28

(E) 29

Question 58–60 refer to the following experimental setups.

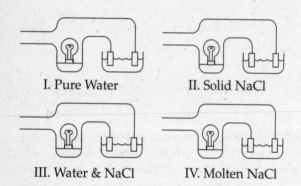

I. Pure Water   II. Solid NaCl

III. Water & NaCl   IV. Molten NaCl

58. Which of the following experimental setups will complete the circuit?

(A) II only

(B) III only

(C) III, and IV

(D) II, III, and IV

(E) I, II, III, and IV

59. What types of bonds are found in solid NaCl?

(A) van der Waals

(B) ionic

(C) covalent

(D) hydrogen

(E) hydrophobic

60. What would happen if the electrodes were put in a saturated solution of glucose dissolved in water?

(A) Light bulb would glow.

(B) Light bulb would remain dark.

(C) Apparatus would combust.

(D) Glucose molecules would dissociate.

(E) None of the above

61. Which of the following statements correctly characterizes a galvanic cell?

I. Oxidation occurs at the anode, which is negative.

II. Oxidation occurs at the anode, which is positive.

III. Reduction occurs at the cathode, which is positive.

(A) II only

(B) III only

(C) I and III only

(D) I, II, and III

(E) II and III

GO ON TO THE NEXT PAGE

62. The heat of combustion of gaseous ammonia is 81 kcal/mole. How much heat is evolved in the reaction of 34 grams of ammonia with excess oxygen?

    (A) 40.5 kcal
    (B) 60.3 kcal
    (C) 75.8 kcal
    (D) 81 kcal
    (E) 162 kcal

63. When there are two electrons in the 3s sub level

    (A) they must occupy different orbitals
    (B) the Heisenberg uncertainty principle predicts that they must periodically jump to the 3p sub level
    (C) the oxidation state of the atom must be 2+
    (D) they must have opposite spins
    (E) they are oppositely charged

64. In order to make a buffer solution, a weak monoprotic acid could be added to

    (A) another acid
    (B) another base
    (C) its conjugate base
    (D) its conjugate acid
    (E) a strong base

65. How will the equilibrium of the following reaction be affected if the temperature is increased?

    $N_2(g) + 3H_2 (g) \rightleftharpoons 2NH_3(g)$ $\Delta H = -30kJ/mole$

    A) It will be shifted to the right.
    (B) It will be shifted to the left.
    (C) It will be unaffected.
    (D) The effect on the equilibrium cannot be determined without more information.
    (E) None of the above

66. For a sample of an ideal gas of fixed weight and at a fixed temperature,

    I. the volume varies directly with the pressure exerted on it
    II. the volume varies inversely with the pressure exerted on it
    III. the pressure varies directly with the density of the gas

    (A) I only
    (B) II only
    (C) III only
    (D) I and II
    (E) II and III

67. When 200 g of $MgCl_2$ is added to 1kg of water, what is the molality of the solution?

    (A) 200/(24+71)
    (B) 0.200/(24+71)
    (C) 0.2(24+71)
    (D) 0.200
    (E) 200

GO ON TO THE NEXT PAGE

68. How many moles of water are formed by a mixture of 100 grams of H_2 and 100 grams of O_2? (Assume the reaction goes to completion.)

(A) $100/32 + 100/64$

(B) $100 + 2(100/32)$

(C) $2(100/32)$

(D) $100(100/32)$

(E) $200(100/32)$

69. An oxide of arsenic contains 65.2% arsenic by weight. What is its simplest formula?

(A) AsO

(B) $As2O_3$

(C) AsO_2

(D) As_2O_5

(E) As_2O

STOP! **END OF TEST.**

Turn the page
for answers and explanations
to Practice Test 1: Diagnostic.

Answer Key

1. A	15. C	29. C	43. C	57. D
2. D	16. A	30. C	44. D	58. C
3. C	17. A	31. A	45. D	59. B
4. B	18. C	32. D	46. D	60. B
5. E	19. B	33. D	47. A	61. C
6. D	20. A	34. C	48. C	62. E
7. C	21. D	35. D	49. B	63. D
8. B	22. C	36. D	50. E	64. C
9. E	23. B	37. E	51. A	65. B
10. A	24. B	38. E	52. D	66. E
11. B	25. D	39. C	53. C	67. A
12. E	26. D	40. D	54. C	68. C
13. D	27. E	41. B	55. C	69. D
14. B	28. A	42. A	56. A	

101. T, F	109. T, T
102. T, F	110. T, T
103. T, T, CE	111. F, F
104. F, F	112. T, F
105. F, T	113. F, F
106. T, F	114. T, F
107. T, T, CE	115. T, F
108. T, F	116. T, T, CE

Answers and Explanations to
Practice Test 1: Diagnostic

1. A

Atomic mass is the averaged mass of the atoms of an element, taking into account the relative abundance of the various isotopes in a naturally occurring substance.

2. D

Electronegativity is a measure of the ability of an atom to attract the electrons in a bond. A common scale is the Pauling scale.

3. C

The atomic radius is defined as the average distance between a nucleus and the outermost electrons. It is usually measured as half of the distance between two nuclei of an element in its elemental form.

4. B

The atomic number is defined as the number of protons in a given element. It defines an element.

5. E

The ionization potential, or ionization energy, is the energy required to remove an electron from the valence shell of a gaseous atom.

6. D

Krypton is the first element with a filled $4p$ orbital so the element must fall after it in the periodic table. Rb is the only element with enough electrons to have a filled $4p$ subshell.

7. C

Of the two ions, Cl^- and Ca^+, Cl^- has the same electron configuration as Argon, a noble gas, and therefore has all paired electrons.

8. B

Fe has 26 electrons, configured as listed.

9. E

An element is a substance that cannot be further broken down by chemical means.

10. A

A cation is an ionic species with a positive charge. An anion is an ionic species with a negative charge.

11. B

Inert gases, elements in Group 0 (or VIII), of the periodic table contain a full octet of valence electrons in their outermost shells; this makes these elements the least reactive.

12. E

All atoms of a given element have the same numbers of protons.

13. D

A calorimeter is an apparatus used to measure the heat released or absorbed by a reaction.

14. B

A barometer is used to measure atmospheric pressure.

15. C

A balance is used to measure weight.

16. A

Centrifuges can spin at very high rotor speeds (e.g. 60,000 rpm) to sediment out particles in suspension.

17. A

The Gibbs free energy is the energy of a system available to do work. The change in Gibbs free energy, ΔG, can be determined for a given reaction with the equation $\Delta G = \Delta H - T\Delta S$. It is used to predict the spontaneity of a reaction.

18. C

Specific heat is the amount of heat required to raise the temperature of one unit mass of a substance of one degree Celsius.

19. B

The heat of formation is the heat absorbed or released during the formation of a pure substance from its elements at constant pressure.

20. A

NaOH completely dissociates in water to form Na^+ and OH^- ions, the latter of which is a very strong base. This results in a solution with a very high pH (>7).

21. D

CH_3COOH does not completely dissociate in water. There would be some undissociated CH_3COOH and some $CH_3COO^- + H^+$; therefore, it would be a weak acid with a pH slightly less than 7.

22. C

HBr completely dissociates into Br^- and H^+ in water; this solution would have a very low pH (<7).

23. B

CsCl is a salt; it completely dissociates in water into Cs^+ and Cl^-. This will form an ionic solution with a neutral pH.

101. T, F

Diamond and graphite are different carbon compounds with different properties because of different bond structures. Diamond has a covalent crystal structure (lattice positions occupied by atoms with covalent bonds) while in graphite, the carbon molecules are in parallel sheets.

102. T, F

Na donates an electron to become Na^+ while Cl picks up an electron to become Cl^-.

103. T, T, CE

In a concentrated NaCl solution, AgCl will not dissociate because of the common ion effect that states that the solubility of one salt is reduced by the presence of another salt having a common ion.

104. F, F

Hydrogen and deuterium are different isotopes of the same element. Hydrogen has one neutron while deuterium has two. Because they correspond to the same element, they have the same number of protons.

105. F, T

In the periodic table, atomic radius decreases from left to right as electrons are added one at a time to the outer electron shell. Therefore, electrons within a shell cannot shield one another from the attractive pull of protons. Since the number of protons is also increasing, there is a greater positive charge pulling the electrons in close to the nucleus, reducing the atomic radius.

106. T, F
The mass number of an element is equal to the total number of protons and neutrons. The atomic number is the number of protons. Therefore, an element with an atomic number of X and a mass number of N has N–X neutrons.

107. T, T, CE
An example of a first order reaction is radioactive decay. In first order reactions, the rate is proportional to the concentration of one reactant. The half life ($\tau_{1/2}$) of a reaction is the time needed for the concentration of a substance to decrease to one half its original value.

108. T, F
Elements in the same group have the same valence electrons and therefore have similar chemical properties. Oxygen and sulfur are both in Group VI and have six valence electrons. A period is a horizontal row in the periodic table; oxygen and sulfur are in different periods.

109. T, T
Nonelectrolytes do not dissolve in water and do not conduct electricity. Strong electrolytes dissociate completely in water. For example, NaCl will conduct electricity very well. Weak electrolytes such as formic acid ($HCHO_2$) will dissociate slightly in water to H^+ and CHO_2^- and will conduct electricity. Statement II, however, is not the correct explanation for statement I. In fact, it is the other way around: A solution of nonelectrolytes does not conduct electricity because no ions are formed.

110. T, T
The kinetic molecular theory predicts that heavier gases will diffuse more slowly that lighter ones. This was proven in 1832 by Thomas Graham. Because of this theory, SO_3 (MW 80) would diffuse more slowly than CO_2 (MW 44). SO_3 has a trigonal planar structure, with a bond angle of 120°. CO_2 is linear, and thus has a bond angle of 180°. This, however, has nothing to do with the rate of diffusion.

111. F, F
Hydrogen bonding is strongest between hydrogen and highly electronegative atoms such as F, Cl, N, O. There is no H bonding between hydrogen and noble gases.

112. T, F
An electron in the 3s subshell may occasionally enter an excited state and jump into the 3p subshell. The Heinsenberg Uncertainty Principle states that it is impossible to simultaneously determine with perfect accuracy both the momentum and position of a particle.

113. F, F
ΔG, the Gibbs free energy, is the energy of a system available to do work. The change in Gibbs free energy can be determined by the equation $\Delta G = \Delta H - T\Delta S$. A negative ΔG indicates a spontaneous reaction, while a positive ΔG indicates a nonspontaneous reaction. Entropy never decreases in an isolated system.

114. T, F
$P_1 V_1 = P_2 V_2$. An increase in pressure leads to a decrease in volume because they are inversely proportional.

115. T, F
An amphoteric compound can act as either an acid or a base because it can react with either H^+ or OH^-, depending on the nature of the reactants.

116. T, T, CE

An indicator is a chemical substance used in low concentration during a titration reaction. It will change color over a certain pH range. The color change, which occurs as the indicator undergoes a dissociation reaction, is used to identify the end point of a titration reaction.

24. B

To answer this question, you need to write the Lewis dot diagrams and then use any one of several formulas to find the formal charge. One such formula is formal charge = valence electrons – [number of bonds + number of non bonding electrons]. The Lewis dot structure of HNO_3 shows that nitrogen is double bonded to one oxygen, single bonded to another oxygen, and single bonded to another oxygen that is single bonded to a hydrogen—there are no lone electron pairs on the nitrogen. Using the formula, formal charge is equal to the valence electrons of nitrogen (5) minus the sum of the number of bonds and the number of non bonding electrons, which in this case is 4. So, the formal charge on the nitrogen is 5 minus 4, or 1.

25. D

In a solution, HCl contributes 1 mol of H^+ ions per mol of HCl. NaOH also contributes 1 mol of OH^- ions per mol of NaOH. Therefore, to neutralize the 80mL of a 0.5M NaOH solution, you must have the same number of moles of HCl. The number of moles of NaOH you have is:

$$(0.80L)(0.5mol/L) = 0.04mol \text{ of NaOH}$$

and therefore 0.04 mol of OH^- ions.

To calculate the volume of a 1M solution of HCl you will need to get 0.04 mol, perform the following calculation:

$$(1mol/L)(X \text{ L}) = 0.04mol$$

$$X \text{ L} = 0.04 \text{ L}$$

$$X \text{ mL} = 40 \text{ mL}$$

26. D

An electrolyte is a substance that ionizes to yield an electrically conducting solution. A strong electrolyte is one that ionizes completely or nearly completely, and a weak electrolyte doesn't ionize very much at all. Examples of strong electrolytes are NaCl, KCl, HCl, HBr, and HI. Examples of weak electrolytes are water, HF, acetic acid, benzoic acid, and ammonia.

27. E

If an element loses an electron, it will have more protons than electrons and with this stronger positive charge can pull the electrons in closer. Therefore, its ionic radius would be less than its atomic radius. Thus, Na, K, Mg, and Ca will all lose electrons and become smaller. Cl, on the other hand, will gain an electron; its positive nucleus cannot hold onto that extra electron as tightly, and its ionic radius is larger than its atomic radius.

28. A

A crystal is a solid whose atom, ions, or molecules are arranged in a regular 3D lattice structure. The basic repeating structure is known as the unit cell.

29. C

The oxidation state of nitrogen in N_2 is 0. In N_2O it is +1, in NH_3 it is –3, in NO_2 it is +4, and in NO_3 it is +6. Oxygen is typically –2 and H is usually +1. Therefore, NH_3 is the correct answer.

30. C

In order to answer this question correctly, you need to know the first law of thermodynamics: The change in internal energy, ΔE, equals the amount of heat transferred, q, minus the amount of work transferred, w. If heat is added to the system, q is positive; if it is transferred to the surrounding, it is negative. If work is done by the system, w is positive; if work is done on the system, w is negative. In this example, 500 joules of heat is added to the system, so q is equal to 500 joules. The system does 75 joules of work on the surrounding, so w is equal to

75 joules. Plugging these into our equation, the change in internal is equal to 500 minus 75, or 425 joules.

31. A

NaOH and H_2SO_4 (a strong base and a strong acid) both dissociate completely in H_2O into their ionic components Na^+, OH^-, $2H^+$, and SO_4^{2-}, which then form a salt and water. The driving force of this double displacement reaction is the formation of water by the OH^- an H^+ ions.

32. D

To answer this question, you simply need to know the definition of ground state and excited state. The ground state of an atom is the state in which all the electrons in the atom are in their lowest energy state. If any electron has absorbed energy and been promoted to a higher energy orbital in the atom without actually leaving the atom, the atom is said to be in an excited state. An atom can have any number of excited states depending on how many electrons have been promoted and what orbitals they end up in. Usually, excited states are unstable and the atom will release energy and return to the ground state. So which of these choices explains the difference between these two definitions? Choice D, when electrons gain energy and change which atomic orbital they're in, they change their quantum numbers which in turn changes the electronic configuration around the nucleus.

33. D

This question deals with solubility constant and the common ion effect. This first step to solving this problem is to express the solubility product constant, which is the ion product of the saturated solution, as the product of the concentration of Mg^{2+} ion and the concentration of the hydroxide ion squared. Second, you must determine the minimum concentration of hydroxide necessary to precipitate the $Mg(OH)_2$. From the equation given, you can solve for the concentration of hydroxide, which is the square root of the K_{sp} divided by the concentration

of Mg^{2+}. Substituting in the values provided in the question, you should find that the concentration is equal to 10^{-3} mol/L. That means that the pOH of the solution will be 3; therefore, the pH must be 11.

34. C

The random motion of a gas holds the most translational kinetic energy.

35. D

This question asks you which of the choices will increase the rate at which ice melts in a closed container. If you lower the temperature, you are tipping the equilibrium toward the ice, so B is wrong. To deal with pressure changes, we must apply Le Chatelier's principle, which says that a system in equilibrium that is subject to stress will shift its equilibrium so as to relieve the stress. If the pressure is lowered, as in choice C, the system will counteract the change in pressure by shifting its equilibrium toward the phase that is less dense. In the case of water, that is the ice. Remember that water has a strange property in that the solid form, ice, at zero degrees Celsius is less dense than the liquid phase, water, at that temperature. Since the reduction of pressure drives the systems to produce ice, choice C is incorrect. However, choice D, an increase in pressure, will have the opposite effect and the water will be produced preferentially, meaning that the ice is melting faster. Choices A and E both favor ice formation.

36. D

The most polar bond is the one that has the greatest difference in the electronegativities of the two elements. Of the choices, H–F is the most polar since hydrogen and fluorine are the farthest apart in the periodic table.

37. E

This question asks you to determine the rate law of the reaction for the formation of ammonia, NH_3. A rate law is an equation that gives the relationship between the rate of the reaction and the concentration of the reactants, each raised to an appropriate power that depends on the exact reaction. For example, the rate of this reaction is equal to $k[N_2]x[H_2]y$, where k is the rate constant and x and y are real numbers. K, x, and y can only be determined experimentally by systematically varying the initial concentration of one reactant. Since there is no experimental data, there is no way for us to know what these values are.

38. E

$10N\ H_2SO_4$ is equal to $5M\ H_2SO_4$ as each molecule has $2H^+$ ions associated with it. Use the equation $M_1V_1 = M_2V_2$. $(100)(5) = M_2(800)$. $M_2 = 5/8$.

39. C

This question asks you which molecule contains both ionic and covalent bonds. Choice C is ammonium sulfate. The bonds between nitrogen and hydrogen in the ammonium ion and between sulfur and oxygen in the sulfate ion are all covalent. However, the bond between the ammonium ion and the sulfate ion is an ionic bond.

40. D

The law of Charles states that $V_1/T_1 = V_2/T_2$, and you must convert Celsius to Kelvin. Rearrange your equation to $V_1/V_2 = T_1/T_2$. $T_1 = 303$ and $T_2 = 293$, and the ratio is $303/293$.

41. B

This question tests your knowledge of Le Châtelier's principle. What will happen to this reaction if more chlorine is added? Adding more chlorine puts a stress on the system, and the system alleviates that stress by using up that added chlorine: The equilibrium shifts to the left.

42. A

This is a redox reaction. We can tell this because the oxidation states of the species change during the reaction. The oxidation state of bromine in the product, molecular bromine, is 0. The oxidation state of bromine in the bromide reactant is –1, and the oxidation state of the bromine in the bromate is +5. The fact that the bromine of one reactant reduces the bromine of the other should not bother you, since it is immaterial. The important thing in redox reactions is the number of electrons. Since bromide is acting as a reducing agent, going form the –1 oxidation state to the 0 oxidation state, it transfers one electron per bromide. Bromate, however, is going from the +5 oxidation state to the 0 oxidation state, requiring 5 electrons. Bromate gets these electrons from bromide, requiring 5 of them. The ratio of bromate to bromide is therefore 1 to 5.

43. C

To figure out how many moles of Na_2CO_3 are in a 120mL of a 1.5M solution, perform the following calculation:

$$(1.5\text{mol/L})(0.12\text{L}) = 0.18\text{mol}$$

To figure out how many grams are in 0.18mol, you need to multiply the number of moles by the formula weight:

$$(106\text{g/mol})(0.18\text{mol}) = 19 \text{ grams}$$

44. D

Thomas Graham stated in 1832 that the rates for two gases diffusing is inverserly proportional to the square root of the molar mass. Therefore for $H_2:O_2 = [32/2]^{1/2}$. Simplify this equation and you have a ratio of diffusion rates of 4:1 for $H_2:O_2$.

45. D

Questions of nuclear chemistry can, as in this case, often be translated into basic arithmetic problems. Based on the law of conservation of mass, the sum of the mass numbers (superscripted) must be the same on each side of the arrow. Similarly, conservation of charge mandates that the sum of the nuclear charges (subscripted) be the same on each side of the arrow. We can thus translate the nuclear question posed into two elementary arithmetic questions:

$$27 + 4 = 30 + ? \text{ and } 13 + 2 = 15 + ?$$

Solving these two problems results in a mass number of 1 and a nuclear charge of 0; this set of values corresponds to the neutron $_0^1 n$. As for the wrong choices, in choice A, the beta particle is a nuclear electron and thus has mass and charge numbers of 0 and –1 respectively. The positron, choice B, has the mass of an electron but the opposite charge; the numbers are thus 0 and +1 respectively. Alpha particles are helium nuclei; choice C corresponds to a mass number of 4 and a nuclear charge of +2. Choice E, gamma ray, is short-wavelength electromagnetic radiation, i.e., light; as such, it has no mass and no charge.

46. D

Cr metal in its elemental state has an oxidation number of 0. In $K_2Cr_2O_7$, O has an oxidation number of –2 and K has an oxidation number of +1.

O	(–2)(7)	= –14
K	(+1)(2)	= 2
Total		–12

Cr must cancel out the –12, so the two Cr molecules must have a charge of +12. 12/2 = 6 for each Cr molecule.

47. A

When a Ca molecules loses an electron to become Ca^+, it goes from 20 to 19 electrons. K in its elemental state has 19 electrons.

48. C

This question requires you to know something about the periodicity of the elements. Basically, all you are asked is why all the halogens behave so similarly in reactions. You know from studying the periodic table that there must be something that repeats or the table would not be periodic. What is it about the elements in a column that are the same? The number of valence electrons. Since the identity of the valence electrons is the same in each column, that is, columns in the s block have valence s electrons, columns in the p block have valence s and p electrons, and columns in the d block have valence s and d electrons, the only thing that can make them act similarly is the number of electrons in these shells. Furthermore, the number of electrons in the valence shell affects the reactivity and stability of the shell. The number of electrons affects the ionization energy and electron affinity, whether the atom will form cations or anions, and even the number of bonds the atom can participate in.

49. B

The atomic weight of an element depends on the number of protons and neutrons in its nucleus; those numbers are always unique to a particular element, regardless of the number of electrons in the species. The ionization potential depends chiefly on the radius of the parent atoms and the effective charge of the nucleus. Since K^+ and Cl^- have different numbers of protons by definition, they will have different effective charges, different atomic radii, and therefore different ionization potential. Choice B, however, says that the two ions have the same electronic configuration. Chlorine belongs to the third period and potassium belongs to the fourth, so in their unionized forms, chlorine's third shell contains seven electrons and potassium's fourth shell contains one electron. If chlorine gains one electron and potassium loses one electron, both will have eight electrons in the third shell, which becomes the valence shell. So, the potassium and chlorine ions described in the question both contain the same number of electrons and the same number of occupied orbitals and thus share the same electronic configuration.

50. E

Cl^- has the largest ionic radius. When an atom gains an electron, positive protons in the nucleus cannot hold on to the electrons as well and therefore the electron shells are able to spread farther out.

51. D

The balanced equation looks like this:

$$8H^+ + MnO_4^- + 5Fe^{2+} \rightarrow Mn^{2+} + 5Fe^{3+} + 4H_2O.$$

One Mn^{2+} and five Fe^{3+} yield a total charge of +17.

52. D

The balanced equation looks like this:

$$1Ag(NH_3)_2^+ \rightarrow 1Ag^+ + 2NH_3$$
$$1 + 1 + 2 = 4$$

53. C

$$1Ag(NH_3)_2^+/2NH_3 = X/11NH_3$$
$$X = 11/2$$

5.5 moles

54. C

$$Ag = 108$$
$$N = (14)2 = 28$$
$$H = (1)6 = 6$$

Total Molecular Weight $108 + 28 + 6 = 142$

Percent Ag: $(108/142) \times 100\% = 76.1\%$

55. C

Balanced equation: $C_3H_8 + 5O_2 \rightarrow 3CO_2 + 4H_2O$

$$160g\ O_2/32g/mol = 5\ moles\ O_2$$

$$88g\ C_3H_8/44g/mole = 2\ moles\ C_3H_8$$

Therefore, oxygen is the limiting reagent; 5 moles of O_2 and one mole of C_3H_8 will form 3 moles of CO_2.

$$3\ moles \times 44\ g/mole = 132\ g\ CO_2$$

56. A

$P_1V_1 = P_2V_2$ is the equation you will use.

Oxygen: $(200)(200) = P_2(500)$
$$P_2 = 80mmHg, the\ new\ pressure$$

Neon: $(300)(100) = P_2(500)$
$$P_2 = 60mmHg$$

57. D

60 is the mass number (the number of protons + the number of neutrons). 27 is the atomic number (the number of protons). In a beta-decay reaction, the atomic number increases by one as a neutron turns into a proton and an electron that is ejected.

58. C

Ionic solutions will conduct electricity. The NaCl will dissociate to become Na^+ and Cl^-. The same occurs when you melt the NaCl. Solid NaCl and pure water are not ionic solutions.

59. B

NaCl is an ionic compound with ionic bonds because it is formed when Na donates an electron to Cl. The bond is formed by the electrostatic interaction between the positive and negative ions.

60. B

Glucose molecules carry no electrical charge, so there are no electrical charges in the solution to provide conduction.

61. C

This roman numeral question asks which of the given statements describe a galvanic cell (remember more than one of the answers may be correct). Galvanic cells are capable of spontaneous reactions. In all electrochemical cells, oxidation occurs at the anode and reduction occurs at the cathode. In addition, the anode in a galvanic cell is negative, meaning that it is a source of electrons. Since a species loses electrons when it is oxidized, this should make sense. There is a trick to remembering these facts. In a galvanic cell, oxidation occurs at the anode, which is negative. Alphabetically, anode comes before cathode, oxidation before reduction, and negative before positive. However, this little trick only works for galvanic cells.

62. E

34 g of NH_3 at 17 g/mol equals 2 mol combusted. Two moles are combusted at 81 kcal/mol. This equation would be (2mol)(81kcal/mol) = 162 kcal.

63. D

The s subshells only contain 2 electrons. Their principal, azimuthal, and magnetic quantum numbers are identical but due to Pauli's Exclusion principle their spin quantum numbers must be +1/2 and −1/2. Therefore, they have opposite spins.

64. C

A buffer is a solution made from a mixture of a weak acid and a salt containing its anion. Buffers resist pH change due to the addition of acid or base. Thus, to make a buffer solution with a weak monoprotic acid, the addition of the corresponding salt is required. This salt must contain the anion, or conjugate base, of the acid.

65. B

If the temperature of an exothermic reaction is increased, the reaction shifts to the left. If the temperature of an endothermic reaction is increased, it shifts to the right. The reaction in question has a negative enthalpy of reaction, meaning that it is an exothermic reaction. If the temperature of this reaction is increased it will shift to the left.

66. E

Boyle's law states that at constant temperatures the volume of a gas is inversely proportional to its pressure. Statement II is therefore correct. Statement III is also correct as density, defined as mass divided by volume, will increase proportionally to pressure since the volume is decreasing.

67. A

Molality equals the number of moles of solute/kg of solvent. 200g/[24 g/mol Mg + 71 g/mole for 2 Cl] is the number of moles of solute. This is also the value of molality since it is added to 1 kg of solvent.

68. C

$$100g\ H_2 = 100g/2g/mole = 50\ mol\ H_2$$
$$100g\ O_2 = 100g/32g/mole = 100/32\ mol$$
$$2H_2 + O_2 \rightarrow 2H_2O$$

Therefore, oxygen is the limiting reagent since it is the compound in the least amount. According to the balanced equation, for every mole of oxygen you will get two moles of water. Therefore, to discover how many moles of water would be produced, multiply the number of moles of oxygen you begin with by two, to account for the fact that two moles of water are produced for every mole of oxygen. Thus, the correct equation would be (2mol H_2O_2/1mol of O_2)(100/32 mol of O_2).

69. D

Perhaps the easiest way to solve such a problem is to imagine a particular sample mass of the compound. For the sake of convenience, choose 100 grams, though any mass will let you arrive at the correct answer. Because the oxide of arsenic contains only arsenic and oxygen, a 100 gram sample would contain 65.2 grams of arsenic and the remainder, 34.8 grams, must be oxygen. To find the ratio between these two elements in the compound, divide the mass of arsenic by the atomic weight of arsenic, 74.7 grams/mole, and the mass of oxygen by the atomic weight of oxygen, 16 grams/mole. This will give the mole ratio between arsenic and oxygen in the compound. To convert to a more easily useful ratio, divide both by the lowest number of the two, in this case arsenic. This gives us a ratio of 1 mole of arsenic to 2.5 moles oxygen, which is better stated by doubling both numbers and getting 2 moles of arsenic to 5 moles of oxygen. This would correspond to the formula in choice D, As_2O_5.

CHAPTER 3

ATOMIC STRUCTURE

The atom is the basic building block of matter, representing the smallest unit of a chemical element. An atom in turn is composed of subatomic particles called protons, neutrons, and electrons. In 1911, Ernest Rutherford provided experimental evidence that an atom has a dense, positively charged nucleus that accounts for only a small portion of the volume of the atom. The protons and neutrons in an atom form the nucleus, the core of the atom. The electrons exist outside the nucleus in characteristic regions of space called orbitals. All atoms of an element show similar chemical properties and cannot be further broken down by chemical means.

Subatomic Particles

There are three kinds of particles found in a typical atom: protons and neutrons, which together make up the nucleus, and electrons, which are found in specific regions of space (known as orbitals) around the nucleus.

Protons

Protons carry a single positive charge and have a mass of approximately one atomic mass unit or amu (see table below). The atomic number (Z) of an element is equal to the number of protons found in an atom of that element. All atoms of a given element have the same atomic number; in other words, the number of protons an atom has defines what kind of element it is.

The atomic number of an element can be found in the Periodic Table (see chapter 4) as an integer above the symbol for the element.

Neutrons

Neutrons carry no charge and have a mass only slightly larger than that of protons. The total number of neutrons and protons in an atom, known as the mass number, determines its mass.

The convention $^{A}_{Z}X$ is used to show both the atomic number and mass number of an X atom, where Z is the atomic number and A is the mass number.

Even though the number of protons must be the same for all atoms of an element, the number of neutrons, and hence the mass number, can be different. Atoms of the same element with different masses are known as isotopes of one another. Isotopes are referred to either by the convention described above or, more commonly, by the name of the element followed by the mass number. For example, carbon-12 is a carbon atom with 6 protons and 6 neutrons, while carbon-14 is a carbon atom with 6 protons and 8 neutrons. Since isotopes have the same number of protons and electrons, they generally exhibit the same chemical properties. Chapter 15 describes nuclear chemistry in greater detail.

Electrons

Electrons carry a charge equal in magnitude but opposite in sign to that of protons. An electron has a very small mass, approximately 1/1837 the mass of a proton or neutron, which is negligible for most purposes. The electrons farthest from the nucleus are known as valence electrons. The farther the valence electrons are from the nucleus, the weaker the attractive force of the positively charged nucleus and the more likely the valence electrons are to be influenced by other atoms. Generally, the valence electrons and their activity determine the reactivity of an atom. In a neutral atom, the number of electrons is equal to the number of protons. A positive or negative charge on an atom is due to a loss or gain of electrons; the result is called an ion. A positively charged ion (one that has lost electrons) is known as a cation; a negatively charged ion (one that has gained electrons) is known as an anion.

Some basic features of the three subatomic particles are summarized in the table below.

Table 1.1

Subatomic Particle	Symbol	Relative Mass	Charge	Location
Proton	$^{1}_{1}H$	1	+1	Nucleus
Neutron	$^{1}_{0}n$	1	0	Nucleus
Electron	e^{-}	0	−1	Electron Orbitals

Don't Mix These Up on Test Day

- Atomic number
  = number of protons

- Mass number
  = number of protons and neutrons

Example: Determine the number of protons, neutrons, and electrons in a Nickel-58 atom and in a Nickel-60^{2+} cation.

Solution: ^{58}Ni has an atomic number of 28 according to the Periodic Table and a mass number of 58. Therefore, ^{58}Ni will have 28 protons, 28 electrons (since it is a neutral atom), and 58 – 28, or 30, neutrons.

In the ^{60}Ni^{2+} species, the number of protons is the same as in the neutral ^{58}Ni atom. However, ^{60}Ni^{2+} has lost two electrons and thus will only have 26 electrons: this is what gives it the +2 charge. Also, the mass number is two units higher than for the ^{58}Ni atom, and this difference in mass must be due to two extra neutrons; thus it has a total of 32 neutrons.

Atomic Weights and Isotopes

To report the mass of something, one generally gives a number together with a unit like pounds, kilograms (kg), grams (g), etcetera. Because the mass of an atom is so small, however, these units are not very convenient, and new ways have been devised to describe how much an atom weighs. A unit that can be used to report the mass of an atom is the atomic mass unit (amu). One amu is approximately the same as 1.66×10^{-24} g. How is this particular value chosen? Why not, for example, have 1 amu be equal to a nice round number like 1.00×10^{-24} g instead? The answer is that it is chosen so that a carbon-12 atom, with 6 protons and 6 neutrons, will have a mass of 12 amu. In other words, the amu is defined as one-twelfth the mass of the carbon-12 atom. It does not convert nicely to grams because the mass of a carbon-12 atom in grams is not a nice round number. In addition, since the mass of an electron is negligible, all the mass of the carbon-12 atom is considered to come from protons and neutrons.

Since the mass of a proton is about the same as that of a neutron, and there are 6 of each in the carbon-12 atom, protons and neutrons are considered to have a mass of $\frac{1}{12} \times 12$ amu = 1 amu each.

While it is necessary to have a way of describing the weight of an individual atom, in real life one generally works with a huge number of them at a time. The atomic weight is the mass in grams of one mole (mol) of atoms. Just like a pair corresponds to two, and a dozen corresponds to twelve, a mole corresponds to about 6.022×10^{23}. The atomic weight of an element, expressed in terms of g/mol, therefore, is the mass in grams of 6.022×10^{23} atoms of that element. This number, roughly 6.022×10^{23}, to which a mole corresponds, is known as Avogadro's number. Why this particular value and not something like 1.0×10^{24}, for example? Once again, the answer lies

Basic Concept

1 mole = 6.022 × 10²³
        = Avogadro's number

in the carbon-12 atom: a mole of carbon-12 atoms weigh exactly 12 g. In other words, a mole is defined as the number of atoms in 12 g of carbon-12. A mole of atoms of an element heavier than carbon-12 (such as oxygen) would have an atomic weight higher than 12 g/mol, while a mole of atoms of an element lighter than carbon-12 (such as helium) would have an atomic weight less than 12 g/mol. Six g of carbon-12 would mean 3.011 × 10²³ cabon-12 atoms, etcetera.

As we have seen, Avogadro's number serves as a conversion factor between one of something and a mole of something. Since 12 amu is the mass of 1 carbon-12 atom while 12 g is the mass of 1 mole of carbon-12 atoms, Avogadro's number also helps to convert between the mass units. Specifically:

$$12 \text{ amu} \times (6.022 \times 10^{23}) = 12 \text{ g}$$

$$1 \text{ amu} = \frac{1 \text{g}}{6.022 \times 10^{23}} = 1.66 \times 10^{-24} \text{ g}$$

which is the conversion factor we gave above. We can now see how this is derived from (or related to) the concept of the mole.

The atomic weight of an element is also found in the Periodic Table, as the number appearing below the symbol for the element. Notice, however, that these numbers are not whole numbers, which is odd considering that a proton and a neutron each have a mass of 1 amu and an atom can only have a whole number of these. Furthermore, even carbon, the element with which we have set the standards, does not have a mass of 12.000 exactly. This is due to the presence of isotopes, as mentioned above. The masses listed in the Periodic Table are weighted averages that account for the relative abundance of various isotopes. The word *weighted* is important: It is not simply the average of the masses of individual isotopes, but takes into account how frequently one encounters that isotope in a common sample of the element. There are, for example, 3 isotopes of hydrogen, with 0, 1, and 2 neutrons respectively. Together with the one proton that makes it hydrogen in the first place, the mass numbers for these isotopes are 1, 2, and 3. The atomic weight of hydrogen, however, is not simply 2 (the average of 1, 2 and 3) but about 1.008, that is, much closer to 1. This is because the isotope with no neutrons is so much more abundant that we count it much more heavily in calculating the average. The following example provides a more concrete illustration of the idea.

Example: Element Q consists of three different isotopes, A, B, and C. Isotope A has an atomic mass of 40 amu and accounts for 60% of naturally occurring Q. The atomic mass of isotope B is 44 amu and accounts for 25% of Q. Finally, isotope C has an atomic mass of 41 amu and a natural abundance of 15%. What is the atomic weight of element Q?

Solution: 0.60(40 amu) + 0.25(44 amu) + 0.15(41 amu)
             = 24.00 amu + 11.00 amu + 6.15 amu = 41.15 amu

The atomic weight of element Q is 41.15 g/mol.

(Incidentally, if you have studied physics, you may be aware of the distinction between mass and weight. As you can see, chemists are a bit sloppier on this matter.)

Bohr's Model of the Hydrogen Atom

In his model of the structure of the hydrogen atom, Bohr postulated that an electron can exist only in certain fixed energy states; the energy of an electron is "quantized." According to this model, electrons revolve around the nucleus in orbits. The energy of the electron is related to the radius of its orbit: The smaller the radius, the lower the energy state of the electron. The smallest orbit (radius) an electron can have corresponds to the ground state of the hydrogen electron. At the ground state level, the electron is in its lowest energy state. The fact that only certain energy values are allowed means that only certain orbit sizes are allowed.

The Bohr model is used to explain the atomic emission spectrum and atomic absorption spectrum of hydrogen. Since the energy of electrons can only take on certain values, the energy an atom can emit or absorb is likewise restrained to values that correspond to differences between these levels. When a hydrogen atom absorbs energy in the form of radiation, for example, its electron moves to a higher energy level. When such a process occurs, a peak shows up in the absorption spectrum, signifying that radiation of that particular energy is being absorbed by the atom. The crucial thing to realize is that there will only be a certain number of sharp peaks, corresponding to energy values that match up with the difference between energy levels. The principle behind the emission spectrum is the same: The atom gives off energy as an electron goes from a higher to a lower energy level, and this will show up as distinct peaks in a spectrum corresponding to transition between different levels. The Bohr model successfully accounted for the precise positionings of these peaks (the precise values of the energy that can be emitted).

Quantum Mechanical Model of Atoms

While the concepts put forth by Bohr offered a reasonable explanation for the structure of the hydrogen atom and ions containing only one electron (such as He^+ and Li^{2+}), they did not explain the structures of atoms containing more than one electron. This is because Bohr's model does not take into consideration the repulsion between multiple electrons surrounding one nucleus. Modern quantum mechanics has led to a more rigorous and generalized study of the electronic structure of atoms. The most important difference between the Bohr model and modern quantum mechanical models is that Bohr's assumption that electrons follow a circular orbit at a fixed distance from the nucleus is no longer considered valid. Rather, electrons are described as being in a state of rapid motion within regions of space around the nucleus, called orbitals. An orbital is a representation of the probability of finding an electron within a given region. In the current quantum mechanical description of electrons, pinpointing the exact location of an electron at any given point in time is impossible. This idea is best described by the Heisenberg uncertainty principle, which states that it is impossible to determine, with perfect accuracy, the momentum and the position of an electron simultaneously. This means that if the momentum of the electron is being measured accurately, its position cannot be pinpointed, and vice versa.

Quantum Numbers

Modern atomic theory states that any electron in an atom can be completely described by four quantum numbers n, l, m_l, and m_s. Furthermore, according to the Pauli exclusion principle, no two electrons in a given atom can possess the same set of four quantum numbers. The position and energy of an electron described by its quantum numbers is known as its energy state. The value of n limits the values of l, which in turn limits the values of m_l. The values of the quantum numbers qualitatively give information about the orbitals: n about the size and the distance from the nucleus, l about the shape, and m_l about the orientation of the orbital. All four quantum numbers are discussed below.

Principal Quantum Number

The first quantum number is commonly known as the principal quantum number and is denoted by the letter n. It is used in Bohr's model and can theoretically take on any positive integer value. The larger the integer value of n, the higher the energy level and the farther away, on average, you will find the electron from the nucleus. The maximum number of electrons in energy level n (electron shell n) is $2n^2$.

Azimuthal Quantum Number

The second quantum number is called the azimuthal quantum number and is designated by the letter l. (It is also known as the angular momentum quantum number.) This quantum number refers to the subshells or sublevels that occur within each principal energy level. For any given n, the value of l can be any integer in the range of 0 to $(n - 1)$. For example, the shell of $n = 3$ can have subshells with $l = 0$, $l = 1$, and $l = 2$, whereas the shell

Basic Concept

4 quantum numbers: n, l, m_l, m_s

Basic Concepts

$l = 0$: s subshell
$l = 1$: p subshell
$l = 2$: d subshell
$l = 3$: f subshell

of $n = 1$ can only have the subshell $l = 0$. The four subshells corresponding to $l = 0, 1, 2$, and 3 are known as the s, p, d, and f subshells, respectively. Based on the restriction on l values just discussed, we can see that every shell has an s subshell, while only shells with $n > 1$ have p subshells, etcetera. For atoms with more than one electron, the greater the value of l, the greater the energy of the subshell. However, the energies of subshells from different principal energy levels may overlap. For example, the $4s$ subshell may have a lower energy than the $3d$ subshell.

Magnetic Quantum Number

The third quantum number is the magnetic quantum number and is designated m_l. An orbital is a specific region within a subshell that may contain no more than two electrons. The magnetic quantum number specifies the particular orbital within a subshell where an electron is highly likely to be found at a given point in time. The possible values of m_l are all integers from l to $-l$, including 0. Therefore, the s subshell, where there is one possible value of m_l (0), will contain 1 orbital; likewise, the p subshell will contain 3 orbitals, the d subshell will contain 5 orbitals, and the f subshell will contain 7 orbitals. The shape and energy of each orbital are dependent upon the subshell in which the orbital is found. For example, s orbitals ($l = 0$, $m_l = 0$), are all spherical in shape. Those with larger principal quantum numbers have a larger radius, implying a larger average distance from the nucleus. A p subshell has three possible m_l values ($-1, 0, +1$). The three dumbbell-shaped orbitals are oriented in space around the nucleus along the x, y, and z axes, and are often referred to as px, py, and pz.

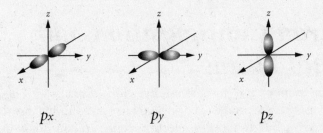

px py pz

The magnetic quantum number can also play a part in determining the shape of an orbital: Orbitals in the d and f subshells, for example, may have different shapes (in addition to just different orientations) depending on the value of ml.

Spin Quantum Number

The fourth quantum number is also called the spin quantum number and is denoted by m_s. Regardless of the shell, subshell, or orbital, any electron can have only one of two values for the spin quantum number.

Basic Concept

The spin quantum number is either $+\frac{1}{2}$ or $-\frac{1}{2}$, regardless of values of other quantum numbers.

The two spin orientations are designated $+\frac{1}{2}$ and $-\frac{1}{2}$.

Whenever two electrons are in the same orbital, they must have opposite spins. This is a consequence of the Pauli exclusion principle; if that were not the case, they would have the identical four quantum numbers. Electrons in different orbitals with the same m_s values are said to have parallel spins. Electrons with opposite spins in the same orbital are often referred to as paired.

For the shell corresponding to $n = 3$, a complete enumeration of the possible sets of quantum numbers follows. Note that the total number of electrons in this shell is $18 = 2\,(3)^2$, which conforms to the formula given above.

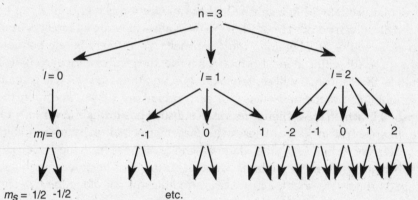

In what order are subshells filled?

(A) lowest to highest energy
(B) highest to lowest energy
(C) randomly

Answer: (A)

Electron Configuration and Orbital Filling

For a given atom or ion, the pattern by which orbitals are filled and the number of electrons within each principal level and subshell are designated by an electron configuration. In electron configuration notation, the first number denotes the principal energy level, the latter designates the subshell, and the superscript gives the number of electrons in that subshell. For example, $2p^4$ indicates that there are four electrons in the second (p) subshell of the second principal energy level.

When writing the electron configuration of an atom, remember the order in which subshells are filled. Subshells are filled from lowest to highest energy, and each subshell will fill completely before electrons begin to enter the next one. The ($n + l$) rule is used to rank subshells by increasing energy. This rule states that the lower the values of the first and second quantum numbers, the lower the energy of the subshell. If two subshells possess the same ($n + l$) value, the subshell with the lower n value has a lower energy and will fill first. The order in which the subshells fill is shown in the following chart, which is arranged so that it is easily remembered: One simply lists the subshells in order, starting each shell with a new line. The order of filling them is found by crossing them with diagonal arrows.

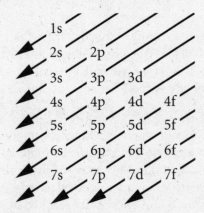

Example: Which will fill first, the 3d subshell or the 4s subshell?

Solution: For 3d, $n = 3$ and $l = 2$, so $(n + l) = 5$. For 4s, $n = 4$ and $l = 0$, so $(n + l) = 4$. Therefore, the 4s subshell has lower energy and will fill first. This can also be determined from the chart by examination.

To determine which subshells are occupied, you must know the number of electrons in the atom. In the case of uncharged atoms, the number of electrons is equal to the atomic number. If the atom is charged, the number of electrons is equal to the atomic number plus the extra electrons if the atom is negative, or the atomic number minus the missing electrons if the atom is positive.

In subshells that contain more than one orbital, such as the 2p subshell with its 3 orbitals, the orbitals will fill according to Hund's rule. Hund's rule states that within a given subshell, orbitals are half-filled so that they each have one electron, all with parallel spins, before any orbital is fully occupied with two electrons of opposite spins. In other words, electrons would tend to avoid pairing as much as possible.

Example: What are the written electron configurations for nitrogen (N) and iron (Fe) according to Hund's rule?

Solution: Nitrogen has an atomic number of 7, thus its electron configuration is $1s^2\, 2s^2\, 2p^3$. According to Hund's rule, the two s orbitals will fill completely, while the three p orbitals will each contain one electron, all with parallel spins.

$$\underset{1s^2}{\underline{\uparrow\downarrow}} \quad \underset{2s^2}{\underline{\uparrow\downarrow}} \quad \underset{2p^3}{\underline{\uparrow\ \ \uparrow\ \ \uparrow}}$$

Basic Concept

Hund's rule: Put one electron in each orbital in the same subshell first before pairing.

Iron has an atomic number of 26, and its $4s$ subshell fills before the $3d$. Using Hund's rule, the electron configuration will be:

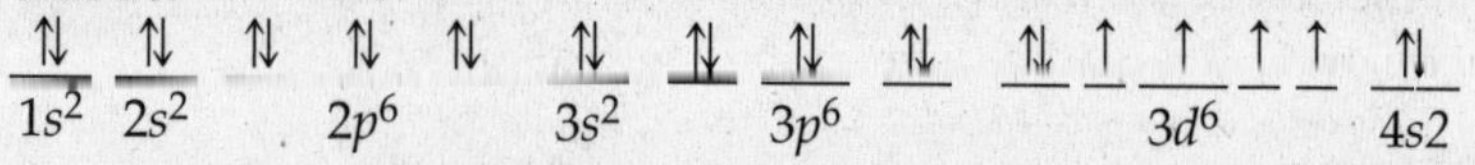

$$\underset{1s^2}{\uparrow\downarrow} \quad \underset{2s^2}{\uparrow\downarrow} \quad \underset{2p^6}{\uparrow\downarrow \ \uparrow\downarrow \ \uparrow\downarrow} \quad \underset{3s^2}{\uparrow\downarrow} \quad \underset{3p^6}{\uparrow\downarrow \ \uparrow\downarrow \ \uparrow\downarrow} \quad \underset{3d^6}{\uparrow\downarrow \ \uparrow \ \uparrow \ \uparrow \ \uparrow} \quad \underset{4s2}{\uparrow\downarrow}$$

Iron's electron configuration is written as $1s^2\ 2s^2\ 2p^6\ 3s^2\ 3p^6\ 3d^6\ 4s^2$. Subshells may be listed either in the order in which they fill (e.g., $4s$ before $3d$) or with subshells of the same principal quantum number grouped together, as shown here. Both methods are correct.

The presence of paired or unpaired electrons affects the chemical and magnetic properties of an atom or molecule. If the material has unpaired electrons, a magnetic field will align the spins of these electrons and weakly attract the atom. These materials are said to be paramagnetic. Materials that have no unpaired electrons and are slightly repelled by a magnetic field are said to be diamagnetic.

Atomic Structure
Review Problems

1. The Mg^{2+} ion has how many electrons?

    A. 12

    B. 10

    C. 14

    D. 24

2. It can be shown using mass spectrometry that the ratio of naturally occurring chlorine-35 to its isotope chlorine-37 is 3:1. Assuming that no other isotopes existed, what would be the atomic weight of chlorine?

3. The maximum number of electrons in a shell with the principal quantum number equal to 4 is

    A. 2

    B. 10

    C. 16

    D. 32

4. If the principal quantum number of a shell is equal to 2, what types of orbitals will be present?

    A. s

    B. s and p

    C. s, p, and d

    D. s, p, d, and f

5. The total number of electrons that could be held in a sublevel with azimuthal quantum number equal to 2 is

    A. 2

    B. 6

    C. 8

    D. 10

6. An element with an atomic number of 26 has how many electrons in the $3d$ orbital?

    A. 0

    B. 2

    C. 6

    D. 8

    E. 10

7. In going from $1s^2\, 2s^2\, 2p^6\, 3s^2\, 3p^6\, 4s^1$ to $1s^2\, 2s^2\, 2p^6\, 3s^2\, 3p^5\, 4s^2$, an electron would

A. absorb energy

B. emit energy

C. relax to the ground state

D. bind to another atom

E. undergo no change in energy

8. Which of the following orbitals has the lowest energy?

A. $2p$

B. $3s$

C. $3d$

D. $4s$

E. $3p$

9. Which of the following correctly represents an excited state of scandium?

A. $1s^2\, 2s^2\, 2p^6\, 3s^2\, 3p^6\, 3d^1\, 4s^2$

B. $1s^2\, 2s^3\, 2p^5\, 3s^2\, 3p^6\, 3d^1\, 4s^2$

C. $1s^2\, 2s^2\, 2p^6\, 3s^2\, 3p^6\, 3d^2\, 4s^1$

D. $1s^2\, 2s^2\, 2p^6\, 3s^2\, 3p^6\, 3d^2\, 4s^2$

E. $1s^2\, 2s^2\, 2p^6\, 3s^2\, 3p^6\, 3d^1\, 4s^0$

KAPLAN

Solutions to Review Problems

1. B

Magnesium has an atomic number of 12, meaning that a neutral atom has 12 protons and 12 electrons. However, the Mg^{2+} ion has a positive charge because it has lost 2 electrons. Therefore, the Mg^{2+} ion has 10 electrons.

2. 35.5

Mass spectrometry shows that 3 out of every 4 chlorine atoms are Cl-35 and 1 is Cl-37. Thus, 75% of all chlorine atoms are Cl-35 and 25% are Cl-37. Using this information and the atomic weights of the isotopes (Cl-35 = 35 g/mol, Cl-37 = 37 g/mol), the atomic weight of chlorine can be determined as follows:

$$(0.75)(35 \text{ g/mol}) + (0.25)(37 \text{ g/mol})$$

$$= (26.25 + 9.25) \text{ g/mol}$$

$$= 35.5 \text{ g/mol}$$

3. D

The maximum number of electrons within a principal energy level is given by the equation $2n^2$. Therefore, a shell with the principal quantum number of 4 will hold a maximum of 32 electrons.

4. B

When the principal quantum number is equal to 2, then the azimuthal quantum number will have values of $l = 0$ and 1. When $l = 0$, the subshell is an s subshell, and when $l = 1$, the subshell is a p subshell.

Therefore, the second principal energy level contains an s subshell and a p subshell.

5. D

The azimuthal quantum number 2 corresponds to the sublevel or subshell d, and the d sublevel, with 5 orbitals ($ml = -2, -1, 0, 1, 2$), is capable of holding a maximum of 5×2 or 10 electrons.

6. C

An element with an atomic number of 26 will have 6 electrons in its $3d$ subshell. This can be determined by writing the electron configuration for the element, $1s^2\, 2s^2\, 2p^6\, 3s^2\, 3p^6\, 3d^6\, 4s^2$. The number of electrons must equal 26; recall that the $4s$ subshell must be filled before the $3d$ because it has the lower energy. Thus, $3d$ will carry 6 electrons.

7. A

The difference between the first and second electron configurations is that in the second configuration one electron has moved from the $3p$ subshell to the $4s$ subshell. Although the $3p$ and $4s$ subshells have the same $(n + l)$ value, the $3p$ subshell fills first because it is slightly lower in energy. In order for an electron to move from the $3p$ subshell to the $4s$ subshell, it must absorb energy.

8.   A

In order to determine which subshell has the lowest energy, the $(n + l)$ rule must be used. The values of the first and second quantum numbers are added together, and the subshell with the lowest $(n + l)$ value has the lowest energy. The sums of the five choices are $(2 + 1) = 3$, $(3 + 0) = 3$, $(3 + 2) = 5$, $(4 + 0) = 4$, $(3 + 1) = 4$. Choices A and B have the same $(n + l)$ value, so the subshell with the lower principal quantum number has the lower energy. This is the $2p$ subshell, choice A.

9.   C

Scandium has 21 electrons. When it is in its excited state, one or more of the electrons will be present in a subshell with a higher energy than the one in which it is usually located. The number of electrons and ordering of subshells will not vary from the ground state electron configuration of scandium. Choice C has one of the $4s$ electrons present in the $3d$ orbital. This represents an excited state because energy is required to cause an electron to jump from $4s$ to $3d$. Note that choice B is not physically possible because the one cannot have three electrons in an orbital.

CHAPTER 4

THE PERIODIC TABLE

The Periodic Table arranges the elements in increasing atomic numbers. Its spatial layout is such that a lot of information about an element's properties can be deduced simply by examining its position. The vertical columns are called groups, while the horizontal rows are called periods. There are seven periods, representing the principal quantum numbers $n = 1$ to $n = 7$, and each period is filled more or less sequentially. The period an element is in tells us the highest shell that is occupied, or the highest principal quantum number. Elements in the same group (same column) have the same electronic configuration in their valence, or outermost shell. For example, both magnesium (Mg) and calcium (Ca) are in the second column; they both have 2 electrons in the outermost s subshell, the only difference being that the principal quantum number is different for Ca ($n = 4$) than for Mg ($n = 3$). Because it is these outermost electrons, or valence electrons, that are involved in chemical bonding, they determine the chemical reactivity and properties of the element. In short, elements in the same group will tend to have similar chemical reactivities.

Valence Electrons and the Periodic Table

The valence electrons of an atom are those electrons that are in its outer energy shell or that are available for bonding. The visual layout of the Periodic Table is convenient for determining the electron configuration of an atom (especially the valence electron configuration); this provides a quick alternative to the methods described in the previous chapter.

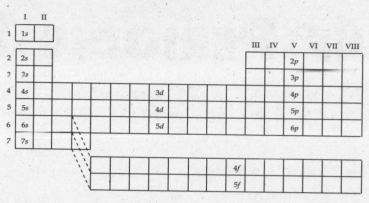

Elements in the leftmost column (Group 1 or IA) all have a single s electron in their valence shell; their electron configuration can therefore be represented as $[X]\ ns^1$, where $[X]$ designates the electron configuration of the noble gas element (see below) that immediately precedes it and is abbreviated this way because these electrons are core electrons that do not generally participate in chemical reactions and are hence uninteresting from a chemical perspective. Elements in the second column (Group 2 or IIA) have electronic configurations $[X]\ ns^2$; their valence electrons are the two electrons in the outermost s subshell.

The next block of elements (elements in the next 10 columns, not including the $4f$ lanthanide and $5f$ actinide series) are all known as transition elements and have electrons in the d subshell; just how many they have depends on exactly which column they are in. Elements in the third column (Group 3 or IIIA), for example, have configurations $[X]\ ns^2\ (n-1)d^1$. (Note that the principal quantum number for the d subshell is one less than that for the s subshell—remember, for example, how after filling the $3p$ subshell, one fills the lower-energy $4s$ orbital first before "going back" to fill the $3d$ subshell.) Their valence electrons are those in the outermost s subshell and in the d subshell of the next-to-outermost energy shell. For the inner transition elements, the valence electrons are those in the s subshell of the outermost energy shell, the d subshell of the next-to-outermost energy shell, and the f subshell of the energy shell two levels below the outermost shell.

The last six columns of the Periodic Table contain elements with s and p valence electrons.

Periodic Trends of the Elements

The properties of the elements exhibit certain trends, which can be explained in terms of the position of the element in the Periodic Table, or in terms of the electron configuration of the element. Elements in general seek to gain or lose valence electrons so as to achieve the stable octet formation possessed by the inert or noble gases of Group VIII (last column of the Periodic Table). Two other important general trends exist. First, as one goes from left to right across a period, electrons are added one at a time; the elec-

Basic Concepts

left → right: atomic radius↓
ionization energy↑
electronegativity↑
(except for noble gases)

top → bottom: atomic radius↑
ionization energy↓
electronegativity↓

trons of the outermost shell experience an increasing amount of nuclear attraction, becoming closer and more tightly bound to the nucleus. Second, as one goes down a given column, the outermost electrons become less tightly bound to the nucleus. This is because the number of filled principal energy levels (which shield the outermost electrons from attraction by the nucleus) increases downward within each group. These trends help explain elemental properties such as atomic radius, ionization potential, electron affinity, and electronegativity.

Atomic Radii

The atomic radius is an indication of the size of an atom. In general, the atomic radius decreases across a period from left to right and increases down a given group; the atoms with the largest atomic radii will therefore be found at the bottom of groups, and in Group I.

As one moves from left to right across a period, electrons are added one at a time to the outer energy shell. Electrons in the same shell cannot shield one another from the attractive pull of protons very efficiently. Therefore, since the number of protons is also increasing, producing a greater positive charge, the effective nuclear charge increases steadily across a period, meaning that the valence electrons feel a stronger and stronger attraction towards the nucleus. This causes the atomic radius to decrease.

As one moves down a group of the periodic table, the number of electrons and filled electron shells will increase, but the number of valence electrons will remain the same. Thus, the outermost electrons in a given group will feel the same amount of effective nuclear charge, but electrons will be found farther from the nucleus as the number of filled energy shells increases. Thus, the atomic radius will increase.

Ionization Energy

The ionization energy (IE), or ionization potential, is the energy required to completely remove an electron from an atom or ion. Removing an electron from an atom always requires an input of energy, since it is attracted to the positively charged nucleus. The closer and more tightly bound an electron is to the nucleus, the more difficult it will be to remove, and the higher the ionization energy will be. The first ionization energy is the energy required to remove one valence electron from the parent atom; the second ionization energy is the energy needed to remove a second valence electron from the ion with +1 charge to form the ion with +2 charge, and so on. Successive ionization energies grow increasingly large; that is, the second ionization energy is always greater than the first ionization energy. For example:

$$Mg\ (g) \rightarrow Mg^+(g) + e^- \quad \text{First Ionization Energy} = 7.646\ eV$$

$$Mg^+\ (g) \rightarrow Mg^{2+}(g) + e^- \quad \text{Second Ionization Energy} = 15.035\ eV$$

Ionization energy increases from left to right across a period as the atomic radius decreases. Moving down a group, the ionization energy decreases as the atomic radius increases. Group I elements have low ionization energies because the loss of an electron results in the formation of a stable octet.

Electron Affinity

Electron affinity is the energy that is released when an electron is added to a gaseous atom, and it represents the ease with which the atom can accept an electron. The stronger the attractive pull of the nucleus for electrons, the greater the electron affinity will be. A positive electron affinity value represents energy release when an electron is added to an atom.

Generalizations can be made about the electron affinities of particular groups in the Periodic Table. For example, the Group IIA elements, or alkaline earths, have low electron affinity values. These elements are relatively stable because their s subshell is filled: They do not particularly "care" to gain an extra electron, even though the process is still favorable. Group VIIA elements, or halogens, have high electron affinities because the addition of an electron to the atom results in a completely filled shell, which represents a stable electron configuration. Achieving the stable octet involves a release of energy, and the strong attraction of the nucleus for the electron leads to a high energy change. The Group VIII elements, or noble gases, have electron affinities on the order of zero since they already possess a stable octet: Gaining an extra electron is really not that favorable and would not result in the release of much energy.

A crude way of describing the difference between ionization energy and electron affinity is that the former tells us how attached the atom is to the electrons it already has, while the latter tells us how the atom feels about gaining another electron.

A Closer Look

Cesium (Cs), at the bottom left of the periodic table, has the largest atomic radius of any naturally occurring atom. It also has the lowest ionization energy, the lowest electron affinity, and the lowest electronegativity of all stable neutral atoms. (Francium is not a stable, naturally occurring element.)

Electronegativity

Electronegativity is a measure of the attraction an atom has for electrons in a chemical bond. The greater the electronegativity of an atom, the greater its attraction for bonding electrons. It is related to ionization energy and electron affinity: Elements with low ionization energies and low electron affinities will have low electronegativities because their nuclei do not attract electrons strongly, while elements with high ionization energies and high electron affinities will have high electronegativities because of the strong pull the nucleus has on electrons. Therefore, electronegativity increases from left to right across periods. In any group, the electronegativity decreases as the atomic number increases, as a result of the increased distance between the valence electrons and the nucleus, i.e., greater atomic radius.

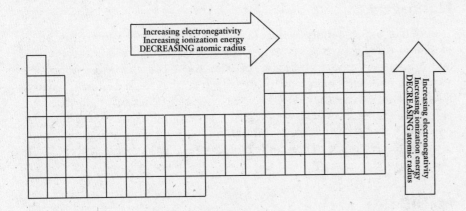

Types of Elements

The elements of the periodic table may be classified into three categories: metals, located on the left side and in the middle of the periodic table; non-metals, located on the right side of the table; and metalloids (semimetals), found along a diagonal line between the other two.

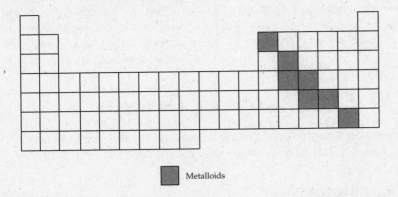

■ Metalloids

Metals

Metals are shiny solids at room temperature (except for mercury, which is a liquid), and generally have high melting points and densities. Metals have the characteristic ability to be deformed without breaking. The ability of a metal to be hammered into shapes is called **malleability** and the ability to be drawn into wires is called **ductility**. Many of the characteristic properties of metals, such as large atomic radius, low ionization energy, and low electronegativity, are due to the fact that the few electrons in the valence shell of a metal atom can easily be removed. Because the valence electrons can move freely, metals are good conductors of heat and electricity. Group IA and IIA represent the most reactive metals. The transition elements are metals which have partially filled d orbitals.

Nonmetals

Nonmetals are generally brittle in the solid state and show little or no metallic luster. They have high ionization energies and electronegativities, and are usually poor conductors of heat and electricity. Most nonmetals share the ability to gain electrons easily (i.e., they tend to form negative ions), but otherwise they display a wide range of chemical behaviors and reactivities. The nonmetals are located on the upper right side of the periodic table; they are separated from the metals by a line cutting diagonally through the region of the periodic table containing elements with partially filled *p* orbitals.

Metalloids

The metalloids, or semimetals, are found along the line between the metals and nonmetals in the periodic table, and their properties vary considerably. Their densities, boiling points, and melting points fluctuate widely. The electronegativities and ionization energies of metalloids lie between those of metals and nonmetals; therefore, these elements possess characteristics of both those classes. For example, silicon has a metallic luster; yet it is brittle and is not an efficient conductor. The reactivity of metalloids is dependent upon the element with which they are reacting. For example, boron (B) behaves as a nonmetal when reacting with sodium (Na) and as a metal when reacting with fluorine (F). The elements classified as metalloids are boron, silicon, germanium, arsenic, antimony, and tellurium.

The Chemistry of Groups

Elements in the same group have the same number of valence electrons, and hence tend to have very similar chemical properties.

Alkali Metals

The alkali metals are the elements of Group IA. They possess most of the physical properties common to metals, yet their densities are lower than those of other metals. The alkali metals have only one loosely bound electron in their outermost shell, giving them the largest atomic radii of all the elements in their respective periods. Their metallic properties and high reactivity are determined by the fact that they have low ionization energies; thus, they easily lose their valence electron to form univalent **cations** (cations with a +1 charge). Alkali metals have low electronegativities and react very readily with nonmetals, especially halogens.

Alkaline Earth Metals

The alkaline earth metals are the elements of Group IIA. They also possess many characteristically metallic properties. Like the alkali metals, these properties are dependent upon the ease with which they lose electrons. The alkaline earths have two electrons in their outer shell and thus have small-

Don't Mix These Up on Test Day

Metalloids

- Boron (B)
- Silicon (Si)
- Germanium (Ge)
- Arsenic (As)
- Antimony (Sb)
- Tellurium (Te)

er atomic radii than the alkali metals. However, the two valence electrons are not held very tightly by the nucleus, so they can be removed to form divalent cations. Alkaline earths have low electronegativities and low electron affinities.

Halogens

The halogens, Group VIIA (second to last column), are highly reactive nonmetals with seven valence electrons (one short of the favored octet configuration). Halogens are highly variable in their physical properties. For instance, the halogens range from gaseous (F_2 and Cl_2) to liquid (Br_2) to solid (I_2) at room temperature. Their chemical properties are more uniform: The electronegativities of halogens are very high, and they are particularly reactive towards alkali metals and alkaline earths, which "want" to donate electrons to the halogens to form stable ionic crystals.

Noble Gases

The noble gases, also called the inert gases, are found in Group VIII. They are fairly nonreactive because they have a complete valence shell, which is an energetically favored arrangement. They thus have high ionization energies. They possess low boiling points and are all gases at room temperature.

Transition Elements

The transition elements are those that are found between the alkaline earth metals and those with valence p electrons (the last six columns). The numbering of the groups can get rather confusing because of the existence of two conventions, but you needn't be too concerned with this. These elements are metals and hence are also known as transition metals. They are very hard and have high melting and boiling points. As one moves across a period, the five d orbitals become progressively more filled. The d electrons are held only loosely by the nucleus and are relatively mobile, contributing to the malleability and high electrical conductivity of these elements. Chemically, transition elements have low ionization energies and may exist in a variety of positively charged forms or oxidation states. This is because transition elements are capable of losing various numbers of electrons from the s and d orbitals of their valence shell. For instance, copper (Cu) can exist in either the +1 or the +2 oxidation state, and manganese (Mn) occurs in the +2, +3, +4, +6, or +7 state. Because of this ability to attain positive oxidation states, transition metals form many different ionic and partially ionic compounds. The dissolved ions can form complex ions either with molecules of water (hydration complexes) or with nonmetals, forming highly colored solutions and compounds, such as $CuSO_4 \cdot 5H_2O$.

Complexes of transition metal ions, called coordination complexes, are an interesting class of species because many of them possess bright colors. This results from the fact that the formation of complexes causes the d orbitals (normally all of the same energy) to be split into two energy sublevels. Many of the complexes can thus absorb certain frequencies of light—those containing the precise amount of energy required to raise electrons from the lower to the higher d sublevel. The frequencies not absorbed give the complexes their characteristic colors.

The Periodic Table
Review Problems

1. Elements in a given period have the same

   A. atomic weight

   B. maximum azimuthal quantum number

   C. maximum principal quantum number

   D. valence electron structure

   E. atomic number

2. Arrange the following species in terms of increasing atomic (or ionic) radius:

   Sr, P, Mg, Mg^{2+}

3. Which of the following elements has the lowest electronegativity?

   A. Cesium

   B. Strontium

   C. Calcium

   D. Barium

   E. Potassium

4. Arrange the following calcium species in terms of increasing size:
   Ca, Ca^+, Ca^{2+}, Ca^{3+}, Ca^-, Ca^{2-}

5. The order of the elements in the Periodic Table is based on

   A. the number of neutrons

   B. the radius of the atom

   C. the atomic number

   D. the atomic weight

   E. the number of oxidation states

6. The elements within each column of the Periodic Table

   A. have similar valence electron configurations

   B. have similar atomic radii

   C. have the same principal quantum number

   D. will react to form stable elements

   E. have no similar chemical properties

7. Arrange the following elements in terms of increasing first ionization energy:

   Ga, Ba, Ru, F, N

8. Arrange the following elements in terms of decreasing electronegativity:

Ca, Cl, Fr, P, Zn

9. Which element has the greatest electronegativity?

A. Chlorine

B. Oxygen

C. Sulfur

D. Phosphorus

E. Fluorine

10. Which of the following elements is most electronegative?

A. S

B. Cl

C. Na

D. Mg

E. P

11. Transition metal compounds generally exhibit bright colors because

A. the electrons in the partially filled d orbitals are easily promoted to excited states

B. the metals become complexed in water

C. the metals conduct electricity, producing colored light

D. the electrons in the d orbitals emit energy as they relax

E. their valence electrons cause them to bind to other metals

12. Identify the following elements as metal, nonmetal, or semimetal (metalloid):

A. Fr

B. Pd

C. I

D. B

E. Sc

F. Si

G. S

Turn the page
for answers and explanations
to the Review Problems.

Solutions to Review Problems

1. C

Refer to the first paragraph of this chapter.

2. $Mg^{2+} < P < Mg < Sr$

The trends in atomic radii are as follows: Going from left to right across a period, the atomic radii decrease because the atomic number increases. The increasing number of protons in the nucleus will have a stronger attraction for the outermost electrons, causing them to be held closer and more tightly to the nucleus. Going down a group, the atomic radius will increase because there are more filled principal energy levels separating the nucleus and the outermost electrons, shielding the attractive force between them. P has a small radius because it lies far to the right and high in a group. The magnesium species will have smaller radii than the strontium species because they are higher in Group II. Finally, positive ions have smaller atomic radii than the corresponding neutral molecules, because the loss of electrons leads to a decrease in electron-electron repulsion within the atom, which in turn allows the electrons to move in closer to the nucleus. Therefore, Mg^{2+} will be smaller than Mg. Mg^{2+} has a smaller radius than P because Mg^{2+} has no electrons in orbitals of the third principal energy level.

3. A

The least electronegative elements are located at the bottom left of the Periodic Table. Cesium has the lowest ionization energy and, likewise, it is the least electronegative. Note that Francium (Fr) would be lower still but is not a stable, naturally occurring element.

4. $Ca^{3+}, Ca^{2+}, Ca^+, Ca, Ca^-, Ca^{2-}$

Positive ions will have smaller radii than the corresponding neutral atoms, and the greater the positive charge, the smaller the ionic radius. Negative ions will have larger radii than the corresponding neutral atoms, and the greater the negative charge, the larger the ionic radius (see answer to question 2, above).

5. C

Refer to the first paragraph of this chapter.

6. A

Refer to the first paragraph of this chapter.

7. $Ba < Ru < Ga < N < F$

Remember two common trends when ordering atoms according to their ionization energies. First, the ionization energy increases toward the right across a period, because the elements are less willing to give up an electron as the attractive pull of the nucleus increases. Second, the ionization energy decreases down a group, because the distance separating the valence electrons from the nucleus increases. Therefore, to order the elements according to their first ionization energy, it is necessary to go from the bottom left of the periodic table, where the lowest values are, across to the top right of the periodic table, where the highest values are.

8.   Cl > P > Zn > Ca > Fr

The two trends to remember with electronegativity are that it increases across a period and decreases down a group. Therefore, chlorine, which is farthest to the top and right, will have the highest value. Francium lies farthest to the left and bottom, so it will have the lowest electronegativity.

9.   E

Fluorine is the most electronegative element among the ones given since it is closest to the upper right hand corner of the periodic table.

10.   B

Chlorine has the greatest electronegativity because, out of all the choices, it lies farthest to the right and top of the periodic table. Chlorine has a great attraction for electrons in a chemical bond because it needs only one more electron to complete a stable octet formation. Therefore, it has a high electronegativity.

11.   A

The closely spaced split d orbitals allow for relatively low energy transitions; these transitions often occur in the visible region of the electromagnetic spectrum. The compound appears to have a color that is complementary to the one that is absorbed. For example, if the transition occurs in the red region of the spectrum, the compound will appear green.

12.

a.   Fr: metal

b.   Pd: metal

c.   I: nonmetal

d.   B: semimetal

Chemical Bonding and Molecular Structure

The atoms of many elements can combine to form molecules. The atoms in most molecules are held together by strong attractive forces called chemical bonds. These bonds are formed via the interaction of the valence electrons of the combining atoms. The chemical and physical properties of the resulting molecules are often very different from those of their constituent elements.

In addition to the very strong forces within a molecule, there are weaker intermolecular forces between molecules. These intermolecular forces, although weaker than the intramolecular chemical bonds, are of considerable importance in understanding the physical properties of many substances. We shall discuss intermolecular forces in greater detail when we look at the different phases of matter later on. Processes that involve the breaking and forming of chemical bonds are generally considered chemical processes, while those that only involve interactions between molecules are generally considered physical processes.

In the formation of chemical bonds, many molecules contain atoms bonded according to the octet rule, which states that an atom tends to bond with other atoms until it has eight electrons in its outermost shell, thereby forming a stable electron configuration similar to that of the noble gas elements. Exceptions to this rule are as follows: hydrogen, which can have only two valence electrons (the configuration of He); lithium and beryllium, which bond to attain two and four valence electrons, respectively; boron, which bonds to attain six, and elements beyond the second row, such as phosphorus and sulfur, which can expand their octets to include more than eight electrons by incorporating d orbitals.

When classifying chemical bonds, it is helpful to introduce two distinct types: ionic bonds and covalent bonds. In ionic bonding, one or more electrons from an atom with a smaller ionization energy are transferred to an atom with a great electron affinity, and the resulting ions are held together by electrostatic forces. In covalent bonding, an electron pair is shared between two atoms. In many cases, the bond is partially covalent and partially ionic; we call such bonds polar covalent bonds.

Basic Concepts

ionic bond: transfer of
                 electron(s)

covalent bond: sharing of
                    electron(s)

Ionic Bonds and Ionic Compounds

When two atoms with large differences in electronegativity react, there is a complete transfer of electrons from the less electronegative atom to the more electronegative atom. The atom that loses electrons becomes a positively charged ion, or cation, and the atom that gains electrons becomes a negatively charged ion, or anion. In general, the elements of Groups I and II (low electronegativities) bond ionically to elements of Group VII (high electronegativities). Elements of Groups I and II (metals) give up their electrons to form cations that have a noble gas configuration, while Group VII elements gain an electron to form anions with the noble gas configuration. For example, a neutral sodium atom has one valence electron in the $3s$ subshell, whereas a neutral chlorine atom has seven valence electrons. If sodium sheds itself of its valence electron, it will possess the same electronic configuration as neon, a noble or inert gas with a filled octet. Conversely, chlorine is one electron short of a stable octet: If it gains an extra electron, it will have the electronic configuration of argon. When the two come together, then, sodium loses an electron to chlorine:

$$Na + Cl \rightarrow Na^+ Cl^-$$

Since opposite charges attract, these two are now held together by electrostatic forces and form the compound we know as sodium chloride, or salt. This force of attraction between the charged ions is called an ionic bond.

As seen from the example above, ionic compounds are formed by the interactions of cations and anions. The nomenclature, or naming, of ionic compounds is based on the names of the component ions. The following are some general guidelines:

1. The cationic species (usually metals) are usually named simply as the element, e.g., NaCl: *sodium* chloride, CaF_2: *calcium* fluoride. For elements that can form more than one positive ion, the charge is indicated by a Roman numeral in parentheses following the name of the element.

   Fe^{2+} Iron (II) Cu^+ Copper (I)

   Fe^{3+} Iron (III) Cu^{2+} Copper (II)

2. An older but still commonly used method is to add the endings *–ous* or *–ic* to the root of the Latin name of the element, to represent the ions with lesser or greater charge respectively.

   Fe^{2+} Ferr*ous* Cu^+ Cupr*ous*

   Fe^{3+} Ferr*ic* Cu^{2+} Cupr*ic*

3. Monatomic anions (single atom anions) are named by dropping the ending of the name of the element and adding *–ide*, as in the examples of sodium *chloride* and calcium *fluoride* above. Also:

   H^- Hydride

   S^{2-} Sulfide

   N^{3-} Nitride

   O^{2-} Oxide

   P^{3-} Phosphide

4. Many polyatomic anions contain oxygen and are called oxyanions. When an element forms two oxyanions, the name of the one with less oxygen ends in *–ite* and the one with more oxygen ends in *–ate*.

   NO_2^- Nitrite SO_3^{2-} Sulfite

   NO_3^- Nitrate SO_4^{2-} Sulfate

5. When the series of oxyanions contains four oxyanions, prefixes are also used. *Hypo–* and *per–* are used to indicate less oxygen and more oxygen, respectively. (Note that these prefixes are used only when there are more than two possibilities for the oxyanion.)

   ClO^- Hypochlorite

   ClO_2^- Chlorite

   ClO_3^- Chlorate

   ClO_4^- Perchlorate

6. Polyatomic anions often gain one or more H^+ ions to form anions of lower charge. The resulting ions are named by adding the word hydrogen or dihydrogen to the front of the anion's name. An older method uses the prefix *bi–* to indicate the addition of a single hydrogen ion.

   HCO_3^- Hydrogen carbonate or bicarbonate

   HSO_4^- Hydrogen sulfate or bisulfate

   $H_2PO_4^-$ Dihydrogen phosphate

Using the above rules, then, one can determine the names of ionic compounds such as the following:

$NaClO_4$ sodium perchlorate

$NaClO$ sodium hypochlorite

$NaNO_3$ sodium nitrate

KNO_2 potassium nitrite

Li_2SO_4 lithium sulfate

$MgSO_3$ magnesium sulfite

Note that the name itself does not explicitly tell us how many ions of each there are; for example, the names lithium sulfite and calcium fluoride do not tell us that there are two lithium ions and two fluoride ions in the respective compound—we have to deduce that from knowing that the positive and negative charges have to balance each other to give a neutral ionic compound. Also, note that in a compound like lithium sulfate, both ionic and covalent bonds exist: The sulfur is bonded covalently to the oxygen atoms, while the sulfate anion as a whole interacts with lithium ions to form ionic bonds.

As described in the previous chapter, metals, which are found in the left part of the periodic table, generally form positive ions, whereas nonmetals, which are found in the right part of the periodic table, generally form negative ions. Note, however, the existence of anions that contain metallic elements, such as MnO_4^- (permanganate) and CrO_4^{2-} (chromate). All elements in a given group tend to form monatomic ions with the same charge. Thus ions of alkali metals (Group I) usually form cations with a single positive charge, the alkaline earth metals (Group II) form cations with a double positive charge, and the halides (Group VII) form anions with a single negative charge.

Ionic compounds have characteristic physical properties. They have high melting and boiling points due to the strong electrostatic forces between the ions. They can conduct electricity in the liquid and aqueous states, though not in the solid state. Ionic solids form crystal lattices consisting of infinite arrays of positive and negative ions in which the attractive forces between ions of opposite charge are maximized, while the repulsive forces between ions of like charge are minimized.

Covalent Bonds

When two or more atoms with similar electronegativities interact, they achieve a noble gas electron configuration by sharing electrons in what is known as a covalent bond. The binding force between the two atoms results from the attraction that each electron of the shared pair has for the two positive nuclei. This sharing of electrons is best envisioned by using dots to represent valence electrons as follows:

$$:\ddot{F}\cdot \quad + \quad \cdot\ddot{F}: \quad\longrightarrow\quad :\ddot{F}:\ddot{F}: \;\text{ or }\; :\ddot{F}-\ddot{F}:$$

Each fluorine atom has seven valence electrons; they are both one short of a stable octet. Unlike the case of ionic bonding, however, there are no willing "electron donors" with low electronegativity around from which they can grab an electron. What they need to do is to each share one electron with its partner: The first structure drawn on the right hand side of the arrow shows how each F atom now has 8 valence electrons; the pair in the middle is shared by both. This pair of electrons is known as a bonding pair of electrons, as opposed to the unshared lone pairs, and is what constitutes the covalent bond between the two F atoms in the F_2 molecule. The bonding nature of these atoms is better indicated by the line between the atoms shown in the second structure.

Sometimes forming an octet requires sharing more than one electron from each atom. The oxygen molecule, O_2, and carbon monoxide, CO, for example, involve two and three pairs of bonding electrons, respectively:

$$\ddot{O}: \quad + \quad :\ddot{O} \quad\longrightarrow\quad \ddot{O}::\ddot{O} \;\text{ or }\; \ddot{O}=\ddot{O}$$

$$\cdot\dot{C}\cdot \quad + \quad :\ddot{O} \quad\longrightarrow\quad :C:::O: \;\text{ or }\; :C\equiv O:$$

When two pairs of electrons are shared, the bond is known as a double bond. When three pairs of electrons are shared, the bond is known as a triple bond. The number of shared electron pairs between two atoms is called the bond order; hence a single bond (as in F_2) has a bond order of one, a double bond has a bond order of two, and a triple bond has a bond order of three.

A covalent bond can be characterized by two features: bond length and bond energy.

Bond Length

Bond length is the average distance between the two nuclei of the atoms involved in the bond. As the number of shared electron pairs increases, the two atoms are pulled closer together, leading to a decrease in bond length. Thus, for a given pair of atoms, a triple bond is shorter than a double bond, which is in turn shorter than a single bond.

Basic Concept

The larger the bond energy, the harder it is to break the bond, and the stronger the bond.

Bond Energy

Bond energy is the energy required to separate two bonded atoms. For a given pair of atoms, the strength of a bond (and therefore the bond energy) increases as the number of shared electron pairs increases. So a triple bond is stronger than a double bond, and a double bond is stronger than a single bond. It is *not* the case, however, that a double bond is twice as strong (that is, needs twice as much energy to break) as a single bond. The reason for this will become clearer as we examine bonding from a slightly different perspective towards the end of this chapter.

Lewis Structures

As mentioned above, the shared valence electrons of a covalent bond are called the bonding electrons, while the valence electrons not involved in the covalent bond are called nonbonding electrons, also called lone electron pairs. A convenient notation, called a Lewis structure, examples of which are already shown above, is used to represent the bonding and nonbonding electrons in a molecule, facilitating chemical "bookkeeping."

The number of valence electrons attributed to a particular atom in the Lewis structure of a molecule is not necessarily the same as the number would be in the isolated atom, and the difference accounts for what is referred to as the formal charge of that atom. Often, more than one Lewis structure can be drawn for a molecule; this phenomenon is called resonance.

Drawing Lewis Structures

A Lewis structure, or Lewis dot symbol, is the chemical symbol of an element surrounded by dots, each representing one of the *s* and/or *p* valence electrons of the atom. The Lewis symbols of the elements found in the second period of the periodic table are shown below.

•Li	Lithium	$\cdot\ddot{N}\cdot$	Nitrogen
•Be•	Beryllium	$\cdot\ddot{O}\colon$	Oxygen
•B•	Boron	$\cdot\ddot{F}\colon$	Fluorine
•Ċ•	Carbon	$\colon\!\ddot{Ne}\colon$	Neon

Just as a Lewis symbol is used to represent the distribution of valence electrons in an atom, it can also be used to represent the distribution of valence electrons in a molecule.

Certain steps must be followed in assigning a Lewis structure to a molecule. These steps are outlined below, using HCN as an example.

- Write the skeletal structure of the compound (i.e., the arrangement of atoms). In general, the least electronegative atom is the central atom. Hydrogen (always) and the halogens F, Cl, Br, and I (usually) occupy the end position.

  In HCN, H must occupy an end position. Of the remaining two atoms, C is the least electronegative, and therefore occupies the central position. The skeletal structure is as follows:

$$H \quad C \quad N$$

- Count all the valence electrons of the atoms. The number of valence electrons of the molecule is the sum of the valence electrons of all atoms present:

  H has one valence electron;

  C has four valence electrons;

  N has five valence electrons; therefore,

  HCN has a total of ten valence electrons.

- Draw single bonds between the central atom and the atoms surrounding it. Place an electron pair in each bond (bonding electron pair).

$$H : C : N$$

  Each bond has two electrons, so $10 - 4 = 6$ valence electrons remain.

- Complete the octets (eight valence electrons) of all atoms bonded to the central atom, using the remaining valence electrons still to be assigned. (Recall that H is an exception to the octet rule since it can have only two valence electrons.) In this example H already has two valence electrons in its bond with C.

$$H : C : \overset{\cdot\cdot}{\underset{\cdot\cdot}{N}} :$$

- Place any extra electrons on the central atom. If the central atom has less than an octet, try to write double or triple bonds between the central and surrounding atoms using the nonbonding, unshared lone electron pairs.

  The HCN structure above does not satisfy the octet rule for C because C possesses only four valence electrons. Therefore, two lone electron pairs from the N atom must be moved to form two more bonds with C, creating a triple bond between C and N.

Finally, bonds are drawn as lines rather than pairs of dots.

$$H - C \equiv N:$$

Now the octet rule is satisfied for all three atoms, since C and N have eight valence electrons and H has two valence electrons.

Formal Charge

The number of electrons officially assigned to an atom in a Lewis structure does not always equal the number of valence electrons of the free atom. The difference between these two numbers is the formal charge of the atom. Formal charge can be calculated using the following formula:

$$\text{Formal charge} = V - \frac{1}{2}N_{bonding} - N_{nonbonding}$$

where V is the number of valence electrons in the free atom, $N_{bonding}$ is the number of bonding electrons, and $N_{nonbonding}$ is the number of nonbonding electrons.

The charge of an ion or molecule is equal to the sum of the formal charges of the individual atoms comprising the ion or molecule. In other words, for a neutral molecule, the formal charges of the individual atoms have to add up to zero.

Example: Calculate the formal charge on the central N atom of NH_4^+.

Solution: The Lewis structure of NH_4^+ is

$$\left[\begin{array}{c} H \\ | \\ H - N - H \\ | \\ H \end{array} \right]^+$$

Nitrogen is in group VA; thus it has 5 valence electrons. In NH_4^+, N has 4 bonds (i.e., 8 bonding electrons and no non-bonding electrons). So, $V = 5$; $N_{bonding} = 8$; $N_{nonbonding} = 0$

Formal charge: $= 5 - \frac{1}{2}(8) - 0$

$= +1$

Thus, the formal charge on the N atom in NH_4^+ is +1.

Resonance Structures

For some molecules, two or more nonidentical Lewis structures can be drawn; these are called resonance structures. The molecule doesn't actually exist as either one of the resonance structures, but is rather a composite, or hybrid, of the two. For example, SO_2 has three resonance structures. Resonance structures are expressed with a double-headed arrow between them; thus,

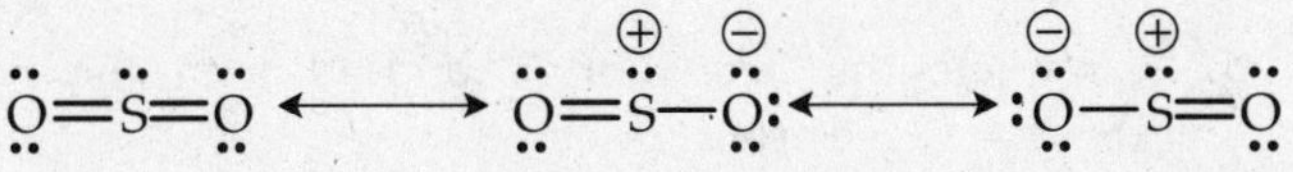

$$\ddot{O}=\ddot{S}=\ddot{O} \longleftrightarrow \ddot{O}=\overset{\oplus}{S}-\overset{\ominus}{\ddot{O}}: \longleftrightarrow :\overset{\ominus}{\ddot{O}}-\overset{\oplus}{S}=\ddot{O}$$

represents the resonance structures of SO_2. The actual molecule is a hybrid of these three structures (the two S–O bonds are actually equivalent: No one bond is stronger than the other).

The last two resonance structures of sulfur dioxide shown above have equivalent energy or stability. Often, nonequivalent resonance structures may be written for a molecule. In these cases, the more stable the structure, the more that structure contributes to the character of the resonance hybrid. Conversely, the less stable the resonance structure, the less that structure contributes to the resonance hybrid. In the example above, it is the structure on the left of the diagram that is the most stable. Formal charges are often useful for qualitatively assessing the stability of a particular resonance structure. The following guidelines are used:

- A Lewis structure with small or no formal charges is preferred over a Lewis structure with large formal charges.

- A Lewis structure in which negative formal charges are placed on more electronegative atoms is more stable than one in which the formal charges are placed on less electronegative atoms.

Resonance structures can differ only in the way the electrons are distributed; the arrangement of the actual atoms cannot change.

Example: Write the resonance structures for NCO^-.

Solution: 1.   C is the least electronegative of the three given atoms, N, C, and O. Therefore the C atom occupies the central position in the skeletal structure of NCO^-.

$$N \quad C \quad O$$

2.   N has 5 valence electrons;

C has 4 valence electrons;

O has 6 valence electrons;

and the species itself has one negative charge, indicating the presence of an extra electron.

Number of valence electrons $= 5 + 4 + 6 + 1$

$= 16$

3. Draw single bonds between the central C atom and the surrounding atoms, N and O. Place a pair of electrons in each bond.

$$N : C : O$$

4. Complete the octets of N and O with the remaining $16 - 4 = 12$ electrons.

$$: \overset{..}{\underset{..}{N}} : C : \overset{..}{\underset{..}{O}} :$$

5. The C octet is incomplete. There are three ways in which double and triple bonds can be formed to complete the C octet: two lone pairs from the O atom can be used to form a triple bond between the C and O atoms;

$$: \overset{..}{\underset{..}{N}} - C \equiv O :$$

or one lone electron pair can be taken from both the O and the N atoms to form two double bonds, one between N and C, and the other between O and C;

$$: \overset{..}{N} = C = \overset{..}{O} :$$

or two lone electron pairs can be taken from the N atom to form a triple bond between the C and N atoms.

$$: N \equiv C - \overset{..}{\underset{..}{O}} :$$

These three are all resonance structures of NCO^-.

Exceptions to the Octet Rule and Lewis Structures

As mentioned above, atoms found in or beyond the third period can have more than eight valence electrons, since some of the valence electrons may occupy d orbitals. (Recall that in order to have electrons in the d subshell, the principal quantum number must be 3 or above. The maximum number of electrons that a shell can hold is $2n^2 \geq 18$ when $n \geq 3$.) These atoms can be assigned more than four bonds in Lewis structures. When drawing the Lewis structure of the sulfate ion, giving the sulfur 12 valence electrons permits three of the five atoms to be assigned a formal charge of zero. The sulfate ion can be drawn in six resonance forms, each with the two double bonds attached to a different combination of oxygen atoms.

KAPLAN

Types Of Covalent Bonding

The nature of a covalent bond depends on the relative electronegativities of the atoms sharing the electron pairs. Covalent bonds are considered to be polar or nonpolar depending on the difference in electronegativities between the atoms.

Polar Covalent Bonds

Polar covalent bonding occurs between atoms with small differences in electronegativity. The bonding electron pair is not shared equally, but pulled more towards the element with the higher electronegativity. Yet the difference in electronegativity is not high enough for complete electron transfer (ionic bonding) to take place. As a result, the more electronegative atom acquires a partial negative charge, δ^-, and the less electronegative atom acquires a partial positive charge, δ^+, giving the molecule partially ionic character. For instance, the covalent bond in HCl is polar because the two atoms have a small difference in electronegativity. Chlorine, the more electronegative atom, attains a partial negative charge and hydrogen attains a partial positive charge. This difference in charge between the atoms is indicated by an arrow crossed (like a plus sign) at the positive end pointing to the negative end, as shown below:

$$\delta^+ \quad \delta^-$$
$$H - Cl$$

This small separation of charge generates what is known as a dipole moment.

Nonpolar Covalent Bonds

Nonpolar covalent bonding occurs between atoms that have the same electronegativities. The bonding electrons are shared equally, with no separation of charge across the bond. Not surprisingly, nonpolar covalent bonds occur in diatomic molecules with the same atoms. Certain elements exist under normal conditions only as diatomic molecules: N_2, O_2, F_2, Cl_2, Br_2, I_2, H_2. Their positions in the Periodic Table form an inverted L-shape towards the top right, excluding the noble gases.

Coordinate Covalent Bonds

In a coordinate covalent bond, the shared electron pair comes from the lone pair of one of the atoms in the molecule. Once such a bond forms, it is indistinguishable from any other covalent bond. Distinguishing such a bond is useful only in keeping track of the valence electrons and formal charges. Coordinate bonds are typically found in Lewis acid-base compounds (see chapter 11 on Acids and Bases). A Lewis acid is a compound that can accept

Quick Quiz

Match each of the following descriptions with the lettered types of bonds below.

1. Bonding between atoms with the same electronegativities

2. Bonding between atoms with small differences in electronegativity

3. Bond in which the shared electron pair comes from the lone pair of one atom

(A) Polar covalent bond
(B) Nonpolar covalent bond
(C) Coordinate covalent bond

Answers: 1. B
2. A
3. C

an electron pair to form a covalent bond; a Lewis base is a compound that can donate an electron pair to form a covalent bond. For example, in the reaction between boron trifluoride (BF_3) and ammonia (NH_3):

$$F\!-\!\underset{\overset{|}{F}}{\overset{\overset{F}{|}}{B}} \quad + \quad :\!\underset{\overset{|}{H}}{\overset{\overset{H}{|}}{N}}\!-\!H \quad \longrightarrow \quad F\!-\!\underset{\overset{|}{F}}{\overset{\overset{F}{|}}{B}}\!-\!\underset{\overset{|}{H}}{\overset{\overset{H}{|}}{N}}\!-\!H$$

Lewis acid Lewis base   Lewis acid-base compound

NH_3 donates a pair of electrons to form a coordinate covalent bond; thus, it acts as a Lewis base. BF_3 accepts this pair of electrons to form the coordinate covalent bond; thus, it acts as a Lewis acid.

Geometry and Polarity of Covalent Molecules

The Lewis structure is not necessarily a good pictorial representation of the three-dimensional appearance of a molecule. The actual geometric arrangement of the bonds and different atoms is obtained by using the VSEPR theory described below. The shape of a molecule can affect its polarity.

The Valence Shell Electron-Pair Repulsion Theory

The valence shell electron–pair repulsion (VSEPR) theory uses Lewis structures to predict the molecular geometry of covalently bonded molecules. It states that the three–dimensional arrangement of atoms surrounding a central atom is determined by the repulsions between the bonding and the nonbonding electron pairs in the valence shell of the central atom. These electron pairs arrange themselves as far apart as possible, thereby minimizing repulsion.

The following steps are used to predict the geometrical structure of a molecule using the VSEPR theory.

- Draw the Lewis structure of the molecule.

- Count the total number of bonding and nonbonding electron pairs in the valence shell of the central atom.

- Arrange the electron pairs around the central atom so that they are as far apart from each other as possible. It is important not to forget to take into consideration nonbonding pairs.

Valence electron arrangements are summarized in the following table:

number of electron pairs	example	geometric arrangement of electron pairs around the central atom	shape	angle between electron pairs
2	$BeCl_2$	X——A——X	linear	180°
3	BH_3		trigonal planar	120°
4	CH_4		tetrahedral	109.5°
5	PCl_5		trigonal bipyramidal	90°,120°,180°
6	SF_6		octahedral	90°,180°

While the number of electron pairs dictate their overall arrangement around the central atom, it is only a starting point in arriving at the actual description of the geometry of the molecule. If one of the X's in the table above is a lone pair of electrons rather than an actual atom or group of atoms, new terms need to be introduced to describe the spatial arrangement of the atoms. The example below illustrates this point.

Example: Predict the geometry of NH_3.

Solution: 1. The Lewis structure of NH_3 is:

$$\begin{matrix} & H & \\ & | & \\ H & —N— & H \\ & \cdot\cdot & \end{matrix}$$

2. The central atom, N, has 3 bonding electron pairs and 1 nonbonding electron pair, for a total of 4 electron pairs.

3. The 4 electron pairs will be farthest apart when they occupy the corners of a tetrahedron.

In describing the shape of a molecule, only the arrangement of atoms (not electrons) is considered. Even though the electron pairs are arranged tetrahedrally, the shape of NH_3 is described as trigonal pyramidal. It is not trigonal planar because the lone pair repels the three bonding electron pairs, causing them to move as far away as possible.

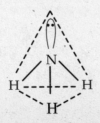

Example: Predict the geometry of CO_2.

Solution: The Lewis structure of CO_2 is O=C=O, with two extra lone pairs of electrons on each oxygen atom so all three atoms have an octet structure.

The double bond behaves just like a single bond for purposes of predicting molecular shape. This compound has two groups of electrons around the carbon. According to the VSEPR theory, the two sets of electrons will orient themselves 180° apart, on opposite sides of the carbon atom, minimizing electron repulsion. Therefore, the molecular structure of CO_2 is linear.

Polarity of Molecules

Basic Concept

A molecule is polar if it has polar bonds and if the dipole moments of these bonds do not cancel one another.

Earlier we talked about the concept of the dipole moment in a polar covalent bond. If a molecule has more than two atoms, there will be more than one bond. Each bond may or may not be a dipole, and in such cases one can talk about the polarity of the molecule as a whole. A molecule is polar if it has polar bonds and if the dipole moments of these bonds do not cancel one another (by pointing in opposite directions, for example). The polarity of a molecule, therefore, depends on the polarity of the constituent bonds and on the shape of the molecule. A molecule with only nonpolar bonds is always nonpolar; a molecule with polar bonds may be polar or nonpolar, depending on the orientation of the bond dipoles. For instance, CCl_4 has four polar C–Cl bonds. According to the VSEPR theory, the shape of CCl_4 is tetrahedral. The four bond dipoles point to the vertices of the tetrahedron and cancel each other, resulting in a nonpolar molecule.

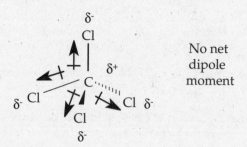

However, if the orientation of the bond dipoles are such that they do not cancel out, the molecules will have a net dipole moment and therefore be polar. For instance, H_2O has two polar O–H bonds. According to the VSEPR model, its shape is angular. The two dipoles add together to give a net dipole moment to the molecule, making the H_2O molecule polar.

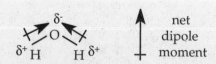

A molecule of two atoms bound by a polar bond must have a net dipole moment and therefore be polar. The two equal and opposite partial charges are localized at the ends of the molecule on the two atoms.

Orbital Hybridization

So far we have ignored the fact that the valence electrons that interact originally occupy different atomic orbitals in the atoms they come from. What does it mean for an electron in the $2p$ subshell of an oxygen atom to be "shared" with another atom, for example? Does it still occupy the same dumbbell–like region of space? Does it matter whether the orbital is $2p_x$, $2p_y$, or $2p_z$? Can it not be a valence electron from the $2s$ subshell instead? All these questions are addressed (or perhaps more correctly, as you shall see, sidestepped) in the orbital hybridization picture of bonding, which "scramble" the atomic orbitals of the central one to form new, hybrid ones that participate in bonding. It is not so much an alternative to VSEPR theory as an extension of it that gives a fuller understanding of the nature of bonding.

sp Hybridization

A molecule such as BeH_2 has a linear geometry. The two valence electrons of Be, originally in the $2s$ orbital, are shared with the hydrogen atoms (which in turn share their $1s$ electrons with Be). In the orbital hybridization picture, each of the two electrons in Be actually occupies an orbital that is a mix of an s and a p orbital, called an sp hybrid orbital.

s orbital p orbital sp orbitals

These two hybrid orbitals are oriented antiparallel to each other, and interact with the $1s$ orbitals of the hydrogen atoms on each end of the molecule. This leads to the 180° bond angle predicted in the VSEPR theory (implied by the linear geometry). The general convention adopted is to call the p orbital that participates in hybridization the p_z orbital, even though such designations are by and large arbitrary. Also, note that we mix two atomic orbitals and end up with two hybrid orbitals: In general, the number of atomic orbitals that "go in" has to equal the number of hybrid orbitals that "come out."

sp² Hybridization

Three sp^2 hybrid orbitals are formed by mixing an s and two p orbitals.

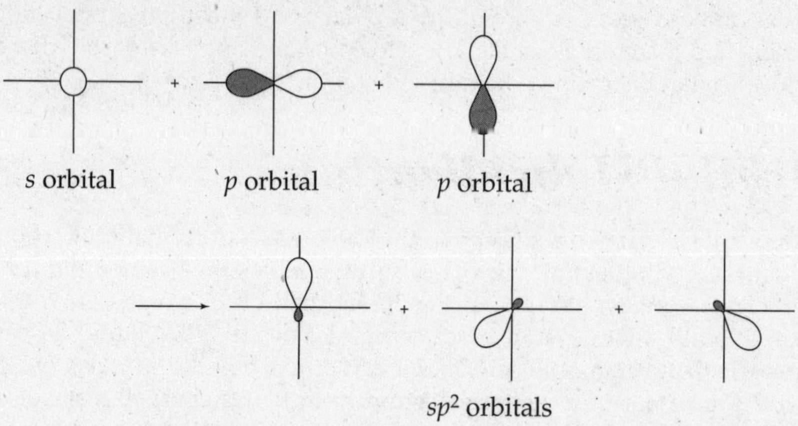

s orbital   p orbital   p orbital

sp² orbitals

Again, three orbitals are mixed to generate three new hybrid orbitals. These are oriented 120° apart from each other and thus the spatial arrangement is in accordance with the trigonal planar geometry of molecules like BF_3.

sp³ Hybridization

Four sp^3 hybrid orbitals are formed by mixing an s and all three of the p orbitals. These hybrid orbitals point towards the four corners of a tetrahedron.

Sigma and Pi Bonds

Going back to the example of BeH_2, one envisions an sp hybridized orbital overlapping with the s orbitals of the two hydrogen atoms to form bonds. These kinds of bonds, which results from end-to-end overlap of orbitals from the two bonded atoms, are known as sigma bonds (σ bonds). When multiple bonds are involved, however, we do not have several σ bonds between two atoms. What we have instead is another kind of bonds known as pi bonds (π bonds), which have a very different spatial arrangement. A useful example to consider is that of the ethene molecule, $H_2C=CH_2$, in which the carbon atoms are double-bonded to each other and each of them is also bonded to two hydrogen atoms. Each carbon atom, then, is bonded to three groups—a CH_2 group and two H atoms, and VSEPR theory would lead us to predict that the three groups are arranged about 120° apart:

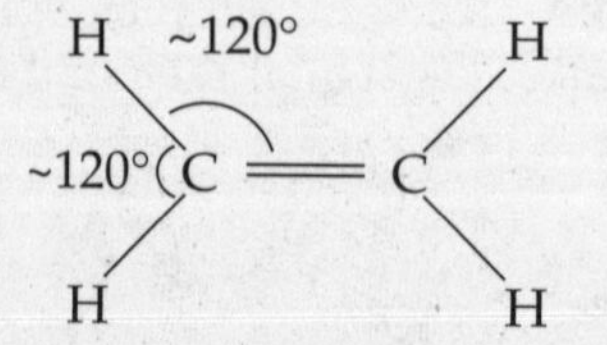

This is in fact the case. The carbon atoms are both sp^2 hybridized. For each of the two carbon atoms, two of the hybrid orbitals are used to form the bonds with the hydrogen atoms, with the remaining one forming a σ bond with the leftover hybrid orbital from the other carbon atom.

Where does one get a double bond? It is *not* the case that each hybrid orbital gives a bond, thus giving two: *Two* hybrid orbitals are needed to interact to give *one* bond. The second bond that is needed to give a double bond comes from the interaction of the p orbitals that are left unused in hybridization (one on each carbon atom). Remember that three sp^2 hybrid orbitals are formed by mixing an s and two p orbitals. Since there are actually three p orbitals, one is left over, sticking out of the plane that contains the hybrid orbitals. (In this case, then, the unhybridized p orbitals are coming out at you.) These two unhybridized orbitals from different carbon atoms can interact and form a weaker π bond. It is weaker because the overlap is not end-to-end but only side-to-side. You can see this by rotating the molecule 90° out of the paper so that the plane of the molecule is now coming out at you:

A single bond then consists of one σ bond, while a double bond consists of a σ and a π bond. Since, as just mentioned, a π bond is weaker, a double bond is not twice as strong as a single bond, even though it is stronger because of the additional interactions.

Similarly, a triple bond consists of a σ and two π bonds: in the sp hybridization scheme, two p orbitals are left unused, one along the x and one along the y direction. Each of these can form a π bond with a parallel p orbital on a neighboring atom. These, together with the σ bond along the z direction formed from the end-on-end overlap of two sp hybridized orbitals, give three bonds total. Again, since π bonds are weaker, a triple bond, though stronger than a single bond, is not three times as strong.

Don't Mix These Up on Test Day

A single bond consists of one σ bond.

A double bond consists of a σ and a π bond.

A triple bond consists of a σ and two π bonds.

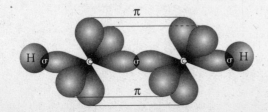

Chemical Bonding and Molecular Structure Review Problems

1. Consider the following reaction:

$$H_2 (g) + F_2 (g) \rightarrow 2\ HF\ (g)$$

Is the H–F bond more or less polar than an H—H bond?

2. Arrange the following compounds in terms of increasing polarity:

HCN, NaCl, Cl_2

3. a. Which has a greater C–C bond distance, C_2H_4 or C_2H_2?

b. Which has a greater C–C bond energy?

4. Which represents the proper Lewis structure of:

a. $CHCl_3$

I.

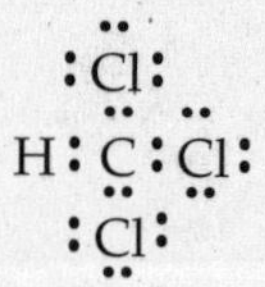

II.

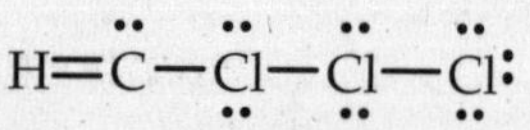

III.

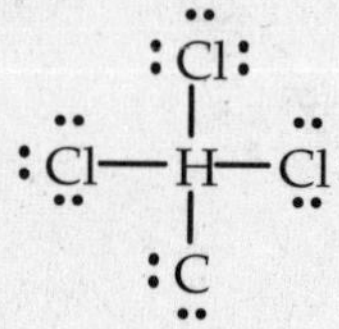

IV.

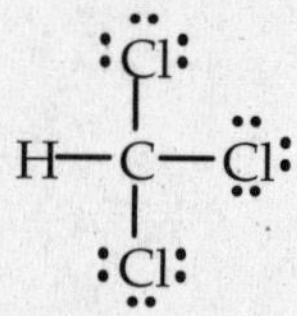

b. N_2

I.

II.

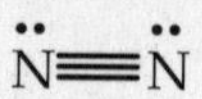

III.

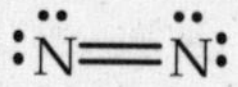

IV.

c. $[ClO_4]^-$

I.

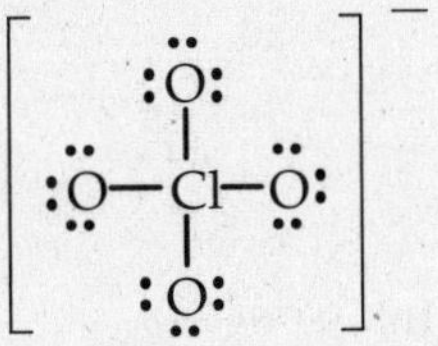

II.

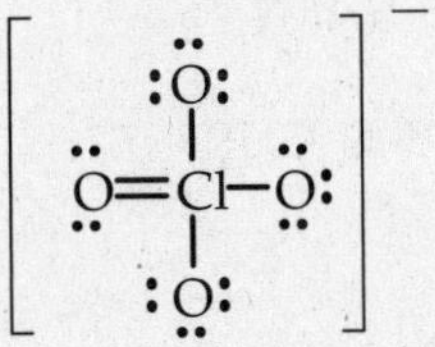

III.

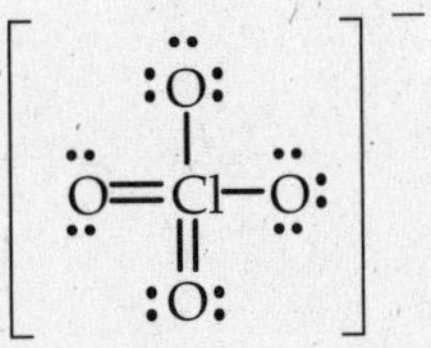

IV.

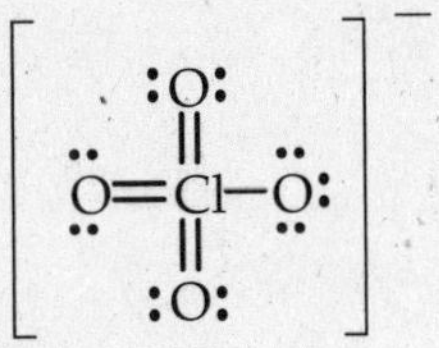

5. Which is not a resonance form of:

a.

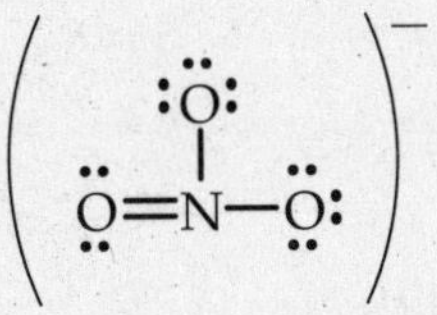

I.

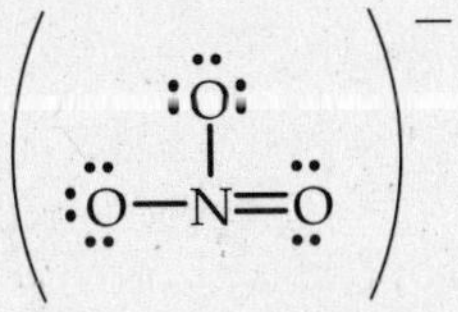

II.

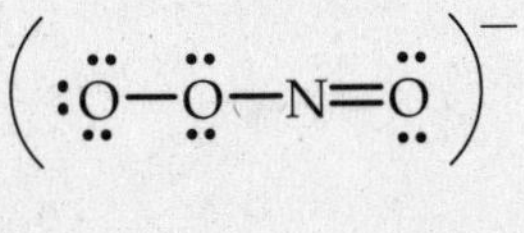

III.

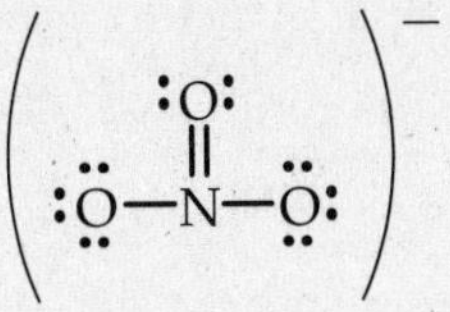

b.

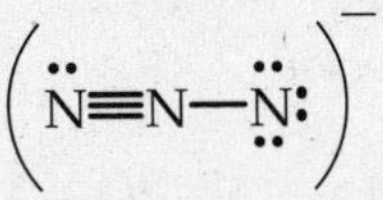

I.

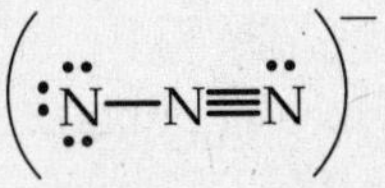

II.

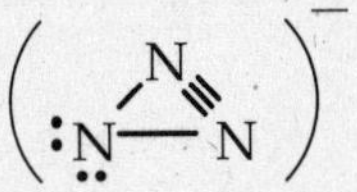

III.

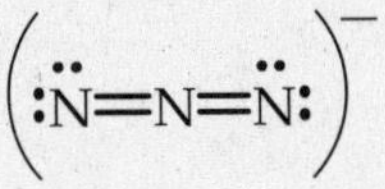

6. Label formal charges and predict which is the most likely resonance structure for N_2O:

a. 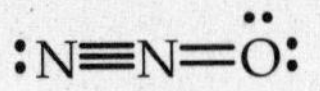 $:N\equiv N=\ddot{O}:$

b. 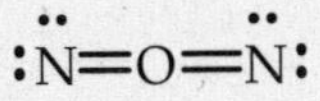 $\ddot{N}=O=\ddot{N}:$

c. 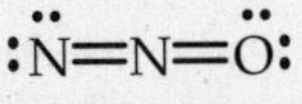 $:\ddot{N}=N=\ddot{O}:$

d. 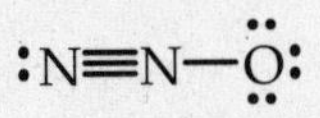 $:N\equiv N-\ddot{O}:$

7. Draw Lewis structures of the most likely ions of the elements from Na to Ca.

8. Draw Lewis structures for each of the following:

   a. nitrate ion ($[NO_3]^-$)

   b. phosphoric acid (H_3PO_4)

   c. aluminum chloride ($AlCl_3$)

   d. sodium phosphate (Na_3PO_4)

9. A hydride is a compound containing hydride ion, H^-. Predict two elements whose hydrides would contain incomplete octets.

10. Which of the following sets of molecules contains only nonpolar species?

    a. BH_3, NH_3, AlH_3

    b. NO_2, CO_2, ClO_2

    c. HCl, HNO_2, $HClO_3$

    d. BH_3, H_2S, BCl_3

    e. BeH_2, BH_3, CH_4

Turn the page
for answers and explanations
to the Review Problems.

Solutions to Review Problems

1.

The HF bond is more polar than an H—H bond.

"Polar" denotes unequal sharing of electrons; H—H must have equal sharing, since the two atoms are the same. H and F are different atoms, with different electronegativities, and so the electrons are unequally shared.

2. $Cl_2 < HCN < NaCl$

Cl_2 is the least polar, because it contains two identical atoms that must share electrons equally. HCN is a linear molecule with a triple bond between C and N; it has a dipole moment pointing from the relatively electropositive H atom toward the rather electronegative N atom. Still, we would expect HCN to be less polar than NaCl; the bond between Na and Cl, a metal and a nonmetal whose electronegativities differ greatly, is completely ionic.

3.

a. C_2H_4

C_2H_4 has greater bond distance because it is a double bond, and is therefore held less closely than C_2H_2, which is a triple bond.

b. C_2H_2

C_2H_2 has a greater bond energy because it is a triple bond, and more energy is needed to break it.

4.

a.   **I and IV**

Choices I and IV are both correct Lewis structures and are, in fact, equivalent. Choice II has an impossible configuration; H can never be double-bonded to anything, since the maximum number of electrons it may possess is 2. Choice III is also impossible, since having four bonds around H would imply 8 electrons.

b.   **II**

Choices I and III must be wrong because, although they satisfy the octet rule, they have the wrong total number of electrons; choice I has 8 valence electrons, while choice III has 12. Given two N atoms, there can be $(2)(5) = 10$ valence electrons, as in correct choice II. Choice IV is doubly wrong because, in addition to having only 8 total electrons, the octet rule is not satisfied, as each nucleus has 7, not 8, valence electrons.

c.   **IV**

Choice IV is the preferred structure, since 4 of the 5 atoms have a formal charge of zero. Since Cl is in the third period, its number of valence electrons can exceed 8.

5.

a. II

In resonance forms, only the electrons change place; atoms are not rearranged. Choices I and III are both resonance structures. Choice II requires rearrangement of the atoms.

b. II

By the same reasoning as above.

6. D

There is no formal charge on the structure of choice A; therefore, this structure should be the more likely resonance structure. However, the expanded octet on N makes the structure impossible. Choice C is incorrect because the negative formal charge is on N, which is not the most electronegative atom. Choice B is incorrect because O, which is the most electronegative atom, has a formal charge of +2.

B. $:\ddot{N}=O=\ddot{N}:$

C. $:\ddot{N}=N=\ddot{O}:$ D. $:N\equiv N-\ddot{\underset{..}{O}}:$

D is most likely because the negative formal charge is on O, the more electronegative element.

7.

Na^+ Mg^{2+} Al^{3+} Si^{4+} $:\overset{..}{\underset{..}{P}}:^{3-}$ $:\overset{..}{\underset{..}{S}}:^{2-}$ $:\overset{..}{\underset{..}{Cl}}:^{-}$
(Ar has none) K^+ Ca^{2+}

Note that a correct ionic Lewis structure must always show the charge on the ion.

8.

a. $[NO_3]-$

		has	needs
charge:		1 electron	
N:		5 electrons	8 electrons
3 O:		18 electrons	24 electrons
		24 electrons	32 electrons

(32 – 24) electrons = 8 electrons = 4 bonds

Place N at the center:

(N and O both have a formal charge.)

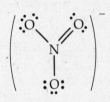

They cannot be reduced because the N octet cannot be expanded. However, since resonance will be present, a better version might be:

b. H_3PO_4

	has	needs
3 H	3 electrons	3(2) = 6 electrons
P	5 electrons	8 electrons (at least)
4 O	24 electrons	32 electrons
	32 electrons	46 electrons

(46 – 32) electrons = 14 electrons = 7 bonds (at least).

Place the P at the center, the 4 oxygens around it, and hydrogens on three of the oxygens, with single bonds between them; this will use all 7 bonds.

Now check the formal charges:

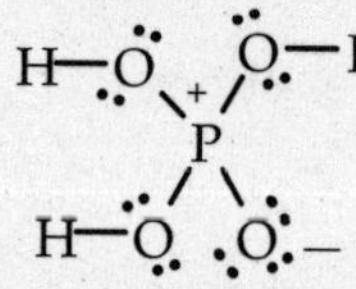

The P has a formal charge of +1 and the O has a formal charge of –1. These can be eliminated by moving a pair of electrons around from the O into a second bond:

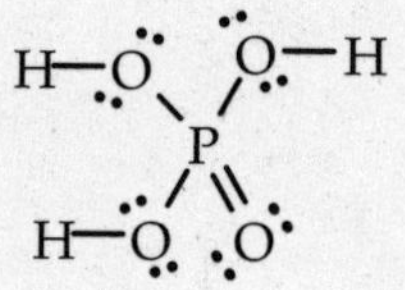

c.

Both aluminum chloride and sodium phosphate are ionic.

$AlCl_3$:

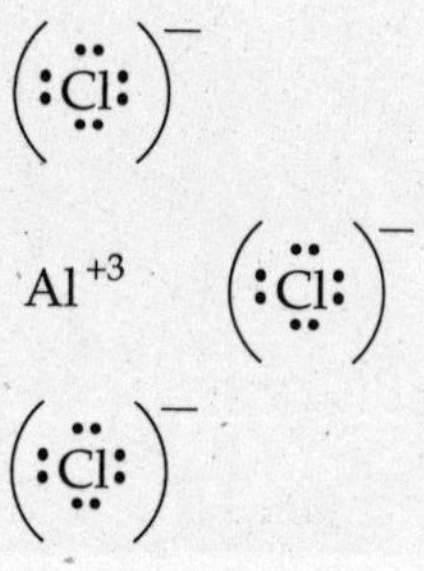

d. Na_3PO_4:

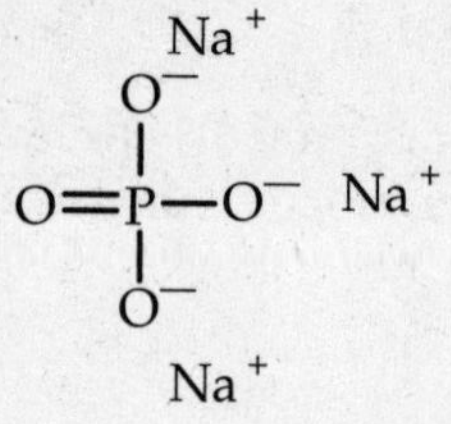

9.

Be and B, because they can join to only two or three hydrogens, respectively, since they have fewer than 4 valence electrons. The elements Mg and Al may also do this, as could Na, Ca, and the other active metals of groups I and II.

10. E

a. NH_3 is polar (positive end at base of pyramid, negative end at N).

b. NO_2 and ClO_2 are both angular molecules, therefore polar.

c. All three are polar.

d. H_2S is angular due to the two lone pairs on S, therefore polar.

CHAPTER 6

CHEMICAL REACTIONS AND STOICHIOMETRY

In the last chapter, we discussed how atoms combine and are held together by bonds that can be either ionic or covalent. After they come together, they may lose some of their individual properties and gain new ones that result from the combination. Water, for example, formed from two hydrogen atoms and an oxygen atom, does not really behave like the elements hydrogen or oxygen.

A compound is a pure substance that is composed of two or more elements in a fixed proportion. Compounds can be broken down chemically to produce their constituent elements or other compounds. All elements, except for some of the noble gases, can react with other elements or compounds to form new compounds. These new compounds can react further to form yet different compounds.

Molecular Weight and Molar Mass

A molecule is a combination of two or more atoms held together by covalent bonds. It is the smallest unit of a compound that displays the properties of that compound. Molecules may contain two atoms of the same element, as in N_2 and O_2, or may be comprised of two or more different atoms, as in CO_2 and $SOCl_2$.

In chapter 3, we discussed the concept of the atomic weight. Like atoms, molecules can also be characterized by their weight. The molecular weight is simply the sum of the weights of the atoms that make up the molecule.

Example: What is the molecular weight of $SOCl_2$?

Solution: To find the molecular weight of $SOCl_2$, add together the atomic weights of each of the atoms.

1S =	1×32 amu	=	32 amu
1O =	1×16 amu	=	16 amu
2Cl =	2×35.5 amu	=	71 amu
molecular weight		=	119 amu

Ionic compounds do not form true molecules. In the solid state they can be considered to be a nearly infinite, three dimensional array of the charged particles of which the compound is composed. Since no actual molecule exists, molecular weight becomes meaningless, and the term *formula weight* is used in its place, although the calculation is the same: We simply add up the atomic masses of the elements in the compound's empirical formula (see below). The formula weight of NaCl, for example, is the atomic weight of Na plus the atomic weight of Cl: (23 + 35.5) amu = 58.5 amu.

Remember that a mole of something is about 6.022×10^{23} of that thing. In addition, the atomic mass of an atom, reported in units of amu, is numerically the same as its mass in grams per mole. For example, one mole of an atom with atomic mass x amu has a mass of x grams. The same relationship holds for molecules: One mole of a compound has a mass in grams equal to the molecular weight of that compound in amu, and contains 6.022×10^{23} molecules of the compound. For example, the molecular weight of carbonic acid, H_2CO_3, is $(2 \times 1 + 12 + 3 \times 16) = 62$ amu. 62 g of H_2CO_3 represents one mole of carbonic acid and contains 6.022×10^{23} H_2CO_3 molecules. In other words, the molar mass of H_2CO_3 is 62 g/mol. This can also be arrived at by simply adding the molar atomic mass of the atoms in the compound: 1 mole of H_2CO_3 contains 2 moles of H atoms, 1 mole of C atoms, and 3 moles of O atoms.

Given the weight of a sample, one can determine the number of moles present with the following formula:

number of moles = weight of sample (g) ÷ molar mass (g/mol)

Example: How many moles are in 9.52 g of $MgCl_2$?

Solution: First, find the molar mass of $MgCl_2$.

1(24.31 g/mol) + 2(35.45 g/mol) = 95.21 g/mol

Now, solve for the number of moles.

$$\frac{9.25 \text{ g}}{95.21 \text{ g/mol}} = 0.10 \text{ mol of } MgCl_2$$

Representation of Compounds

The formula for a chemical compound gives us information about the relative proportions of the different elements that constitute it. Conversely, knowledge of the composition of a compound enables us to determine its (empirical) formula. Knowing how to represent chemical compounds, and knowing how to determine a compound's formula, is very important for the SAT II: Chemistry Test.

Law of Constant Composition

The law of constant composition states that any sample of a given compound will contain the same elements in the identical mass ratio. For instance, every sample of H_2O will contain two atoms of hydrogen for every atom of oxygen, or, in other words, one gram of hydrogen for every eight grams of oxygen. This is hardly surprising since we already know that atoms prefer an octet structure and would combine with other atoms in predictable ways to achieve this.

Empirical and Molecular Formulas

There are two ways to express a formula for a compound. The empirical formula gives the simplest whole number ratio of the elements in the compound. The molecular formula gives the exact number of atoms of each element in a molecule of the compound, and is a multiple of the empirical formula (including a multiple of 1—that is, same as the empirical formula). For example, benzene is a molecule where six carbon atoms are joined together in a ring, with a hydrogen atom attached to each of them. Its molecular formula is therefore C_6H_6, but its empirical formula is just CH. For some compounds, the empirical and molecular formulas are the same, as in the case of H_2O. An ionic compound, such as NaCl or $CaCO_3$, will have only an empirical formula since there are no real molecules in the solid state in these cases, as discussed above.

Don't Mix These Up on Test Day

The molecular formula is either the same as the empirical formula or a multiple of it.

Given a molecular formula, you can always write the empirical formula just by looking to see whether the numbers of atoms are already in the smallest whole number ratio. If not, you can factor out the common factor among them. C_2H_4 is not an empirical formula because you can factor out a two from the subscripts to get CH_2. CH_4, on the other hand, is already an empirical formula. If you are given an empirical formula, however, you need to know the molecular weight (or molar mass) of the compound to find out the actual molecular formula.

Example: A compound with the empirical formula CH_2O has a weight of 180 g/mol. What is the molecular formula?

Solution: Let us first find what the formula weight is from the empirical formula:

$1 \times$ mass of carbon atom $+ 2 \times$ mass of hydrogen atom $+ 1 \times$ mass of oxygen atom $= (1 \times 12 + 2 \times 1 + 1 \times 16)$ g/mol $= 30$ g/mol

The actual molecular weight is 6 times this; therefore the molecular formula must be 6 times the empirical formula: $C_6H_{12}O_6$.

Percent Composition

The percent composition by mass of an element is the weight percent of the element in a specific compound. To determine the percent composition of an element X in a compound, the following formula is used.

$$\% \text{ composition} = \frac{\text{mass of X in formula}}{\text{formula weight of compound}} \times 100\%$$

The percent composition of an element may be determined using either the empirical or molecular formula.

Example: What is the percent composition of chromium in $K_2Cr_2O_7$?

Solution: The formula weight of $K_2Cr_2O_7$ is:

$$2(39 \text{ g/mol}) + 2(52 \text{ g/mol}) + 7(16 \text{ g/mol}) = 294 \text{ g/mol}$$

$$\% \text{ composition of Cr} = \frac{2 \times 52}{294} \times 100\% = 35.4\%$$

Example: What are the empirical and molecular formulas of a compound which contains 40.9% carbon, 4.58% hydrogen, 54.52% oxygen, and has a molecular weight of 264 g/mol?

Solution: First, assume that we have a sample that weighs 100 g total. The percentage then translates directly into the weight of that element in the sample (e.g., 40.9% by weight means 40.9 g in a 100-g sample). Then convert grams to moles by dividing the weight of each element by its molar atomic mass:

$$\# \text{ mol C} = \frac{40.9 \text{ g}}{12 \text{ g/mol}} = 3.41 \text{ mol}$$

$$\# \text{ mol H} = \frac{4.58}{1 \text{ g/mol}} = 4.58 \text{ mol}$$

$$\# \text{ mol O} = \frac{54.52 \text{ g}}{16 \text{g/mol}} = 3.41 \text{ mol}$$

Next, find the simplest whole number ratio of the elements by dividing the number of moles by the smallest number obtained in the previous step.

$$\text{C: } \frac{3.41}{3.41} = 1 \qquad \text{H: } \frac{4.58}{3.41} = 1.33 \qquad \text{O: } \frac{3.41}{3.41} = 1$$

Finally, the empirical formula is obtained by converting the numbers obtained into whole numbers (multiplying them by an integer value). In this case, we want to turn 1.33 into an integer; the smallest number we can multiply it by to make it an integer is 3:

$$1.33 \times 3 = 4$$

The empirical formula is therefore $3 \times C_1H_{1.33}O_1 = C_3H_4O_3$.

This method gives the empirical formula because the elements are always in their smallest whole number ratio. A molecular formula of $C_6H_8O_6$, which is a multiple of two of the empirical formula, would be entirely consistent with the percent compositions given above: You cannot distinguish between the two, or any multiple of the empirical formula, just by percent compositions alone. Incidentally, this is how the term *empirical formula* gets its name: The word *empirical* means experimental, and the values of percent compositions, obtained experimentally through simple analytical techniques, only allow us to determine the empirical formula. (Of course, nowadays with modern technology, we are not limited to experimental techniques that would only give us percent compositions.)

For the second part of the question on the molecular formula, we can use the same approach discussed earlier: Divide the molecular weight by the weight represented by the empirical formula. The resultant value is the number of empirical formula units in the molecular formula. The empirical formula weight of $C_3H_4O_3$ is:

$$3(12 \text{ g/mol}) + 4(1 \text{ g/mol}) + 3(16 \text{ g/mol}) = 88 \text{ g/mol}$$

The molecular weight is given to be 264 g/mol. Therefore:

$$\frac{264}{88} = 3$$

$C_3H_4O_3 \times 3 = C_9H_{12}O_9$ is the molecular formula.

Types of Chemical Reactions

There are many ways in which elements and compounds can react to form other species; memorizing every reaction would be impossible, as well as unnecessary. However, nearly every inorganic reaction can be classified into at least one of four general categories.

Combination Reactions

Combination reactions are reactions in which two or more reactants form one product. The formation of sulfur dioxide by burning sulfur in air is an example of a combination reaction.

$$S (s) + O_2 (g) \rightarrow SO_2 (g)$$

The letters in parentheses designate the phase of the species: *s* for solid, *g* for gas, *l* for liquid, and *aq* for aqueous solution.

Decomposition Reactions

A decomposition reaction is defined as one in which a compound breaks down into two or more substances, usually as a result of heating. An example of a decomposition reaction is the breakdown of mercury (II) oxide (the sign Δ here represents the addition of heat).

$$2HgO(s) \xrightarrow{\Delta} 2Hg(l)O_2(g)$$

Single Displacement Reactions

Single displacement reactions occur when an atom (or ion) of one compound is replaced by an atom of another element. For example, zinc metal will displace copper ions in a copper sulfate solution to form zinc sulfate.

$$Zn\ (s) + CuSO_4\ (aq) \rightarrow Cu\ (s) + ZnSO_4\ (aq)$$

Single displacement reactions are often further classified as redox reactions. (These will be discussed in more detail in chapter 14.)

Double Displacement Reactions

In double displacement reactions, also called metathesis reactions, elements from two different compounds displace each other to form two new compounds. For example, when solutions of calcium chloride and silver nitrate are combined, insoluble silver chloride forms in a solution of calcium nitrate.

$$CaCl_2\ (aq) + 2\ AgNO_3\ (aq) \rightarrow Ca(NO_3)_2\ (aq) + 2\ AgCl\ (s)$$

Neutralization reactions are a specific type of double displacements which occur when an acid reacts with a base to produce a solution of a salt and water. For example, hydrochloric acid and sodium hydroxide will react to form sodium chloride and water.

$$HCl\ (aq) + NaOH\ (aq) \rightarrow NaCl\ (aq) + H_2O\ (l)$$

This type of reaction will be discussed further in the chapter on Acids and Bases, chapter 13.

Net Ionic Equations

Because reactions such as displacements often involve ions in solution, they can be written in ionic form. In the example where zinc is reacted with copper sulfate, the ionic equation would be:

$$Zn\ (s) + Cu^{2+}\ (aq) + SO_4^{2-}\ (aq) \rightarrow Cu\ (s) + Zn^{2+}\ (aq) + SO_4^{2-}\ (aq)$$

When displacement reactions occur, there are usually spectator ions that do not take part in the overall reaction but simply remain in solution throughout. The spectator ion in the equation above is sulfate, which does not undergo any transformation during the reaction. A net ionic reaction can be written showing only the species that actually participate in the reaction:

Quick Quiz

Match the type of reaction with the lettered description below.

1. Combination reaction

2. Decomposition reaction

3. Single displacement reaction

4. Double displacement reaction

(A) Elements from two different compounds displace each other to form two new compounds.

(B) A compound breaks down into two or more substances.

(C) Two or more reactants form one product.

(D) An atom of one compound is replaced by an atom of another element.

Answers: 1. C
2. B
3. D
4. A

$$Zn \ (s) + Cu^{2+} \ (aq) \rightarrow Cu \ (s) + Zn^{2+} \ (aq)$$

Net ionic equations are important for demonstrating the actual reaction that occurs during a displacement reaction.

Balancing Equations

Chemical equations express how much and what type of reactants must be used to obtain a given quantity of product. From the law of conservation of mass, the mass of the reactants in a reaction must be equal to the mass of the products. More specifically, chemical equations must be balanced so that there are the same number of atoms of each element in the products as there are in the reactants. Stoichiometry is essentially the study of how the quantities of reactants and products are related in a chemical reaction. Stoichiometric coefficients are numbers used to indicate the number of moles of a given species involved in the reaction. For example, the reaction for the formation of water is:

$$2 \ H_2 \ (g) + O_2 \ (g) \rightarrow 2 \ H_2O \ (g)$$

The coefficients indicate that two moles of H_2 gas must be reacted with one mole of O_2 gas to produce two moles of water. In general, stoichiometric coefficients are given as whole numbers.

Given the identities of the compound participating in a reaction (the reactants and products), you need to balance the equation for the reaction before you can deduce any stoichiometric information from it. When balancing an equation, the important thing to realize is that you can only change the number in front of the compound, the one that tells you how many molecules (or moles) of that compound are needed for the reaction to occur. You may not change the subscripts—that would change the nature or the identity of the compound, and hence the reaction itself.

For example, in the above reaction, if we are just given the information H_2 and O_2 react to form H_2O and are told to balance the equation to determine the molar relationships, we may *not* write $H_2 + O_2 \rightarrow H_2O_2$. Yes, all the elements are balanced, but the reaction has changed: We have written the formation of hydrogen peroxide instead of water!

Example: Balance the following reaction.

$$C_4H_{10} \ (l) + O_2 \ (g) \rightarrow CO_2 \ (g) + H_2O \ (l)$$

Test Strategy

When balancing equations, it is in general more effective to focus on the least represented elements (often the heavier ones).

Solution: First, we can balance the carbons in reactants and products.

$$C_4H_{10} + O_2 \rightarrow 4\,CO_2 + H_2O$$

Second, balance the hydrogens in reactants and products.

$$C_4H_{10} + O_2 \rightarrow 4\,CO_2 + 5\,H_2O$$

Third, balance the oxygens in the reactants and products.

$$2\,C_4H_{10} + 13\,O_2 \rightarrow 8\,CO_2 + 10\,H_2O$$

Finally, check that all of the elements, and the total charges, are balanced correctly. We could have balanced the elements in a different order, although in general it is easier to tackle the least represented atoms first.

Applications of Stoichiometry

Once an equation has been balanced, the ratio of moles of reactant to moles of products is known, and that information can be used to solve many types of stoichiometry problems.

Example: How many grams of calcium chloride are needed to prepare 72 g of silver chloride according to the following equation?

$$CaCl_2\ (aq) + 2AgNO_3\ (aq) \rightarrow Ca(NO_3)_2\ (aq) + 2AgCl\ (s)$$

Solution: Noting first that the equation is balanced, 1 mole of $CaCl_2$ yields 2 moles of AgCl when it is reacted with two moles of $AgNO_3$. The molar mass of $CaCl_2$ is 110 g, and the molar mass of AgCl is 144 g. As the first step in our calculations, we find out how many moles of AgCl it is that we want:

$$\#\ mol\ AgCl = \frac{72g}{144g/mol} = 0.5\ mol$$

Based on the stoichiometric relationship between AgCl and $CaCl_2$, we know that we need $\frac{0.5}{2} = 0.25$ mol $CaCl_2$. The mass of $CaCl_2$ needed is therefore 0.25 mol $\times$ 110 g/mol = 27.5 g.

This line of reasoning can be quite time-consuming to go through. A powerful technique in handling such problems is that of dimensional analysis, in which we arrange the numbers and quantities so that the units cancel to give us the right one that we want. For this problem, we could have done the calculations in one step:

$$72\ g\ AgCl \times \frac{1\ mol\ AgCl}{144g\ AgCl} \times \frac{1\ mol\ CaCl_2}{2\ mol\ AgCl} \times \frac{110g\ CaCl_2}{1\ mol\ CaCl_2} = 27.5\ g\ CaCl_2$$

Note that we start with the value of 72 g AgCl given in the question and multiply it by three fractions that have in common the property that the numerator is equivalent to (or "corresponds to") the quantity in the denominator: 1 mol of AgCl is equivalent to 144 g of AgCl; 1 mol $CaCl_2$ gives us 2 mol AgCl in this reaction; 1 mol $CaCl_2$ is equivalent to 110 g $CaCl_2$. Because of the equivalence between numerator and denominator, we can switch the two and not affect the fraction as a whole. The way we have decided which to use as numerator and which to use as denominator is dictated by which units we want to cancel. For example, we want to get rid of the weight of AgCl and obtain the number of moles instead, and so we have chosen to put 144 g AgCl in the denominator to cancel the unit of (g AgCl) in the starting value:

$$72 \ \cancel{\text{g AgCl}} \times \frac{1 \text{ mol AgCl}}{144 \ \cancel{\text{g AgCl}}} = \frac{72}{144} \text{ mol AgCl} = 0.5 \text{ mol AgCl}$$

You can verify that all the units do cancel to yield at the end "g $CaCl_2$," which is what we want. The way the equation has to be set up to give the right unit tells us how to manipulate the numbers, without having to spend too much time trying to recall, "Should I divide by the molar mass or multiply?"

Limiting Reactants

When reactants are mixed, they are seldom added in the exact stoichiometric proportions as shown in the balanced equation. Therefore, in most reactions, one of the reactants will be used up first. This reactant is known as the limiting reactant (or limiting reagent) because it limits the amount of product that can be formed in the reaction. The reactant that remains after all of the limiting reactant is used up is called the excess reactant.

Example: If 28 g of Fe react with 24 g of S to produce FeS, what would be the limiting reactant? How many grams of excess reactant would be present in the vessel at the end of the reaction?

Solution: First, the balanced equation needs to be determined. We are told that Fe and S come together to form FeS:

$$Fe + S \rightarrow FeS$$

This is already balanced.

Next, the number of moles for each reactant must be determined.

$$28 \text{ g Fe} \times \frac{1 \text{ mol Fe}}{56 \text{ g}} = 0.5 \text{ mol Fe}$$

$$24 \text{ g S} \times \frac{1 \text{ mol S}}{32 \text{ g}} = 0.75 \text{ mol S}$$

Test Strategy

If the quantities of two reactants are given, be on the lookout for the possibility that one of them is a limiting reactant.

Since one mole of Fe is needed to react with one mole of S, and there are 0.5 mol Fe versus 0.75 mol S, the limiting reagent is Fe. Thus, 0.5 mol Fe will react with 0.5 mol S, leaving an excess of 0.25 mol S in the vessel. The mass of the excess reactant will be:

$$0.25 \text{ mol S} \times \frac{32 g}{1 \text{ mol S}} = 8 \text{ g S}$$

Note that the limiting reactant is not necessarily the one with the smallest mass. It depends also on the molecular (or atomic) weights of all the reactants and also the stoichiometric relationship. In the example above, for example, there is a higher mass of Fe than S, yet as we have seen Fe is the limiting reactant.

Yields

The yield of a reaction, which is the amount of product predicted or obtained when the reaction is carried out, can be determined or predicted from the balanced equation. There are three distinct ways of reporting yields. The theoretical yield is the amount of product that can be predicted from a balanced equation, assuming that all of the limiting reagent has been used, that no competing side reactions have occurred, and all of the product has been collected. The theoretical yield is seldom obtained; therefore, chemists speak of the actual yield, which is the amount of product that is isolated from the reaction experimentally.

The term percent yield is used to express the relationship between the actual yield and the theoretical yield, and is given by the following equation:

$$\text{percent yield} = \frac{\text{actual yield}}{\text{theoretical yield}} \times 100\%$$

Example: What is the percent yield for a reaction in which 27 g of Cu is produced by reacting 32.5 g of Zn in excess $CuSO_4$ solution?

Solution: The balanced equation is as follows:

$$Zn \ (s) + CuSO_4 \ (aq) \rightarrow Cu \ (s) + ZnSO_4 \ (aq)$$

Calculate the theoretical yield for Cu. We are told that $CuSO_4$ is in excess, and so the theoretical yield will be dictated by the amount of Zn:

$$32.5 \text{ g Zn} \times \frac{1 \text{ mol Zn}}{65.4 \text{ g Zn}} \times \frac{1 \text{ mol Cu}}{1 \text{ mol Zn}} \times \frac{63.5 \text{ g Cu}}{1 \text{ mol Cu}} = 31.6 \text{ g Cu}$$

31.6 g Cu = theoretical yield

This is the most one can ever hope to get. The actual yield, we are told, is 27 g. The percent yield is therefore $\frac{27}{31.6} \times 100\% = 85\%$.

Chemical Reactions and Stoichiometry Review Problems

1. What is the sum of the coefficients of the following equation when it is balanced?

$$C_6H_{12}O_6 + O_2 \rightarrow CO_2 + H_2O$$

A. 20

B. 38

C. 21

D. 19

E. 18

2. Determine the molecular formula and calculate the percent composition of each element present in nicotine, which has an empirical formula of C_5H_7N and a molecular weight of 162 g/mol.

3. Acetylene, used as a fuel in welding torches, is produced in a reaction between calcium carbide and water:

$$CaC_2 + 2 H_2O \rightarrow Ca(OH)_2 + C_2H_2$$

How many grams of C_2H_2 are formed from 0.400 moles of CaC_2?

A. 0.400

B. 0.800

C. 4.00

D. 10.4

E. 26.0

4. $CH_3CO_2Na + HClO_4 \rightarrow CH_3CO_2H + NaClO_4$

The above reaction is classified as a

A. double displacement reaction.

B. combination reaction.

C. decomposition reaction.

D. single displacement and decomposition reaction.

E. combination and decomposition reaction.

5. Aspirin ($C_9H_8O_4$) is prepared by reacting salicylic acid ($C_7H_6O_3$) and acetic anhydride ($C_4H_6O_3$):

$$C_7H_6O_3 + C_4H_6O_3 \rightarrow C_9H_8O_4 + C_2H_4O_2$$

How many moles of salicylic acid should be used to prepare six 5-grain aspirin tablets? (1 g = 15.5 grains)

A. 0.01

B. 0.1

C. 1.0

D. 2.0

E. 31.0

6. The percent composition of an unknown element X in CH_3X is 32%. Which of the following is element X?

   A. H

   B. F

   C. Cl

   D. Na

   E. Li

7. 27.0 g of silver was reacted with excess sulfur, according to the following equation:

   $2Ag + S \rightarrow Ag_2S$

   25.0 g of silver sulfide was collected. What are the theoretical yield, actual yield, and percent yield?

8. What is the mass in grams of a single chlorine atom? Of a single molecule of O_2?

The following reaction should be used to answer questions 9–11.

$$Ag(NH_3)_2+ \rightarrow Ag+ + 2NH_3$$

9. How many moles of Ag are required for the production of 11 moles of ammonia (NH_3)?

10. If 5.8 g of $Ag(NH_3)_2{}^+$ yields 1.4 g of ammonia, how many moles of silver are produced?

    A. 4.4

    B. 5.8

    C. 0.041

    D. 0.054

    E. 7.2

11. What are the percent compositions of Ag, N, and H in $Ag(NH_3)_2{}^+$?

12. A hydrocarbon (which by definition contains only C and H atoms) is heated in an excess of oxygen to produce 58.67 g of CO_2 and 27 g of H_2O. What is the empirical formula of the hydrocarbon?

13. Balance the following reactions:

    A. $I_2 + Cl_2 + H_2O \rightarrow HIO_3 + HCl$

    B. $MnO_2 + HCl \rightarrow H_2O + MnCl_2 + Cl_2$

    C. $BCl_3 + P_4 + H_2 \rightarrow BP + HCl$

    D. $C_3H_5(NO_3)_3 \rightarrow CO_2 + H_2O + N_2 + O_2$

    E. $HCl + Ba(OH)_2 \rightarrow BaCl_2 + H_2O$

Turn the page
for answers and explanations
to the Review Problems.

Solutions to Review Problems

1.   D

In order to answer this question, the equation must first be balanced. Starting with carbon, it can be seen that there are six carbons on the reactant side and only one on the product side, so a coefficient of six should be placed in front of the carbon dioxide. For the hydrogen, there are 12 atoms on the left and only two on the right; thus, a coefficient of six should go in front of water. Now, for oxygen, there are eight atoms on the left and 18 on the right. In order to balance the oxygen, ten more atoms of oxygen must be added to the left side. The best way to do this is to put a coefficient of six in front of oxygen, since putting a stoichiometric coefficient in front of the glucose molecule would unbalance the equation in terms of carbon and hydrogen. Therefore, the final balanced equation is:

$$C_6H_{12}O_6 + 6\,O_2 \rightarrow 6\,CO_2 + 6\,H_2O$$

2.   C₁₀H₁₄N₂, 74.1% C, 8.6% H, 17.3% N

To determine the molecular formula of nicotine, the empirical weight of the compound must be calculated.

$$5(C) + 7(H) + 1(N) = \text{empirical weight}$$

$$5(12 \text{ g/mol}) + 7(1 \text{ g/mol}) + 14 \text{ g/mol} = 81 \text{ g/mol}$$

The empirical weight (81 g/mol) is then divided into the molecular weight (162 g/mol) to determine the number by which each subscript in the empirical formula must be multiplied to obtain the molecular formula.

$$\frac{162}{81} = 2$$

$$2(C_5H_7N) = C_{10}H_{14}N_2 = \text{molecular formula}$$

To find the percent composition of each element, the following calculations are carried out.

$$\% \text{ C} = \frac{10 \times 12}{162} \times 100\% = 74.1\%$$

$$\% \text{ H} = \frac{14 \times 1}{162} \times 100\% = 8.6\%$$

$$\% \text{ N} = \frac{2 \times 14}{162} \times 100\% = 17.3\%$$

The same percentages would be obtained if we use the empirical formula for this calculation.

3.   D

According to the balanced equation, one mole of CaC_2 yields one mole of C_2H_2. Therefore, if 0.400 moles of CaC_2 were used, 0.400 moles of C_2H_2 would be produced. The molecular weight of C_2H_2 is 2(12 g/mol) + 2(1 g/mol) = 26 g/mol. Thus, the mass of C_2H_2 is 26 g/mol × 0.400 mol = 10.4 g.

4.   A

Since the only change is that the Na from CH_3CO_2Na exchanges with the H from $HClO_4$, this is a double displacement reaction. Alternatively, this reaction could be classified as a neutralization, since it is an acid and a base that react.

5. A

According to the balanced equation, one mole of salicylic acid will yield one mole of aspirin. Therefore, to solve this question, the number of moles of aspirin in six 5-grain tablets, or 30 grains of aspirin, must be determined, using the following relationship.

$$\frac{1 \text{ g}}{15.5 \text{ grains}} = \frac{x}{30 \text{ grains}}$$

$$x \approx 2 \text{ g}$$

Therefore, the weight of the aspirin produced is about 2 grams, which must be converted to moles. The molecular weight of aspirin is $9(C) + 8(H) + 4(O) = 9(12 \text{ g/mol}) + 8(1 \text{ g/mol}) + 4(16 \text{ g/mol}) = 180 \text{ g/mol}$. Then, the number of moles in two grams of aspirin is calculated.

$$\frac{2 \text{ g}}{180 \text{ g/mol}} = 0.01 \text{ mol}$$

6. E

Let n be the atomic mass of element X. The expression for the percent composition of X in CH_3X is:

$$\frac{1 \times n}{1 \times 12 + 3 \times 1 + 1 \times n} \times 100\% = \frac{n}{15 + n} \times 100\%$$

This, we are told, is equal to 32%. Therefore

$$\frac{n}{15 + n} = 0.32$$

$$n = 4.8 + 0.32\,n$$

$$0.68\,n = 4.8$$

$$n = 7$$

Answer choice E, Li, has an atomic weight of 7.

7. 31 g, 25 g, 81%

According to the balanced equation, two moles of silver should react with one mole of sulfur to form one mole of silver sulfide. The theoretical yield is the amount of product that would be collected if all of the limiting reagent reacts. Using a stoichiometric calculation, the theoretical yield of silver sulfide if 27.0 g of silver is used would be as follows:

$$27 \text{ g Ag} \times \frac{1 \text{ mol Ag}}{108 \text{ g Ag}} \times \frac{1 \text{ mol Ag}_2\text{S}}{2 \text{ mol Ag}}$$

$$\times \frac{248 \text{ Ag}_2\text{S}}{1 \text{ mol Ag}_2\text{S}} = 31 \text{ g Ag}_2\text{S}$$

The actual yield is the amount of product that is obtained from the experiment. It is usually less than the theoretical yield since the reagents may not react completely, and the product may be difficult to collect. In this experiment, the actual yield is 25.0 g of silver sulfide. Finally, the percent yield represents the percentage of product actually collected in reference to the theoretical yield. Thus, the percent yield for this experiment would be

$$\frac{25.0}{31.0} \times 100\% = 81\%$$

8. 5.81×10^{-23} g/atom, 5.31×10^{-23} g/molecule

The mass of a single atom is determined by dividing the atomic weight by Avogadro's number. Therefore, the mass of a chlorine atom is

$$\frac{35 \text{ g}}{6.02 \times 10^{23} \text{ atoms}} = 5.81 \times 10^{-23} \text{ g/atom}$$

The mass of an oxygen molecule (O_2) is similarly determined by dividing the molecular weight of oxygen by Avogadro's number.

$$\frac{32 \text{ g}}{6.02 \times 10^{-23} \text{ molecules}} = 5.31 \times 10^{-23} \text{ g/molecule}$$

9.   5.5 mol

From the balanced equation, it can be seen that for every mole of ammonia produced, 0.5 mol $Ag(NH_3)_2^+$ is needed (a ratio of 2:1). So, to produce 11 mol ammonia, 5.5 mol $Ag(NH_3)_2^+$ is required and since a mole of $Ag(NH_3)_2^+$ contains 1 mole of Ag, 5.5 moles of Ag are required.

10.   C

In order to answer this question, you must use the law of conservation of mass, which says that the mass of the products must be equal to the mass of the reactants. Therefore, if 5.8 g of $Ag(NH_3)_2^+$ are allowed to dissociate to form 1.4 g of ammonia, 5.8 – 1.4 or 4.4 g of silver must be formed. The following calculation is used to determine the number of moles of silver that are formed.

$$\# \text{ mol Ag} = \frac{4.4 \text{ g}}{108 \text{ g/mol}} = 0.041 \text{ mol}$$

11.   76.1%, 19.7%, 4.2%

The percent composition of elements in a compound or formula is determined by dividing the mass of the element by the total formula weight of the compound. Therefore, in the complex ion, $Ag(NH_3)_2^+$, which has a formula weight of 142 g/mol, the percent compositions of Ag, N, and H are as follows:

$$\% \text{ Ag} = \frac{108 \text{ g/mol}}{142 \text{ g/mol}} \times 100\% = 76.1\%$$

$$\% \text{ N} = \frac{2 \times 14 \text{ g/mol}}{142 \text{ g/mol}} \times 100\% = 19.7\%$$

$$\% \text{ H} = \frac{6 \times 1 \text{ g/mol}}{142 \text{ g/mol}} \times 100\% = 4.2\%$$

Note that $(NH_3)_2$ implies N_2H_6: 2 atoms of N and 6 atoms of H.

12.   C4H9

The reaction is of the form:

$$\text{hydrocarbon} + O_2 \rightarrow CO_2 \text{ and } H_2O$$

We can see that all the carbon and all the hydrogen on the product side have to come from the hydrocarbon originally, and so from the amounts of CO_2 and H_2O given, the moles of carbon and hydrogen in the hydrocarbon can be calculated.

$$\frac{58.67 \text{ g } CO_2}{44 \text{ g/mol}} = 1.33 \text{ mol CO2}$$

Since each mole of CO_2 contains one mole of carbon, 1.33 moles of CO_2 contains 1.33 moles of carbon. Therefore, the hydrocarbon contains 1.33 moles of carbon.

$$\frac{27 \text{ g } H_2O}{44 \text{ g/mol}} = 1.5 \text{ mol } H_2O$$

Since one mole of H_2O contains two moles of hydrogen atoms, 1.5 moles of H_2O contains 3.0 moles of hydrogen. Therefore, the hydrocarbon contains 3 moles of hydrogen.

Using these calculations, the simplest formula that can be written is $C_{1.33}H_3$. However, empirical and molecular formulas are not expressed with decimals or fractions, so these coefficients should be multiplied by their least common multiple to get whole number coefficients. Both 1.33 and 3 are multiplied by 3 to give an empirical formula of C_4H_9.

13.

The following are the correct balanced equations.

A. $I_2 + 5 Cl_2 + 6 H_2O \rightarrow 2 HIO_3 + 10 HCl$

B. $MnO_2 + 4 HCl \rightarrow 2 H_2O + MnCl_2 + Cl_2$

C. $4 BCl_3 + P_4 + 6 H_2 \rightarrow 4 BP + 12 HCl$

D. $4 C_3H_5(NO_3)_3 \rightarrow 12 CO_2 + 10 H_2O + 6 N_2 + O_2$

E. $2 HCl + Ba(OH)_2 \rightarrow BaCl_2 + 2 H_2O$

CHAPTER 7

THE GAS PHASE

Among the different phases of matter, the gaseous phase is the simplest to understand and to model, since all gases, to a first approximation, display similar behavior and follow similar laws regardless of their identity. The atoms or molecules in a gaseous sample move rapidly and are far apart. In addition, intermolecular forces between gas particles tend to be weak; this results in certain characteristic physical properties, such as the ability to expand to fill any volume and to take on the shape of a container. Furthermore, gases are easily, though not infinitely, compressible.

The state of a gaseous sample is generally defined by four variables: **pressure** (P), **volume** (V), **temperature** (T), and **number of moles** (n), though as we shall see, these are not all independent. The pressure of a gas is the force per unit area that the atoms or molecules exert on the walls of the container through collision. The SI unit for pressure is the pascal (Pa), which is equal to 1 newton per meter squared. (The SI units are those that are based on the simple metric units of kilogram, meter, second, etcetera. A newton, for example, is a $kg \bullet m/s^2$.) In chemistry, however, gas pressures are more commonly expressed in units of atmospheres (atm) or millimeters of mercury (mmHg or torr), which are related as follows:

$$1 \text{ atm} = 760 \text{ mmHg} = 760 \text{ torr} = 101325 \text{ Pa}$$

Volume is generally expressed in liters (L) or milliliters (mL). The temperature of a gas is usually given in Kelvin (K, not °K), and its value, also known as the absolute temperature, is related to the temperature in degrees Celsius by the expression $T(K) = T(°C) + 273.15$. Gases are often discussed in terms of standard temperature and pressure (STP), which refers to conditions of 273.15 K (0°C) and 1 atm.

Note that it is important not to confuse STP with standard conditions encountered in thermochemistry—the two standards involve different temperatures and are used for different purposes. STP (0°C or 273 K) is generally used for gas law calculations; standard conditions (usually 25°C or 298 K) is used when measuring standard enthalpy, entropy, Gibbs's free energy, and cell potential.

Measurement of Gas Pressures

As stated above, the pressure of a gas is the force it exerts per unit area. This force can push a column of liquid up a tube: the higher the pressure, the larger the force for a given area, and consequently the more liquid that is pushed up. This behavior is exploited in the measuring of pressure. There are two different kinds of setups or devices: the barometer and the manometer.

The simplest form of barometer consists of a container of liquid open to the atmosphere, with an inverted tube placed in it. Care is taken so that there is no air trapped within the inverted tube. The pressure exerted by the air in the environment (i.e., the atmospheric pressure) pushes the liquid up the tube. The pressure is reported as the height to which the liquid rises. A value of 760 mm Hg, for example, which is a typical reading for atmospheric pressure, means that liquid mercury (Hg) rises to a height of 760 mm.

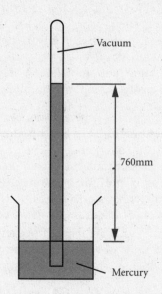

If, instead of measuring just the atmospheric pressure, we wish to measure the pressure of a gas generated in a reaction, a slightly different setup is needed, although it is one that uses the same principle. In an open-tube manometer, the liquid (usually mercury) is placed in a U-tube with one end exposed to the atmosphere and the other end connected to the closed vessel holding the gas whose pressure we want to measure:

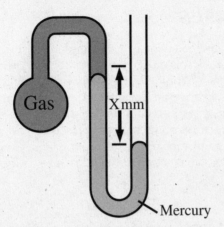

Mercury

It is the difference in heights of the two ends of the mercury that enables us to calculate the pressure of the gas in the vessel. If the end exposed to the atmosphere is x mm lower than the other end (as in the diagram above), then the pressure of the gas is $(760 - x)$ mmHg: The pressure is lower than atmospheric pressure, hence the mercury is pushed down further on the open-end side. Conversely, if the exposed end is x mm higher than the other end, the pressure of the gas is higher than atmospheric pressure, more exactly $(760 + x)$ mmHg.

A close-tube manometer setup, considered a hybrid between the barometer and the open-end manometer, is also possible. The end that used to be exposed is now sealed off as well, with any trapped air again evacuated. This time, the difference in heights is read off directly as the pressure of the gas, without having to reference the atmospheric pressure. In the diagram below, then, the pressure of the gas is simply x mmHg.

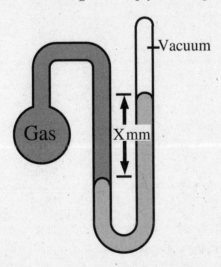

Vacuum

This setup is generally used to measure only pressures that are lower than atmospheric pressure.

Ideal Gases

When examining the behavior of gases under varying conditions of temperature and pressure, it is most convenient to treat them as ideal gases. An ideal gas represents a hypothetical gas whose molecules have no intermolecular forces (that is, they do not interact with each other) and occupy no volume. Although gases in reality deviate from this idealized behavior, at relatively low pressures (atmospheric pressure) and high temperatures many gases behave in a nearly ideal fashion. Therefore, the assumptions used for ideal gases can be applied to real gases with reasonable accuracy.

Boyle's Law

Experimental studies on the relationship between the pressure and the volume of a gas performed by Robert Boyle in 1660 led to the formulation of Boyle's law. His work showed that for a given gaseous sample held at constant temperature (isothermal conditions), the volume of the gas is inversely proportional to its pressure:

$$PV = k \text{ or } P_1V_1 = P_2V_2$$

where k is a proportionality constant and the subscripts 1 and 2 represent two different sets of conditions. A plot of volume versus pressure for an ideal gas is shown below.

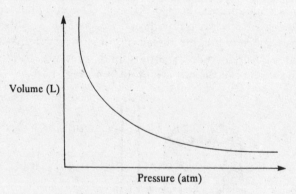

Volume (L)

Pressure (atm)

Example: Under isothermal conditions, what would be the volume of a 1 L sample of helium after its pressure is changed from 12 atm to 4 atm?

Solution: $P_1 = 12 \text{ atm } P_2 = 4 \text{ atm}$

$V_1 = 1 \text{ L} \quad V_2 = X$

$P_1V_1 = P_2V_2$

$12 \text{ atm } (1 \text{ L}) = 4 \text{ atm } (X)$

$X = 3 \text{ L}$

Dalton's Law of Partial Pressures

When two or more ideal gases are found in one vessel without chemical interaction, each gas will behave independently of the other(s). Therefore, the pressure exerted by each gas in the mixture will be equal to the pressure that gas would exert if it were the only one in the container. The pressure exerted by each individual gas is called the partial pressure of that gas. In 1801, John Dalton derived an expression, now known as Dalton's law of partial pressures, which states that the total pressure of a gaseous mixture is equal to the sum of the partial pressures of the individual components:

$$P_T = P_A + P_B + P_C + \ldots$$

The mole fraction of a gas A in a mixture of gases is defined as:

$$X_A = \frac{\text{number of moles of A}}{\text{total number of moles of gases}}$$

The partial pressure of a gas is related to its mole fraction and can be determined using the following equation:

$$P_A = P_T X_A$$

Example: A vessel contains 0.75 mol of nitrogen, 0.20 mol of hydrogen, and 0.05 mol of fluorine at a total pressure of 2.5 atm. What is the partial pressure of each gas?

Solution: First calculate the mole fraction of each gas.

$$X_{N_2} = \frac{0.75\text{mol}}{1.00\text{mol}} = 0.75,$$

$$X_{H_2} = \frac{0.20\text{mol}}{1.00\text{mol}} = 0.20,$$

$$X_{F_2} = \frac{0.05\text{mol}}{1.00\text{mol}} = 0.05$$

Then calculate the partial pressure:

$$P_A = X_A P_T$$

$$P_{N_2} = (2.5\text{ atm})(0.75) = 1.875\text{ atm}$$

$$P_{H_2} = (2.5\text{ atm})(0.20) = 0.50\text{ atm}$$

$$P_{F_2} = (2.5\text{ atm})(0.05) = 0.125\text{ atm}$$

Test Strategy

Look for opposing answers in the answer selections. If two answers are close in wording or if they contain opposite ideas, there is a strong possibility that one of them is the correct answer.

Law of Charles and Gay-Lussac

The Law of Charles and Gay-Lussac, more simply known as Charles's Law, states that at constant pressure, the volume of a gas is directly proportional to its absolute temperature;

$$\frac{V}{T} = k \text{ or } \frac{V_1}{T_1} = \frac{V_2}{T_2}$$

where k is a constant and the subscripts 1 and 2 represent two different sets of conditions. A plot of volume versus temperature is shown below.

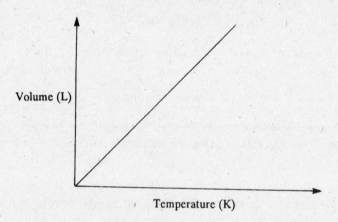

Volume (L)

Temperature (K)

Example: If the absolute temperature of 2 L of gas at constant pressure is changed from 283.15 K to 566.30 K, what would be the final volume?

Solution: $T_1 = 283.15 \text{ K}$ $\qquad V_1 = 2 \text{ L}$

$T_2 = 566.30 \text{ K}$ $\qquad V_2 = X$

$$\frac{V_1}{T_1} = \frac{V_2}{T_2}$$

$$\frac{2L}{283.15K} = \frac{X}{566.30K}$$

$$X = 2L \times \frac{566.30K}{283.15K} = 2L \times 2 = 4L$$

Avogadro's Law

For all gases at a constant temperature and pressure, the volume of the gas will be directly proportional to the number of moles of gas present; therefore, all gases at the same temperature and pressure have the same number of moles in the same volume.

$$\frac{n}{V} = k \text{ or } \frac{n_1}{V_1} = \frac{n_2}{V_2}$$

The subscripts 1 and 2 once again apply to two different sets of conditions with the same temperature and pressure.

Ideal Gas Law

The ideal gas law combines the relationships outlined in Boyle's law, Charles's law, and Avogadro's law to yield an expression which can be used to predict the behavior of a gas. The ideal gas law shows the relationship between four variables that define a sample of gas—pressure (P), volume (V), temperature (T), and number of moles (n)—and is represented by the equation

$$PV = nRT$$

The constant R is known as the (universal) gas constant. Under STP conditions (273.15 K and 1 atmosphere), 1 mole of gas was shown to have a volume of 22.4 L. Substituting these values into the ideal gas equation gave $R = 8.21 \times 10^{-2}$ L • atm/(mol • K).

The gas constant may be expressed in many other units; another common value is 8.314 J/(K.mol), which is derived when the SI units of pascals (for pressure) and cubic meters (for volume) are substituted into the ideal gas law. When carrying out calculations based on the ideal gas law, it is important to choose a value of R that matches the units of the variables.

Example: What volume would 12 g of helium occupy at 20°C and a pressure of 380 mmHg?

Solution: The ideal gas law can be used, but first, all of the variables must be converted to yield units that will correspond to the expression of the gas constant as 0.0821 L • atm/(mol • K).

P = 0.5 atm

T = 293.15K

$n = 12 \text{ g} \times (1 \text{ mol He}/4 \text{ g}) = 3$ mol

Substituting into the ideal gas equation:

Basic Concept

All the information in Boyle's law, Charles's law, and Avogadro's law is contained in the ideal gas equation:

$$PV = nRT$$

$$PV = nRT$$

$$(0.5 \text{ atm})(V) = (3 \text{ mol})(0.0821 \text{ L} \bullet \text{atm}/(\text{mol} \bullet \text{K}))(293.15\text{K})$$

$$V = 144.4 \text{ L}$$

Basic Concept

density = mass/volume

Further Uses of the Ideal Gas Law

In addition to standard calculations to determine the pressure, volume, or temperature of a gas, the ideal gas law may be used to determine the density and molar mass of a gas.

Calculating Density.

Density is defined as the mass per unit volume of a substance and, for gases, is usually expressed in units of g/L. By rearrangement, the ideal gas equation can be used to calculate the density of a gas.

$$PV = nRT$$

n, the number of moles, is also equal to m/M where m = total mass (in g) and M = molar mass (in g/mol).

$$\therefore PV = \frac{m}{M}RT$$

$$\text{density} = \frac{m}{V} = \frac{PM}{RT}$$

Another way to find the density of a gas is to start with the volume of a mole of gas at STP, 22.4 L, calculate the effect of pressure and temperature on the volume, and finally calculate the density by dividing the mass by the new volume. The following equation, derived from Boyle's and Charles's Laws, is used to relate changes in the temperature, volume and pressure of a gas:

$$\frac{P_1V_1}{T_1} = \frac{P_2V_2}{T_2}$$

where the subscripts 1 and 2 refer to the two states of the gas (at STP and under the actual conditions). Rearranging it gives:

$$V2 = V_1\left(\frac{P_1}{P_2}\right)\left(\frac{T_2}{T_1}\right)$$

V_2 is then used to find the density of the gas under nonstandard conditions by d = m/V_2. Since we started by assuming 1 mole of gas (22.4 L at STP), the mass used will be the molar mass.

If you try to anticipate how the changes in pressure and temperature affect the volume of the gas, this can serve as a check to be sure you have not accidentally confused the pressure or temperature value that belongs in the numerator with the one that belongs in the denominator.

Example: What is the density of HCl gas at 2 atm and 45°C?

Solution: At STP, a mole of gas occupies 22.4 liters. Since the increase in pressure to 2 atm decreases volume, 22.4 L must be multiplied by 1/2. And since the increase in temperature increases volume, the temperature factor will be 318/273.

$$V_2 = 22.4 \text{ L/mol} \times (1/2) \times (318/273) = 13.0 \text{ L/mol}$$

$$d = (36 \text{ g/mol})/(13.0 \text{ L/mol}) = 2.77 \text{ g/L}$$

Calculating Molar Mass

Sometimes the identity of a gas is unknown, and the molar mass must be determined in order to identify it. The density of the gas at STP is calculated using the gas laws and experimentally determined values of mass and volume. The molecular weight is then found by multiplying the density at STP by 22.4 liters, the volume of one mole of gas at STP. Alternatively, the number of moles of the gas can be found using the ideal gas law, and then the molar mass can be found by dividing the total mass of the gas by the number of moles contained.

Example: What is the molar mass of a 2.8 g sample of gas that occupies a volume of 3L at a temperature of 546 K and a pressure of 1.5 atm?

Solution: $V_{STP} = 3L\left(\dfrac{273K}{546K}\right)\left(\dfrac{1.5\text{atm}}{1\text{atm}}\right) = 2.25L$

$\dfrac{2.8g}{2.25L} = 1.24g/L$ at STP

molar mass $= 1.24 \text{ g/L} \times 22.4 \text{ L/mol} = 28 \text{ g/mol}$

The gas is therefore most likely N_2.

Kinetic Molecular Theory of Gases

As indicated by the gas laws, all gases show similar physical characteristics and behavior. A theoretical model to explain why gases behave the way they do was developed during the second half of the 19th century. The combined efforts of Boltzmann, Maxwell, and others led to the **kinetic molecular theory of gases**, which gives us an understanding of gaseous behavior on a microscopic, molecular level. Like the gas laws, this theory was developed in reference to ideal gases, although it can be applied with reasonable accuracy to real gases as well.

The assumptions of the kinetic molecular theory of gases are as follows:

1. Gases are made up of particles whose volumes are negligible compared to the container volume.

2. Gas atoms or molecules exhibit no intermolecular attractions or repulsions.

3. Gas particles are in continuous, random motion, undergoing collisions with other particles and with the container walls.

4. Collisions between any two gas particles are elastic, meaning that no energy is dissipated or, equivalently, that kinetic energy is conserved.

5. The average kinetic energy of gas particles is proportional to the absolute temperature of the gas, and is the same for all gases at a given temperature.

Average Molecular Speeds

According to the kinetic molecular theory of gases, the average kinetic energy of a gas is proportional to the absolute temperature of the gas; more specifically, the kinetic energy of one mole of gas is 3/2 RT. Since the kinetic energy is related to the speed ($KE = 1/2\ mv^2$), this also means that the higher the temperature, the faster the gas molecules are moving. However, because the large number of rapidly and randomly moving gas particles do not all move at the same speed, the speed of an individual gas molecule is nearly impossible to define and is not a very useful concept. There will be molecules that move faster and those that move slower than the average value. A Maxwell-Boltzmann distribution curve shows the distribution of speeds of the gas particles at a given temperature. The figure below shows a distribution curve of molecular speeds at two temperatures, T_1 and T_2, where $T_2 > T_1$. Notice that the bell-shaped curve flattens and shifts to the right as the temperature increases, indicating that at higher temperatures more molecules are moving at high speeds. The area under the curve is the total number of gas molecules and thus has to remain constant even as the shape of the curve changes.

Basic Concept

For gases of the same molecules, the higher the temperature, the higher the kinetic energy, and thus the higher the average speed of the molecules.

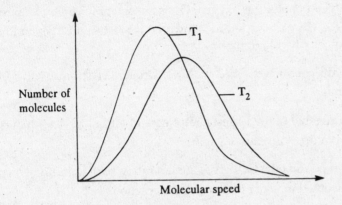

Graham's Law of Diffusion and Effusion

The typical speed of a gas molecule in room temperature is quite high. If, however, a bottle of perfume at the center of a room is opened while you are standing in a corner, the time lapse between the opening of the bottle and when you can detect the smell may be longer than what would be expected from the molecular speed. This is because the perfume molecules are constantly colliding with air molecules that change their course in a random fashion. This process in which gas molecules move through a mixture is known as diffusion (although this term is not limited to the gaseous phase). Effusion, on the other hand, is the flow of gas particles under pressure from one compartment to another through a small opening. The kinetic molecular theory of gases predicted that heavier gas molecules move more slowly than lighter ones under the same conditions. Under isothermal (same temperature) and isobaric (same pressure) conditions, the rates at which two gases diffuse and effuse follow the same mathematical rule and are inversely proportional to the square root of their molar masses:

$$\frac{r_1}{r_2} = \left(\frac{M_2}{M_1}\right)^{1/2} = \sqrt{\frac{M_2}{M_1}}$$

where r_1 and M_1 represent the diffusion/effusion rate and molar mass of gas 1, and r_2 and M_2 represent the diffusion/effusion rate and molar mass of gas 2.

Descriptive Chemistry of Some Common Gases

There are certain miscellaneous facts on the properties of some common gases that one needs to be aware of for the SAT II: Chemistry Test. These properties are exploited in qualitative tests designed to detect their presence:

A Closer Look

One common application of Graham's Law is in the separation of isotopes, for example in the enrichment of uranium for nuclear reactors. Most of the naturally occurring atoms of uranium are U-238, while a few are the U-235 isotopes used as fuel for nuclear fission reactors. This lighter isotope is collected and concentrated by letting a sample of uranium (compounded with fluorine) effuse into an evacuated container. The fluoride compound with the lighter isotope would diffuse faster and thus be present in a higher concentration in the second container. This procedure is repeated several times to attain the desired isotopic purity.

- Oxygen: molecular oxygen, O_2, is a reactant in combustion reactions. If a glowing splint is lowered into a test tube containing oxygen, it will reignite.

- Hydrogen: when ignited in air, hydrogen, H_2, burns with a blue flame.

- Nitrogen: N_2, the largest component of air (a little less than 80% by volume) is relatively inert.

- Carbon dioxide: CO_2 gives a moderately acidic solution when dissolved in water because of the reaction:

$$CO_2 \ (g) + H_2O \ (l) \rightleftharpoons H_2CO_3 \ (aq)$$

When carbon dioxide is passed through limewater, $Ca(OH)_2$, the solution turns cloudy from the formation of insoluble calcium carbonate:

$$CO_2 + Ca(OH)_2 \rightarrow CaCO_3 + H_2O$$

The precipitation of calcium carbonate, however, does not go on indefinitely. As just mentioned, water containing CO_2 is slightly acidic, and this causes the calcium carbonate to dissolve:

$$CaCO_3 \ (s) + H_2O \ (l) + CO_2 \ (g) \rightarrow Ca^{2+} \ (aq) + 2HCO_3^- \ (aq)$$

The Gas Phase
Review Problems

1. Boyle's Law can be used for which of the following?

   A. Predicting the expected volumes of two party balloons

   B. Predicting the relative pressures inside a hot air balloon

   C. Predicting the change in volume of an inflatable toy from summer to winter

   D. Predicting the height of a mercury barometer column in a low-pressure system

   E. Predicting the change in volume of a party balloon inside a bell jar as a vacuum is being drawn

2. A sample of argon occupies 50 L at standard temperature. Assuming constant pressure, what volume will the gas occupy if the temperature is doubled?

   A. 25 L

   B. 50 L

   C. 100 L

   D. 200 L

   E. 2500 L

3. What is the molecular weight of an unknown gas if 2.5 g of it occupies 2 L at 630 torr and a temperature of 600 K?

4. Explain the conditions that define an ideal gas.

5. If a 360 mL sample of helium contains 0.25 mol of the gas, how many molecules of chlorine gas would occupy the same volume at the same temperature and pressure?

   A. 1.2×10^{24}

   B. 6.022×10^{23}

   C. 3.01×10^{23}

   D. 1.51×10^{23}

   E. 7.55×10^{22}

6. In the kinetic molecular theory of gases, which of the following statements concerning average speeds is true?

   A. Most of the molecules are moving at the average speed.

   B. Any given molecule moves at the average speed most of the time.

   C. When the temperature increases, more of the molecules will move at the new average speed.

   D. When the temperature increases, fewer molecules will move at the new average speed.

   E. When the temperature increases, the average speed decreases.

7. What is the density of a gas at 76 torr and 37°C (molar mass = 25 g/mol)?

   A. 0.1 g/L

   B. 0.8 g/L

   C. 22.4 g/L

   D. 75 g/L

   E. 633 g/L

8. The following reaction represents the production of hydrogen chloride gas.

   $$H_2 + Cl_2 \rightarrow 2\ HCl$$

   How many grams of chlorine gas are needed to produce 3 L of HCl gas at a pressure of 2 atm and a temperature of 19°C?

9. A student performing an experiment has a bulb containing 14 g of N_2, 64 g of O_2, 8 g of He, and 35 g of Cl_2, at a total pressure of 380 torr. What are the partial pressures of each gas?

10. All of the following statements underlie the kinetic molecular theory of gases EXCEPT:

    A. Gas molecules have no intermolecular forces.

    B. Gas particles are in random motion.

    C. The collisions between gas particles are elastic.

    D. Gas particles have no volume.

    E. The average kinetic energy is proportional to the temperature (°C) of the gas.

11. What will be the final pressure of a gas that expands from 1 L at 10°C to 10 L at 100°C, if the original pressure was 3 atm?

12. In the reaction $N_2 + 2O_2 \rightarrow 2\ NO_2$, what volume of NO_2 is produced from 7 g of nitrogen gas at 27°C and 0.9 atm?

Use the following information to answer questions 13 and 14.

Water undergoes electrolysis to produce hydrogen and oxygen gas at 14°C, with the products collected above water. The vapor pressure of water at 14°C is 12 mmHg. The total pressure is 740 mmHg.

13. What is the pressure due to the electrolysis products (hydrogen gas plus oxygen gas)?

14. If the partial pressure of oxygen is 242.7 mm Hg, what is the mole fraction of hydrogen gas?

Turn the page
for answers and explanations
to the Review Problems.

Solutions to Review Problems

1. E

Boyle's Law states that when a gas is held at constant temperature, its pressure and volume are inversely proportional. This means that as the pressure increases, the volume decreases, and vice versa. Of the answer choices, the only one that involves both pressure and volume—in addition to a controlled variation of one of the variables—is choice E. When a balloon is placed in a bell jar, the volume of the balloon will increase as a vacuum is being drawn in the jar. Boyle's Law can be used to predict this behavior.

2. C

This question is an application of Charles's Law, which states that at constant pressure, the volume and temperature of a gas will vary in direct proportion to each other. If a 50 L volume of gas is heated from standard temperature, which is 273 K, to two times standard temperature, 576 K, the volume will double as well. Therefore, the volume of the gas will increase from 50 L to 100 L.

3. 74 g/mol

Using the ideal gas law, we can determine the number of moles of the gas:

$$n = \frac{PV}{RT} = \frac{\left(630 \text{ torr} \times \frac{1\text{atm}}{760\text{torr}}\right)(2\text{L})}{(0.082\text{L}\bullet\text{atm/mol}\bullet\text{K})(600\text{K})} = 0.034 \text{ mol}$$

The molecular weight, or molar mass, of the gas can then be found by dividing the mass of the gas by the number of moles it contains: 2.5 g / 0.034 mol = 74 g/mol.

4.

The conditions that define an ideal gas are low pressure and high temperature. Under these conditions, the ideal gas assumption that gas molecules have no intermolecular forces and occupy no volume are most valid.

5. D

This question is an application of Avogadro's Principle, which states that at a constant temperature and pressure, all gases will have the same number of moles in the same volume. This is true regardless of the identity of the gas. Thus, if there is 0.25 mol of He gas under one set of conditions, there will likewise be 0.25 mol of chlorine gas under the same set of conditions. The number of chlorine gas molecules is therefore 0.25 times Avogadro's number:

$$(0.25 \text{ mol}) (6.022 \times 10^{23} \text{ molecules/mol}) = 1.51 \times 10^{23} \text{ molecules}$$

Note that the actual numerical value of the volume does not even come into play.

6.   D

The average speed of a gas is defined as the mathematical average of all the speeds of the gas particles in a sample. To answer this question, you must understand the Maxwell-Boltzmann distribution curve, which shows the distribution of speeds of all the gas particles in a sample at a given temperature. The distribution curve is a bell-shaped curve that flattens and shifts to the right as the temperature increases. The flattening of the curve means that gas particles within the sample are traveling at a greater range of speeds. As a result, a smaller proportion of the molecules will move at exactly the new average speed.

7.   A

A gas weighing 25 g/mol will have a density of (25 g/mol) / (22.4 L/mol) = (25/22.4) g/mol at STP. The density at 76 torr and 37°C is found by calculating the change in volume of a mole of gas under these conditions:

$$V_2 = V_1 \left(\frac{P_1}{P_2} \right) \left(\frac{T_2}{T_1} \right)$$

$$= (22.4 \text{L/mol}) \left(\frac{760 \text{torr}}{76 \text{torr}} \right) \left(\frac{310 \text{K}}{273 \text{K}} \right)$$

$$= 254 \text{L / mol}$$

Therefore the density of the gas is 25g / 254L = 0.1 g/mol. Alternatively, we can use the equation given earlier: density $= \dfrac{m}{V} = \dfrac{PM}{RT}$.

8.   8.88 g

First, find out the volume of one mole of gas at the pressure and temperature given:

$$22.4 \text{ L/mol} \times (1 \text{ atm} / 2 \text{ atm})$$
$$\times (292 \text{K} / 273 \text{ K}) = 12.0 \text{ L}$$

Since one mole of HCl occupies 12 L at this temperature and pressure, 3 L HCl then corresponds to 0.25 mol. Since 2 mol of HCl are produced from each mol of Cl_2, 0.25 mol HCl would be produced from 0.125 mol of Cl_2. The molecular weight of Cl_2 is 71, so the answer is (71 g/mol of Cl_2) (0.125 mol) = 8.88 g.

9.   38 torr, 152 torr, 152 torr, 38 torr

According to Dalton's law of partial pressures, the sum of the partial pressures of the gases in a mixture is equal to the total pressure of the mixture. Therefore, the partial pressures of nitrogen, oxygen, helium, and chlorine will add up to 380 torr. The partial pressure of a gas A, P_A, is calculated using the equation $P_A = P_T X_A$ where X is the mole fraction of the gas A and P_T is the total pressure. First, then, one must calculate the number of moles of each gas present by dividing the mass of each gas by its molar mass, and then determine the mole fraction of each. It can be easily verified that $X_{N_2} = 0.1$, $X_{O_2} = 0.4$, $X_{He} = 0.4$, and $X_{Cl_2} = 0.1$. Now the partial pressures may be calculated:

$P_{N_2} =$	(380 torr)(0.1) =	38 torr
$P_{O_2} =$	(380 torr)(0.4) =	152 torr
$P_{He} =$	(380 torr)(0.4) =	152 torr
$P_{Cl_2} =$	(380 torr)(0.1) =	38 torr

10.   E

The average kinetic energy of a gas is proportional to its temperature in K, not °C.

11. 0.4 atm

Rearranging the equation:

$$\frac{P_1}{V_1} = \frac{P_2 V_2}{T_2}$$

gives:

$$P_2 = P_1 \left(\frac{V_1}{V_2}\right)\left(\frac{T_2}{T_1}\right) = (3atm)\left(\frac{1L}{10L}\right)\left(\frac{373K}{283K}\right) = 0.4atm$$

12. 13.7 L

7g of nitrogen corresponds to 0.25 mol, and hence from the balanced reaction, one would expect that 0.50 mol of NO_2 will be produced. Therefore, the volume at STP will be (0.5 mol NO_2) (22.4 L/mol at STP) = 11.2 L NO_2

Now find the volume under the conditions given:

$$(11.2L)\left(\frac{1atm}{0.9atm}\right)\left(\frac{300K}{2734K}\right) = 13.7L$$

13. 728 mm Hg

There are three different gases in this system: hydrogen, oxygen, and water vapor. Therefore, the pressure due to hydrogen plus oxygen will be equal to the total pressure minus the vapor pressure of the water, which is 12 mm Hg.:

740 mm Hg – 12 mm Hg = 728 mm Hg

14. 0.66

From Dalton's law of partial pressures, the sum of the partial pressures is equal to the total pressure. In Question 13 it was determined that the pressure due to H_2 and O_2 is 728 mm Hg. If the partial pressure of oxygen is 242.7 mm Hg, then the partial pressure of hydrogen is 728 mm Hg – 242.7 mm Hg = 485.3 mm Hg Now, the mole fraction can be calculated using the equation

$$P_A = P_T X_A$$
$$485.3 \text{ mm Hg} = (X_{H_2})(740 \text{ mm Hg})$$
$$X_{H_2} = 0.66$$

CHAPTER 8

THE CONDENSED PHASES AND PHASE CHANGES

When the attractive forces between molecules overcome the random thermal kinetic energy that keeps molecules apart in the gas phase, the molecules cluster together so that they can no longer move about freely, and enter the liquid or solid phase. Because of their smaller volume relative to gases, liquids and solids are often referred to as the condensed phases.

General Properties of Liquids

In a liquid, atoms or molecules are held close together with little space between them. As a result, liquids, unlike gases, have definite volumes and cannot easily be expanded or compressed. However, the molecules can still move around and are in a state of relative disorder. Consequently, a liquid can change shape to fit its container, and its molecules are able to diffuse and evaporate.

One of the most important properties of liquids is their ability to mix, both with each other and with other phases, to form solutions. The degree to which two liquids can mix is called their miscibility. Oil and water are almost completely immiscible because of their polarity difference. Oil and water normally form separate layers when mixed, with oil on top because it is less dense. Under extreme conditions, such as violent shaking, two immiscible liquids can form a fairly homogeneous mixture called an emulsion. Although they look like solutions, emulsions are actually mixtures of discrete particles too small to be seen distinctly.

A Closer Look

Although they look like solutions, emulsions are actually mixtures of discrete particles too small to be seen distinctly.

General Properties of Solids

In a solid, the attractive forces between atoms, ions, or molecules are strong enough to hold them rigidly together; thus, the particles' only motion is vibration about fixed positions, and the kinetic energy of solids is predominantly vibrational energy. As a result, solids have definite shapes and volumes.

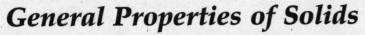

Crystalline Solids

A solid may be crystalline or amorphous. A crystalline solid, such as NaCl, possesses an ordered structure; its atoms exist in a specific three-dimensional geometric arrangement with repeating patterns of atoms, ions, or molecules. Because of this long-range order, crystalline solids have highly regular shapes.

The repeating units of crystals are represented by the unit cell, the smallest structural unit that contains all the information about the spatial arrangement of the particles. Unit cells can be thought of as building blocks of crystalline solids: Repeating them in all directions will reproduce the crystal. There are many types of unit cells, differing in geometry, size, etc. Below are the structures of the three cubic unit cells: simple cubic, body-centered cubic, and face-centered cubic.

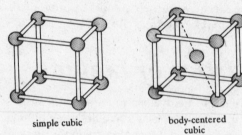 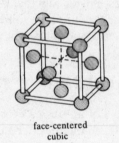

simple cubic   body-centered cubic   face-centered cubic

The structure of crystalline solids can be determined using the technique of X-ray crystallography. When electromagnetic radiation (of which X-ray is a type) passes through matter, it interacts with the electrons and is scattered in different directions. If the matter is made up of a regularly-spaced array of structural units (as is the case for crystals), the X-ray is scattered in such a way that it produces a characteristic pattern of spots. This phenomenon is known as diffraction. The arrangement of the spots, known as the diffraction pattern, is determined by the structure of the crystal (the dimensions and shape of the unit cell) and the arrangement of particles within it. Obtaining the diffraction pattern, then, allows one to deduce the crystal structure.

Most solids are crystalline in structure, and can be further categorized by the nature of the forces that hold the particles together.

Ionic Crystals

Ionic solids are generally formed between atoms with significantly different electronegativities. The atoms exist in their ionic form in the solid, with the positive and negative ions occupying alternate sites within the unit cell. Examples include NaCl and CaF_2. Ionic solids are brittle and are characterized by high melting points, high boiling points, and poor electrical conductivity in the solid phase. These properties are due to the compounds' strong electrostatic interactions, which also cause the ions to be relatively immobile. Because they are aggregates of positively and negatively charged ions, there are no discrete molecules, and their formulas are empirical formulas that describe the ratio of ions in the lowest possible whole numbers. For example, the empirical formula $BaCl_2$ gives the ratio of Ba^{2+} to Cl^- ions within the crystal.

Covalent Crystals

Covalent crystals are comprised of atoms linked by strong covalent bonds into a large three-dimensional structure that can be thought of as one giant molecule. Because of the strong forces binding the atoms together, covalent crystals have high melting points and are hard. The most common example of covalent crystals is diamond, which is a network of carbon atoms, each of which is bonded to four others in a tetrahedral arrangement.

Metallic Crystals

Metallic solids consist of metal atoms packed together. They have high melting and boiling points as a result of their strong covalent attractions. They differ from normal covalent solids in that the valence orbitals of the constituent atoms interact to form molecular orbitals delocalized over the entire crystal and through which electrons can move. Because of the ability of electrons to move freely within the such solids, metallic crystals conduct electricity and heat very well.

Molecular Crystals

The molecules or atoms making up these solids are held together by intermolecular forces that tend to be weaker than ionic and covalent forces. They thus tend to have low melting points and are easily deformable. The structures of a lot of biological molecules (such as proteins) are determined by performing X-ray crystallography on the molecular crystals that they form.

Amorphous Solids

An amorphous solid, such as glass, does not possess the long-range, periodic structure that crystalline solids do, although the molecules are also fixed in place.

Liquid Crystals

Not all material falls neatly into one of the above categories. Metalloids, for example, possess properties that are intermediate between those of metallic and covalent solids. A more astounding example is that of liquid crystals, a state of matter in between liquids and solids.

Most of the substances that can form liquid crystals are long, rodlike organic molecules. Above a certain temperature, they behave like normal liquids,

A Closer Look

Liquid crystals are a state of matter in between liquids and solids. Above a certain temperature, they behave like normal liquids, oriented randomly and moving relatively freely among themselves. As they are cooled, they pass through several intermediate phases, each of increasing order. In particular, these rod-like molecules tend to line up parallel to one another. The precise orientation these molecules choose to line up in is affected by external electric and magnetic fields. It is this property that makes liquid crystals useful in electronic displays such as certain computer screens.

oriented randomly and moving relatively freely among themselves. As they are cooled, they pass through several intermediate phases, each of increasing order. In particular, these rodlike molecules tend to line up parallel to one another. The precise orientation these molecules choose to line up in is affected by external electric and magnetic fields. It is this property that makes liquid crystals useful in electronic displays such as certain computer screens.

Basic Concept

From strongest to weakest:

- Hydrogen bonding
- Dipole-dipole interactions
- Dispersion (London) forces

Intermolecular Forces

Molecular crystals, as mentioned above, and most liquids (except those that are molten forms of ionic solids) are held together by intermolecular attractions that are generally weaker than ionic and covalent interactions. These attractive forces are electrostatic in nature, and affect the atoms or molecules even though they are neutral because of the asymmetric distribution of charge density. They can be roughly categorized into three types: dipole-dipole interactions, London dispersion forces, and hydrogen bonding. These intermolecular forces are sometimes referred to collectively as van der Waals forces.

Dipole-Dipole Interactions

Polar molecules tend to orient themselves so that the positive region of one molecule is close to the negative region of another molecule. This arrangement is energetically favorable because of the electrostatic attraction between unlike charges. The magnitude of this kind of interaction increases with increasing polarity of the molecules.

Dipole-dipole interactions are present in the solid and liquid phases when the constituent molecules are polar, but often become negligible in the gas phase because the molecules are generally much farther apart. Polar species tend to have higher boiling points than nonpolar species of comparable molecular weight.

Dispersion Forces

Even though atoms or nonpolar molecules have no dipole moment, they experience attractive forces among themselves. This is because at any particular point in time, the electron density will be distributed randomly throughout the orbital; that is, the electron density fluctuates with time. This leads to rapid polarization and counterpolarization of the electron cloud and thus the formation of short-lived dipoles. These dipoles interact with the electron clouds of neighboring molecules, inducing the formation of more dipoles. The attractive interactions of these short-lived dipoles are called dispersion or London forces.

Dispersion forces are generally weaker than other intermolecular forces. They do not extend over long distances and are therefore most important when molecules are close together. The strength of these interactions within a given substance depends directly on how easily the electrons in the molecules can move (i.e., be polarized). Large molecules in which the electrons are far from the nucleus are relatively easy to polarize and therefore experience greater dispersion forces among themselves. If it were not for dispersion forces, the noble gases would not liquefy at any temperature since no other intermolecular forces exist between the noble gas atoms. The low temperature at which the noble gases liquefy is indicative of the relatively small magnitude of dispersion forces between the atoms.

Hydrogen Bonding

Hydrogen bonding is a specific, unusually strong form of dipole-dipole interaction. When hydrogen is bound to a highly electronegative atom such as fluorine, oxygen, or nitrogen, the hydrogen atom carries little of the electron density of the covalent bond, most of which is shifted over to the electronegative atom. This positively charged hydrogen atom interacts with the partial negative charge located on the electronegative atoms of nearby molecules, causing the two molecules to experience an attraction for each other. Substances which display hydrogen bonding tend to have unusually high boiling points compared with compounds of similar molecular formula that do not participate in hydrogen bonding. The difference derives from the energy required to break the hydrogen bonds. Hydrogen bonding is particularly important in the behavior of water, alcohols, amines, and carboxylic acids. In fact, if it were not for the hydrogen bonding ability of water, life as we know it would not be possible on Earth.

Phase Equilibria and Phase Changes

The different phases of matter interchange upon the absorption or release of energy, and more than one of them may exist in equilibrium under certain conditions. For example, at 1 atm and 0°C, an ice cube floating in water is a system in which the liquid and the solid phases coexist in equilibrium. On the microscopic level, however, the two phases are constantly interconverting while in this seemingly static state. Some of the ice may absorb heat and melt, but an equal amount of water will release heat and freeze. That is, individual H_2O molecules are going between the solid and liquid phases constantly, but in a way such that the relative amounts of ice and water remain constant. This condition in which two opposing processes occur such that the net change in the outcome is zero is known as dynamic equilibrium.

Gas-Liquid Equilibrium

The temperature of a liquid is related to the average kinetic energy of the liquid molecules; however, the kinetic energy of the individual molecules will vary (just as there is a distribution of molecular speeds in a gas). A few molecules near the surface of the liquid may have enough energy to leave the liquid phase and escape into the gaseous phase. This process is known as **evaporation** (or **vaporization**). Each time the liquid loses a high-energy particle, the average kinetic energy of the remaining molecules decreases, which means that the temperature of the liquid decreases. Evaporation is thus a cooling process. Given enough kinetic energy, the liquid will completely evaporate.

If a cover is placed on a beaker of liquid, the escaping molecules are trapped above the solution. These molecules exert a countering pressure, which forces some of the gas back into the liquid phase; this process is called **condensation**. Atmospheric pressure acts on a liquid in a similar fashion as a solid lid. As evaporation and condensation proceed, an equilibrium is reached in which the rates of the two processes become equal; that is, the liquid and the vapor are in dynamic equilibrium. The pressure that the gas exerts when the two phases are at equilibrium is called the vapor pressure. Vapor pressure increases as temperature increases, because more molecules will have sufficient kinetic energy to escape into the gas phase. The temperature at which the vapor pressure of the liquid equals the external (most often atmospheric) pressure is called the boiling point. In general, then, the temperature at which a liquid boils is dependent on the pressure surrounding it. We know water to boil at 100°C because it is at this temperature that its vapor pressure (or the pressure exerted by the gas phase H_2O molecules) is equal to one atmosphere. At places of high elevation, the surrounding pressure is lower than 1 atm and so water boils at a lower temperature. By controlling the ambient pressure, then, we can change the temperature at which water boils. This is the principle behind the pressure cooker: By maintaining a high pressure, water can reach a temperature higher than 100°C before it vaporizes, thus making it more effective at heating things.

Liquid-Solid Equilibrium

The liquid and solid phases can also coexist in equilibrium (such as in the ice-water mixture discussed above). Even though the atoms or molecules of a solid are confined to definite locations, each atom or molecule can undergo motions about some equilibrium position. These motions (vibrations) increase when energy (most commonly in the form of heat) is supplied. If atoms or molecules in the solid phase absorb enough energy in this fashion, the solid's three-dimensional structure breaks down and the liquid phase begins. The transition from solid to liquid is called **fusion** or **melting**. The reverse process, from liquid to solid, is called **solidification**, **crystallization**, or freezing. The temperature at which these processes occur is called the melting point or freezing point, depending on the direction of the transition. Whereas pure crystals have distinct, very sharp melting points, amorphous solids, such as glass, tend to melt over a larger range of temperatures, due to their less-ordered molecular distribution.

Basic Concept

• Triple Point

In a phase diagram, the point at which all three lines intersect is called the triple point.

Gas-Solid Equilibrium

A third type of phase equilibrium is that between a gas and a solid. When a solid goes directly into the gas phase, the process is called **sublimation**. Dry ice (solid CO_2) sublimes under atmospheric pressure; the absence of the liquid phase makes it a convenient refrigerant. The reverse transition, from the gaseous to the solid phase, is called deposition.

Phase Diagrams

Implicit in the discussion above is the fact that the phase in which a substance finds itself is a function of external conditions. On a plot of pressure versus temperature, one can imagine dividing the area of the quadrant into three sections, one corresponding to each of the three phases. The x and y values falling within each section are all the combinations of the pressure and temperature values at which the substance will be in that phase. In general, the gas phase is found at high temperature and low pressure; the solid phase is found at low temperature and high pressure; and the liquid phase is found at high temperature and high pressure. A typical phase diagram is shown below:

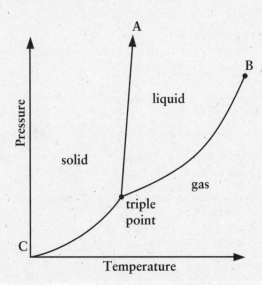

The three phases are demarcated by lines indicating the temperatures and pressures at which two phases are in equilibrium. Along line A, the solid and liquid phases are in equilibrium; along line B, liquid and gas; and along line C, solid and gas. Crossing one of these lines represents a phase change process: Crossing line B, for example, denotes either evaporation or condensation, depending on the direction of travel. The intersection of the three lines is called the triple point. At this temperature and pressure, unique for a given substance, all three phases are in equilibrium. Each substance has its own characteristic phase diagram that describes its physical properties. The reason why dry ice sublimes rather than melts, for example, is because the triple point of carbon dioxide lies at a pressure above 1 atm. The process of raising its temperature in open air (atmospheric pressure) thus occurs in the lower portion of the plot and the phase transition takes the substance across line C, bypassing the liquid phase. If the external pressure is 8 atm, then heating a block of dry ice would cause it to melt into the liquid state.

The liquid-gas equilibrium curve, line B, terminates at a point known as the critical point, beyond which there are no distinct liquid and gas phases. Instead, the substance exists in a form known as a supercritical fluid. On the other hand, the boundary between the solid and liquid phases continues indefinitely (hence the arrowhead on line A), and for almost all substances leans to the right, which means that as the pressure increases, a higher and higher temperature is needed to cause melting (solid to liquid) to occur. This is because high pressure favors the typically denser solid phase over the liquid one. H_2O is unique in that its solid form is generally less dense than its liquid form (the reason ice floats on water). As a result, the phase diagram for H_2O has a solid-liquid equilibrium curve that slopes to the left.

The Condensed Phases and Phase Changes Review Problems

1. Which of the following indicates the relative randomness of molecules in the three states of matter?

   A. Solid > liquid > gas

   B. Liquid < solid < gas

   C. Liquid > gas > solid

   D. Gas > liquid > solid

   E. None of the above

2. What factors determine whether or not two liquids are miscible?

   A. Molecular size

   B. Molecular polarity

   C. Density

   D. Both B and C

3. Discuss the physical properties of ionic crystals.

4. Alloys are mixtures of pure metals in either the liquid or solid phase. Which of the following is usually true of alloys?

   A. The melting/freezing point of an alloy will be lower than that of either of the component metals, because the new bonds are stronger.

   B. The melting/freezing point of an alloy will be lower than that of either of the component metals, because the new bonds are weaker.

   C. The melting/freezing point of an alloy will be greater than that of either of the component metals, because the new bonds are weaker.

   D. The melting/freezing point of an alloy will be greater than that of either of the component metals, because the new bonds are stronger.

Refer to the phase diagram below for questions 5–7.

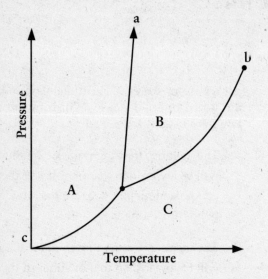

5. What is the typical form of a substance in state B?

    A. Pure crystalline solid

    B. Amorphous solid

    C. Gas

    D. Liquid

6. What is the typical form of a substance in state C?

    A. Pure crystalline solid

    B. Amorphous solid

    C. Gas

    D. Liquid

7. What is the triple point?

Turn the page
for answers and explanations
to the Review Problems.

Solutions to Review Problems

1.   D

Because gas molecules have the greatest freedom to move around, gases have the greatest disorder. Liquids are denser than gases and therefore the molecules experience stronger intermolecular attractions and are less free to move around. The arrangement of molecules in solids is the least random. Thus, melting and boiling are accompanied by an increase in entropy, i.e., $\Delta S > 0$. (See chapter 11.)

2.   B

The miscibility of two liquids strongly depends on their polarities. In general, polar and nonpolar liquids are not miscible, while a polar liquid can usually be mixed with another polar liquid, and a nonpolar liquid with another nonpolar liquid. Choice A, molecular size, and choice C, the density of a liquid, do not directly affect the miscibility (although choice C should remind you that two immiscible liquids will form separate layers, with the denser liquid on the bottom). Thus, B is the only correct choice.

3.

Ionic crystals contain repeating units of cations and anions. Because of the strong electrostatic attraction between the ions, these crystals have high melting points. Since the charges in these crystals are tightly fixed in the lattice, ionic solids are poor conductors of electricity. In the liquid or solution phase the charged particles can move around, and thus liquid ionic compounds will conduct electricity, as do solutions of such salts.

4.   B

The bonds between different metal atoms in an alloy are much weaker than those between the atoms in pure metals. Therefore, breaking these bonds requires less energy than does breaking the bonds in pure metals. Since melting and freezing points increase as the stability of bonds, they tend to be lower for alloys than for pure metals.

5.   D

Refer to the discussion of phase diagrams.

6.   C

Refer to the discussion of phase diagrams.

7.

The unique combination of temperature and pressure at which the solid, liquid, and gas phases coexist at equilibrium.

CHAPTER 9

SOLUTION CHEMISTRY

Solutions are homogeneous (that is, everywhere the same) mixtures of substances that combine to form a single phase, generally the liquid phase. Many important chemical reactions, both in the laboratory and in nature, take place in solution (including almost all reactions in living organisms).

A solution consists of a **solute** (e.g., NaCl, NH_3, or $C_{12}H_{22}O_{11}$) dispersed (dissolved) in a **solvent** (e.g., H_2O or benzene). The solvent is the component of the solution whose phase remains the same after mixing. For example, a solid cube of sugar dissolved in water (a liquid) yields a liquid mixture of water and sugar: Water is therefore the solvent and sugar the solute. If the two substances are already in the same phase, the solvent is generally taken to be the component present in greater quantity. Solute molecules move about freely in the solvent and can interact with other molecules or ions; consequently, chemical reactions occur easily in solution.

Solvation

The interaction between solute and solvent molecules is known as solvation or dissolution; when water is the solvent, it is also known as hydration and the resulting solution is known as an aqueous solution. Solvation is possible when the attractive forces between solute and solvent are stronger than those between the solute particles, i.e., when solute-solvent interactions overcome solute-solute interactions. It then becomes more energetically favorable for the solute particles to be surrounded each by the solvent rather than stay close together. For example, when NaCl dissolves in water, its component ions (Na^+ and Cl^-) dissociate from one another and become surrounded by water molecules. Because water is polar, ion-dipole interactions can occur between the Na^+ and Cl^- ions and the water molecules. For nonionic solutes, solvation involves van der Waals forces between the solute and solvent molecules. The general rule is that like dissolves like: Ionic and polar solutes are soluble in polar solvents, and nonpolar solutes are soluble in nonpolar solvents.

Basic Concept

solute = substance being dissolved (often solid)

solvent = substance doing the dissolving (often liquid)

solution = solvent + dissolved solute

This same generalization also applies to cases where the solute is not a solid. Water and oil do not mix, for example, because they do not "dissolve each other." The interactions between water molecules and the nonpolar molecules of oil are nowhere near as strong as the hydrogen bonding among water molecules themselves which would need to be disrupted if mixing were to occur. The two liquids therefore stay separate from each other.

Solubility and Concentration Units

The solubility of a substance is the maximum amount of that substance that can be dissolved in a particular solvent at a particular temperature. When this maximum amount of solute has been added, the solution is said to be saturated. In this state, the solution is in a state of dynamic equilibrium: The two opposite processes of dissolution and precipitation (or crystallization) are taking place at the same rate.

If more solute is added to a saturated solution, it will not dissolve. For example, at 18°C, a maximum of 83 g of glucose ($C_6H_{12}O_6$) will dissolve in 100 mL of H_2O. Thus, we can say that the solubility of glucose is 83 g/100 mL. If more glucose is added, it will remain in solid form, precipitating to the bottom of the container. Occasionally more solid can be dissolved than is allowed, in which case the solution is said to be supersaturated. This is, however, an unstable system, and often the slightest disturbance would cause the excess solute to precipitate out.

A solution in which the proportion of solute to solvent is small is said to be dilute, and one in which the proportion is large is said to be concentrated. Of course, very often we need to specify the amount of solute dissolved in a solvent more exactly, and this is expressed by the quantity of concentration. Solubility, then, can be thought of as the maximum possible concentration for the given solute-solvent pair in question. In the example above, we mentioned that the maximum concentration of glucose in water at 18°C is 83 g/100 mL. More commonly, however, the concentration of a solution is expressed as percent composition by mass, mole fraction, molarity, molality, or normality.

Percent Composition by Mass

The percent composition by mass of a solution is the mass of the solute divided by the mass of the solution (solute plus solvent), multiplied by 100.

Example: What is the percent composition by mass of a salt water solution if 200 g of the solution contains 0.3 mol of NaCl?

Solution: First we need to convert the amount of NaCl into mass:

0.3 mol NaCl $= 0.3$ mol $\times$ (atomic mass of Na^+ + atomic
                        mass of Cl^-)

                    $= 0.3$ mol $\times (23 + 35.5)$ g/mol

                    $= 17.6$ g NaCl

percent by mass $= (17.6$ g $/ 200$ g$) \times 100\% = 8.8\%$

Mole Fraction

The mole fraction (X) of a compound is equal to the number of moles of the compound divided by the total number of moles of all species within the system. The sum of the mole fractions in a system will always equal 1.

Example: If 92 g of glycerol is mixed with 90 g of water, what will be the mole fractions of the two components? (MW of $H_2O = 18$; MW of $C_3H_8O_3 = 92$).

Solution: 90 g water $= 90$ g $\times (1$ mol $/ 18$ g$) = 5$ mol

92 g glycerol $= 92$ g $\times (1$ mol $/ 92$ g$) = 1$ mol

total number of moles $= 5 + 1 = 6$

$X_{water} = 5$ mol $/ 6$ mol $= 0.833$

$X_{glycerol} = 1$ mol $/ 6$ mol $= 0.167$

Since these are the only two components in the system, one can verify that the two mole fractions add up to be 1.

Molarity

The **molarity** (M) of a solution is the number of moles of solute per liter of solution. Solution concentrations are usually expressed in terms of molarity. Molarity depends on the volume of the solution, not on the volume of solvent used to prepare the solution. In other words, mixing 1 mol of solute with 1 L of solvent will not in general give a 1 M solution since the final volume after mixing may be different from 1 L. You may recall from your laboratory experience that in order to produce a solution of a particular molarity, you add solvent to a container (volumetric flask) that already has the solute weighed out in it, until the total volume of the mixture reaches a specific value (marked by a ring around the narrow neck of the flask). One generally is not interested in, and does not keep track of, the volume of solvent actually added.

Example: If enough water is added to 11 g of $CaCl_2$ to make 100 mL of solution, what is the molarity of the solution?

Solution: 11 g $CaCl_2$ = 0.10 mol $CaCl_2$

100 mL = 0.10 L

∴ molarity = 0.10 mol / 0.10 L = 1.0 M

Molality

Don't Mix These Up on Test Day

Do not confuse molarity and molality!

Molarity = moles of solute/volume of solution

Molality = moles of solute/mass of solvent

The **molality** (m) of a solution is the number of moles of solute per kilogram of solvent. For dilute aqueous solutions at 25°C the molality is approximately equal to the molarity, because the density of water at this temperature is 1 kilogram per liter and the volume of the solution is presumed to be approximately the same as that of the solvent (i.e., water) added. But note that this is an approximation and true only for dilute aqueous solutions.

Example: If 10 g of NaOH are dissolved in 500 g of water, what is the molality of the solution?

Solution: 10 g NaOH = 0.25 mol NaOH

500 g = 0.5 kg

∴ molality = 0.25 mol solute / 0.5 kg solvent = 0.50 m

Normality

Basic Concept

Normality is best thought of as "molarity of the stuff of interest" in a solution.

The **normality** (N) of a solution is equal to the number of equivalents of solute per liter of solution. An equivalent is a measure of the reactive capacity of a molecule, and is defined according to the type of reaction being considered.

To calculate the normality of a solution, then, we must know for what purpose the solution is being used, because it is the concentration of the reactive species with which we are concerned. For example, a 1 molar solution of sulfuric acid would be 2 normal for acid-base reactions (because each mole of sulfuric acid, H_2SO_4, provides 2 moles of H^+ ions) but is only 1 normal for a sulfate precipitation reaction (because each mole of sulfuric acid only provides 1 mole of sulfate ions). Normality is always a whole-number multiple of molarity: for example, a 3M solution may be 3N, 6N, etcetera, but not 2N or 5N.

Dilution

A solution is diluted when solvent is added to a solution of high concentration to produce a solution of lower concentration. The concentration of a solution after dilution can be conveniently determined using the equation below:

$$M_i V_i = M_f V_f$$

where M is molarity, V is volume, and the subscripts i and f refer to initial and final values, respectively. Note that the product MV gives the number of moles of solute, and the equation is just a statement on the conservation of matter: The amount of solute dissolved in the solution remains constant after a dilution.

Example: How many mL of water must be added to 65 mL of a 5.5 M solution of NaOH in order to prepare a 1.2 M NaOH solution?

Solution: The first step is to find the final volume of the solution:

$$5.5 \text{ M} \times 0.065 \text{ L} = 1.2 \text{ M} \times V_f$$

$$V_f = 5.5 \times 0.065 / 1.2 = 0.3 \text{ L} = 300 \text{ mL}$$

The volume of water that needs to be added is therefore $(300 - 65) \text{ mL} = 235 \text{ mL}$

Electrolytes and Conductivity

The ability of a substance to conduct electrical currents depends on how easily charges can move through it. Metallic solids, as we have seen, conduct electricity because the structure of the electronic orbitals allows electrons to migrate freely. In the case of aqueous solutions, electrical conductivity is governed by the presence and concentration of ions in solution. The movement of these ions in response to an electric field is what makes up a current. Therefore, pure water does not conduct an electrical current well since the concentrations of hydronium and hydroxide ions are very small. Solutes whose solutions are conductive are called electrolytes. A solute is considered a strong electrolyte if it dissociates completely into its constituent ions. Examples of strong electrolytes include ionic compounds, such as NaCl and KI, and molecular compounds with highly polar covalent bonds that dissociate into ions when dissolved, such as HCl in water. A weak electrolyte, on the other hand, ionizes or hydrolyzes incompletely in aqueous solution and only some of the solute is present in ionic form. Examples include acetic acid and other weak acids, ammonia and other weak bases, and $HgCl_2$. Many compounds do not ionize at all in aqueous solution, retaining their molecular structure in solution. These compounds are called nonelectrolytes and include many nonpolar gases and organic compounds, such as oxygen and sugar.

Colligative Properties

The presence of solute particles can make the physical properties (such as boiling point and freezing point) of the solution different from those of the pure solvent. Such effects are more easily studied systematically in cases where the solution is relatively dilute and the solute is nonvolatile (negligible presence in the gas phase; does not exert vapor pressure of its own). The more numerous these solute particles are in solution, the more pronounced the changes on the physical properties. Physical properties that depend on the number of dissolved particles in the solution but not on their chemical identity or nature are known as colligative properties.

Vapor-Pressure Lowering (Raoult's Law)

A Closer Look

Raoult's Law is the basis of the technique of distillation, used to separate substances with different boiling points or volatilities.

When solute B is added to pure solvent A, the vapor pressure of A above the solvent decreases. If the vapor pressure of A above pure solvent A is designated by $P°_A$ and the vapor pressure of A above the solution containing B is P_A, the change in vapor pressure, defined as $\Delta P = P°_A - P_A$, is

$$\Delta P = X_B P°_A$$

where X_B is the mole fraction of the solute B in solvent A. For a two-component system (that is, no other kind of solute present), $X_B = 1 - X_A$, and so by substituting this into the equation, together with the definition $\Delta P = P°_A - P_A$, one obtains Raoult's law:

$$P_A = X_A P°_A$$

The more solute particles there are in solution, the lower the mole fraction of the solvent would be, and hence the lower the vapor pressure. One limiting case is the trivial scenario where only the solute is present; without any solvent, X_A is zero, and so the vapor pressure of A is zero by the equation, which certainly makes sense because there simply isn't any solvent around to exert a vapor pressure. In the other extreme, when no solute is present, the system is comprised entirely of solvent A; its mole fraction is therefore one and its vapor pressure would be the same as that of pure A. In between these two cases, Raoult's law states that the vapor pressure is linearly proportional to the mole fraction of the solvent.

It should be pointed out that even though we have introduced Raoult's law in the study of colligative properties, it is not limited to the solvent-non-volatile solute systems on which we have focused. In a solution with several volatile components (a mixture of benzene and toluene, for example), Raoult's law states that the vapor pressure of *each* component is proportional to its mole fraction in the solution:

$$P_A = X_A P°_A$$

$$P_B = X_B P°_B$$

$$P_C = X_C P°_C, \text{ etcetera}$$

The sum of all the mole fractions has to equal one. The total vapor pressure over the solution, then, is the sum of the partial vapor pressures of each component:

$$P_{tot} = P_A + P_B + P_C + ... = X_A P°_A + X_B P°_B + X_C P°_C + ...$$

This last result is simply an application of Dalton's law of partial pressures.

Raoult's law is actually only an idealized description of the behavior of solutions, and holds only when the attraction between molecules of the different components of the mixture is equal to the attraction between the molecules of any one component in its pure state. When this condition does not hold, the relationship between mole fraction and vapor pressure will deviate from Raoult's Law. Solutions that obey Raoult's law are called ideal solutions, much in the same way that gases obeying PV = nRT are called ideal gases.

A Closer Look

Solutions that obey Raoult's law are called ideal solutions.

Freezing-Point Depression

Pure water (H_2O) freezes at 0°C at 1 atm; however, for every mole of solute particles dissolved in 1 L of water, the freezing point is lowered by 1.86°C. This is because the solute particles interfere with the process of crystal formation that occurs during freezing; the solute particles lower the temperature at which the molecules can align themselves into a crystalline structure.

The formula for calculating this freezing-point depression is:

$$\Delta T_f = K_f m$$

where ΔT_f is the freezing-point depression (the number of degrees or Kelvin the freezing point is lowered by), K_f is a proportionality constant characteristic of a particular solvent, and m is the molality of the solution (mol solute/kg solvent). Each solvent has its own characteristic K_f. The larger the value, the more sensitive its freezing point is to the presence of solutes.

A Closer Look

Freezing-point depression is the principle behind spreading salt on ice: The freezing point of water is lowered by the presence of the salt, and so the ice melts. Antifreeze (mostly ethylene glycol) also operates by the same principle.

Note that the molality in question is the total molality of all particles present. A 1m aqueous solution of NaCl, for example, would correspond to 2m in solute particles since it dissociates to give 1m of Na^+ ions and 1m of Cl^- ions. It would lead to a freezing-point depression that is twice the magnitude of that of a 1m aqueous solution of sugar.

Freezing-point depression is the principle behind spreading salt on ice: The freezing point of water is lowered by the presence of the salt, and so the ice melts. Antifreeze (mostly ethylene glycol) also operates by the same principle.

Boiling-Point Elevation

A liquid boils when its vapor pressure equals the atmospheric pressure. Since, as we have seen above, the vapor pressure of a solution is lower than that of the pure solvent, more energy (and consequently a higher temperature) will be required before its vapor pressure equals atmospheric pressure. In other words, the boiling point of a solution is higher than that of the pure solvent. The extent to which the boiling point of a solution is raised relative to that of the pure solvent is given by the following formula:

$$\Delta T_b = K_b m$$

where ΔT_b is the boiling-point elevation, K_b is a proportionality constant characteristic of a particular solvent, and m is the molality of the solution. Note how similar this equation is in form to that for freezing-point depression: The only difference is that a solvent will have different values for Kf and Kb, and that it is important to keep in mind that in one case the temperature is raised ($\Delta T > 0$), while in the other case the temperature is lowered ($\Delta T < 0$).

Osmotic Pressure

Consider a container separated into two compartments by a semipermeable membrane (which, by definition, selectively permits the passage of certain molecules). One compartment contains pure water, while the other contains water with dissolved solute. The membrane allows water but not the solute molecules to pass through. Because it is more favorable for the two compartments to equalize their concentration, water will diffuse from the compartment containing pure water to the compartment containing the water-solute mixture. This net flow will cause the water level in the compartment containing the solution to rise above the level in the compartment containing pure water.

Because the solute cannot pass through the membrane, the concentrations of solute in the two compartments can never be equal. The pressure exerted by the water level in the solute-containing compartment will eventually oppose the influx of water, and thus the water level will rise only to the point at which it exerts a sufficient pressure to counterbalance the tendency of water to flow across the membrane. This pressure is defined as the osmotic pressure (Π) of the solution, and is given by the formula:

$$\Pi = MRT$$

where M is the molarity of the solution, R is the ideal gas constant, and T is the temperature on the Kelvin scale. This equation clearly shows that molarity and osmotic pressure are directly proportional; that is, as the concentration of the solution increases, the osmotic pressure also increases.

The setup behind the concept of osmotic pressure, as described above, may seem at first glance artificial and contrived, but actually is very important in cellular biology, because the cell membrane is a semipermeable membrane that allows only certain types of molecules to diffuse through. The solute concentration in the cytoplasm of a cell relative to that of its environment determines the net direction of the flow of water, and may lead to either shrinking or swelling (maybe even bursting, or lysing) of the cell.

A Closer Look

The solute concentration in the cytoplasm of a cell relative to that of its environment determines the net direction of the flow of water, and may lead to either shrinking or swelling (maybe even bursting, or lysing) of the cell.

Solution Chemistry Review Problems

1. Which of the following choices correctly describes the solubility behavior of potassium chloride (KCl)?

   A. Solubility in CCl_4 > Solubility in CH_3CH_2OH > Solubility in H_2O

   B. Solubility in H_2O > Solubility in CH_3CH_2OH > Solubility in CCl_4

   C. Solubility in CH_3CH_2OH > Solubility in CCl_4 > Solubility in H_2O

   D. Solubility in H_2O > Solubility in CCl_4 > Solubility in CH_3CH_2OH

2. How much NaOH must be added to make 200 mL of a 1M NaOH solution?

   A. 8 g

   B. 16 g

   C. 40 g

   D. 80 g

3. To what volume must 10.0 mL of 5.00 M HCl be diluted to make a 0.500 M HCl solution?

   A. 1 mL

   B. 50 mL

   C. 100 mL

   D. 500 mL

   E. 1,000 mL

4. What is the normality of a 2M solution of phosphoric acid, H_3PO_4, for an acid-base titration?

   A. 0.67

   B. 2

   C. 3

   D. 6

5. Given that the molecular weight of ethyl alcohol, CH_3CH_2OH, is 46, and that of water is 18, how many grams of ethyl alcohol must be mixed with 100 mL of water for the mole fraction (X) of ethyl alcohol to be 0.2?

6. Which of the following will be the most electrically conductive?

A. Sugar dissolved in water

B. Salt water

C. Salt dissolved in an organic solvent

D. An oil and water mixture

7. A semipermeable membrane separates a container of fresh water from one of salt water. If the volume of fresh water decreases significantly, what must be true of the semipermeable membrane? Assume that evaporation is negligible.

8. Once equilibrium is reached, if the temperature in Question 7 is suddenly increased, the osmotic pressure

A. will decrease

B. will increase

C. will remain the same

D. cannot be determined

9. What is the freezing point of a solution containing 0.5 mol of glucose dissolved in 200 g of H_2O? (The K_f for water is $1.86°Cm^{-1}$.)

10. The osmotic pressure at STP of a solution made from 1 L of NaCl (*aq*) containing 117 g of NaCl is

A. 44.77 atm

B. 48.87 atm

C. 89.54 atm

D. 117 atm

11. At 18°C, the vapor pressure of pure water is 0.02 atm, and the vapor pressure of pure ethyl alcohol (MW = 46) is 0.50 atm. For a water-alcohol mixture with the alcohol present in a mole fraction of 0.2, find the vapor pressure due to the alcohol and the vapor pressure due to water. (Assume solution to be ideal.)

Solutions to Review Problems

1. B

KCl is an ionic salt, and therefore should be soluble in polar solvents and insoluble in nonpolar solvents. Water is a highly polar liquid. The carbon atom in carbon tetrachloride, CCl_4, is bonded to four atoms, so the molecule is tetrahedral. This geometry means that the individual dipole moments of the bonds cancel and CCl_4 is nonpolar. Ethanol (CH_3CH_2OH) has two carbon atoms in tetrahedral arrangement; most of the dipole moments associated with the bonds are the same, but the C—C and C—OH bonds are different, so ethanol is somewhat polar. Thus, the polarities of the three solvents decrease in the following sequence: $H_2O > CH_3CH_2OH > CCl_4$, with the solubility of KCl decreasing along that sequence.

2. A

A 1M NaOH solution means that there is 1 mol of NaOH for every liter of the solution. We are interested, however, in a final volume of 200 mL, which is a fifth of a liter. 0.2 mol of NaOH is therefore needed. The formula weight of NaOH is 40, and so the amount of NaOH needed is 0.2 mol $\times$ 40 g/mol = 8 g.

3. C

When a solution is diluted, more solvent is added, yet the number of moles of solute remains the same. To solve a dilution problem, the following equation is used:

$$M_i V_i = M_f V_f$$

where i represents the initial conditions and f represents the final conditions. Therefore, the calculation to solve for the final volume is:

$$(5.0 \text{ M})(0.01 \text{ L}) = (0.50 \text{ M})(V_f)$$

$$V_f = 0.100 \text{ L}$$

$$= 100 \text{ mL}$$

4. D

Each mole of H_3PO_4 contains 3 moles of hydrogen and (since this is an acid) three equivalents. A 2M solution of this acid is thus 2 M $\times$ 3 N/M = 6N.

5. 64.4 g

The number of moles of water is found by estimating the density of water to be 1 g/mL.

$$\text{mol } H_2O = (100 \text{ mL } H_2O)(1 \text{ g/mL}) / (18 \text{ g/mol})$$
$$= 5.6 \text{ mol}$$

If the mole fraction of ethyl alcohol is to be 0.2, then the mole fraction of water must be 0.8. If n equals the total number of moles:

$$5.6 = 0.8 \text{ n}$$

$$n = 7 \text{ moles.}$$

Then

$$\text{mol ethyl alcohol} = (.2)(7) = 1.4 \text{ mol}$$

and

$$(1.4 \text{ mol ethyl alcohol}) (46 \text{ g/mol})$$
$$= 64.4 \text{ g ethyl alcohol}$$

6. B

Only ionic compounds (electrolytes) dissolved in polar solvents will conduct electricity. Sugar is a covalent solid, and therefore is not an electrolyte even when dissolved in water. Choice C is incorrect because salt will not dissolve appreciably in an organic solvent and so no ions will be present. Choice D is incorrect because oil and water are immiscible, and neither on its own contains a significant amount of electrolytes. Salt water is in essence an aqueous solution of NaCl, which dissociates to generate Na^+ and Cl^- ions, and so choice B is correct.

7.

The membrane must be permeable to water, but not to salt. If it were permeable to salt, the salt would diffuse across the membrane into the freshwater container (down its concentration gradient) until the molarities of the two containers were the same. However, the volume of the salt water container increased, indicating that fresh water diffused across the membrane from a region of low solute concentration to one of high solute concentration. The water level rose until it exerted enough pressure to counterbalance the tendency to diffuse; this pressure is known as the osmotic pressure.

8. B

Using the formula $\Pi = MRT$, we see that osmotic pressure and temperature are directly proportional; i.e., if temperature increases, osmotic pressure will also increase. Thus, choice B is the correct answer.

9. $-4.65°C$

This question applies the concept of freezing-point depression. If 0.5 mol of a nonelectrolyte solute such as glucose is dissolved in 200 g of H_2O, then the molality of the solution is

0.5 mol / 0.200 kg solvent = 2.5 m

Using the equation $\Delta T_f = K_f m$, the freezing point depression is $2.5 \times 1.86°C$, or $4.65°C$, and the new freezing point is thus original freezing point $- 4.65°C = (0 - 4.65)°C = -4.65°C$.

10. C

The osmotic pressure (Π) of a solution is given by $\Pi = MRT$. At STP, T = 273K. The formula weight of NaCl is 58.5. The number of moles in the solution described is:

117 g / (58.5 g/mol) = 2 moles of NaCl

But since NaCl is a strong electrolyte, it dissociates in aqueous solution and there are actually 4 moles of particles present in the solution, i.e., 2 moles of Na^+ and 2 moles of Cl^-. The volume of the solution is 1 L, and so the total molarity of solutes is 4 M.

$$P = (4 \text{ M})(8.2 \times 10{-2} \text{ L} \bullet \text{atm}/(\text{K} \bullet \text{mol}))(273 \text{ K})$$

$$= 89.54 \text{ atm}$$

11.

Use Raoult's Law to answer this question, with A = H_2O and B = ethyl alcohol.

$$P_A = X_A P°_A$$

$$P_B = X_B P°_B$$

Since we are told that $X_B = 0.2$, then $X_A = 1 - X_B = 1 - 0.2 = 0.8$

$$P°_A = 0.02 \text{ atm}$$

$$P°_B = 0.50 \text{ atm}$$

$$P_A = (0.8)(0.02 \text{ atm}) = 0.016 \text{ atm}$$

$$P_B = (0.2)(0.50 \text{ atm}) = 0.10 \text{ atm}$$

Thus the vapor pressure due to water is 0.016 atm and the vapor pressure due to the alcohol is 0.10 atm.

CHAPTER 10

CHEMICAL EQUILIBRIUM

Dynamic Equilibrium

When we write a balanced chemical equation of the form $2NO_2 \rightarrow N_2O_4$, the meaning seems clear: Two molecules of NO_2 come together in a combination reaction to yield a molecule of N_2O_4. From our discussions earlier on stoichiometry, we know that we can also look at it as 2 moles of NO_2 coming together to form one mole of N_2O_4, or as 5 moles of NO_2 coming together to form two and a half moles of N_2O_4, etcetera. In real life, however, if we put a certain amount of gaseous NO_2 in a vessel, we will most likely not end up with half that number of moles of N_2O_4; in other words, the reaction is not seen to go to *completion*. Instead, what we would have is a mixture of both gases. What we have failed to take into account is that just as two molecules of nitrogen dioxide can combine to form a molecule of N_2O_4, a molecule of N_2O_4 can also undergo decomposition to give back two molecules of nitrogen dioxide. In the beginning, because there is no N_2O_4 around, only the combination reaction takes place; however, as the product of this reaction, N_2O_4, accumulates, the decomposition reaction starts to "kick in" and works to undo what the combination reaction has done. In general, for every reaction there is a reverse reaction that takes place simultaneously in opposition to it. In the long run, a state is eventually reached where the two reactions, while still going on, reach a stalemate so that no one side is gaining any net ground. From a macroscopic perspective (to our naked eye, if you will) no change is occurring in the composition of the system. This state is known as a dynamic equilibrium and will persist until the surrounding conditions are disturbed.

Law of Mass Action

For a general reaction of the form

$$a\,A + b\,B \rightarrow c\,C + d\,D,$$

the law of mass action states that the equilibrium condition is expressed by the equation:

$$K_c = \frac{[C]^c[D]^d}{[A]^a[B]^b}$$

where K_c is a constant known as the concentration equilibrium constant. Its value changes depending on the reaction in question and on the temperature at which the reaction is carried out, [C] denotes the concentration of C in moles per liter (molarity), etcetera. (Even though molarity is a concept most commonly encountered when talking about aqueous solutions, there is no reason why one cannot talk about the molarity of a gas in a mixture—the number of moles of that gas divided by the volume of the container in liters.) Note that the exponents of the concentrations are the same as the stoichiometric coefficients in the balanced chemical equation.

Example: What is the expression for the equilibrium constant for the following reaction?

$$3\,H_2\,(g) + N_2\,(g) \rightarrow 2\,NH_3\,(g)$$

Solution: $K_c = \dfrac{[NH_3]^2}{[H_2]^3[N_2]}$

It is important to understand what the law of mass action is saying. It is always possible to write an expression like the right hand side in the example above; in fact, this expression involving the concentrations of the species is known generally as the reaction quotient. Its value changes as the reaction progresses: It will be very small in the beginning since the system is composed only of reactants (small numerator, large denominator), but will increase as products accumulate (numerator gets bigger as denominator decreases). When its value has reached that of K_c, the system or the reaction has reached equilibrium and the concentrations of each species will no longer change. The concentrations that appear in the law of mass action are therefore the equilibrium concentrations of the species. A large K_c ($K_c \gg 1$) means that the reaction goes almost to completion: at equilibrium, there is a high concentration of products (high [C] and [D]), and we say that the equilibrium "lies to the right." A small K_c ($K_c \ll 1$), on the other hand, means that the reaction is not very favorable: At equilibrium there are still a lot of reactants around (high [A] and [B]); we say that the equilibrium "lies to the left."

Equilibrium can be reached from either direction. In other words, in the generic equation written above, we can start with a container filled with C and D. At equilibrium, the relative concentrations of the species would be the same. The reaction would now technically be written as

$$c\,C + d\,D \rightarrow a\,A + b\,B,$$

and the mass action expression would be

$$K_c = 1 \div ([C]^c[D]^d / [A]^a[B]^b) = \frac{[A]^a[B]^b}{[C]^c[D]^d}$$

Note that the equilibrium constant for this reverse reaction, K_c, is simply the reciprocal of the equilibrium constant for the forward reaction. That is:

$$K_c = \frac{1}{K_c} \text{ since} = 1 \div ([C]^c[D]^d / [A]^a[B]^b) = \frac{[A]^a[B]^b}{[C]^c[D]^d}$$

The more favorable a reaction is, the less favorable its reverse will be, and vice versa.

Gas-Phase Equilibria

The concentrations of the species in the law of mass action are expressed in moles per liter (M). If, however, the reaction is in the gas phase (if the reactants and products are gases), the more common approach is to formulate the law of mass action in terms of the partial pressures of the species. For the reaction

$$a\,A\,(g) + b\,B\,(g) \rightarrow c\,C\,(g) + d\,D\,(g),$$

the law of mass action is usually written in the form:

$$K_p = \frac{P_C{}^c P_D{}^d}{P_A{}^a P_B{}^b}$$

where P_C is the partial pressure of gas C, etcetera. All partial pressures should be in units of atmospheres. For the reaction mentioned at the beginning of this chapter and the one in the example problem above, then, one can write the following mass action expressions:

$$\frac{P_{N_2O_4}}{P_{NO_2}{}^2} = K_{P_1} \qquad\qquad \frac{P_{NH_3}{}^2}{P_{H_2}{}^3 P_{N_2}} = K_{P_2}$$

The numerical value of K_p would in general be different from that of K_c, the equilibrium constant written with molarities. What is unchanged, however, is the essence of the law of mass action: At equilibrium, a certain relationship will always prevail among the amount of reactants and products, no matter how you choose to report these amounts.

Heterogeneous Equilibria

Reaction equilibria that involve different physical states of matter (different phases) are known as heterogeneous equilibria. For example, for a reaction of the type

$$a A (s) + b B (g) \rightarrow c C (s) + d D (g),$$

one would expect to write the law of mass action as on the previous page:

$$K_c = \frac{[C]^c[D]^d}{[A]^a[B]^b}$$

But the molarity of a solid compound like A is just its density divided by its molar mass, which is a constant that is characteristic of A. The numerical values of [A] and [C] (and consequently $[A]^a$ and $[C]^c$) can therefore be subsumed into the equilibrium constant:

$$K_c = \frac{K_c[A]^a}{[C]^c} = \frac{[D]^d}{[B]^b}$$

Or, keeping in mind what we just discussed above, we would also write

$$K_P = \frac{P_D{}^d}{P_B{}^b}$$

The same holds for reactants and solids that appear as pure liquids. In short, concentrations of pure solids and liquids do not appear in the equilibrium constant expression.

One final note on equilibrium constants: One might expect, from examining the law of mass action, that the units for equilibrium constants would change depending on the reaction, since the mass action expression involves different powers of concentration or pressure depending on the stoichiometry. However, the equilibrium constant is always dimensionless, with the understanding that we are sticking to the units of molars and atmospheres for concentrations and pressures respectively.

Finally, to emphasize the equilibrium nature of a reaction, we often use two arrows, pointing in opposite directions, to separate the reactants and the products. The reaction above could therefore be written as an equilibrium of the form:

$$a A + b B \rightleftharpoons c C + d D$$

Le Châtelier's Principle

The French chemist Henry Louis Le Châtelier stated that a system at equilibrium to which a stress is applied tends to change so as to relieve the applied stress. This rule, known as Le Châtelier's principle, is used to determine the direction in which a reaction at equilibrium will proceed when subjected to a stress, such as a change in concentration, pressure, temperature, or volume.

Changes in Concentration

Increasing the concentration of a species once the system has reached equilibrium will tend to shift the equilibrium away from the species that is added, in order to reestablish its equilibrium concentration, and vice versa. For example, in the reaction

$$A + B \rightleftharpoons C + D,$$

if the concentration of A and/or B is increased, the equilibrium will shift towards (or favor production of) C and D. Conversely, if the concentration of C and/or D is increased, the equilibrium will shift away from the production of C and D, favoring production of A and B. Similarly, decreasing the concentration of a species will tend to shift the equilibrium towards the production of that species. For example, if A and/or B is removed from the above reaction, the equilibrium will shift so as to favor increasing concentration of A and B.

All of this could be read from the law of mass action. If, for example, the concentration of A is increased by the injection of more A into the system, the reaction quotient would have a lower value than the equilibrium constant. Equilibrium is reestablished only by having more reactants go on to yield products: lowering [A] and [B] and raising [C] and [D] so that the value of the reaction quotient would once again be equal to the equilibrium constant.

Note that this new equilibrium does not contain the same amounts of A, B, C, and D; in fact, we know that [C] and [D] will be higher than before, and [B] will be lower (since it is depleted by reaction with the excess A that the system was trying to get rid of). Yet the value of the equilibrium constant is not changed: the concentrations adjust themselves to new values *while maintaining the equality stated in the law of mass action.* A more concrete example may help: Suppose the concentrations of A, B, C and D are 3 M, 4 M, 2 M and 2 M respectively at equilibrium. The injection of more A brings its concentration up to 5 M, at which point the system is no longer at equilibrium. After we have waited for the system to react to this stress and settle into the new equilibrium, we find that the concentrations are now 4.6 M, 3.8 M, 2.4

Basic Concept

Increase reactant concentration: favors products

Increase product concentration: favors reactants

M, and 2.4 M. They are all different from before, yet both sets of values correspond to equilibrium, and both sets satisfy the law of mass action:

$$\frac{2 \times 2}{3 \times 4} = \frac{2.4 \times 2.4}{4.6 \times 3.8} = K_c = 0.33$$

Graphically, a plot of the concentrations of the species as a function of time would look as follows:

Basic Concept

Higher pressure: favors side with fewer moles of gases

Large volume: favors side with more moles of gases

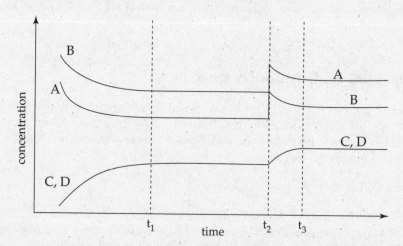

The system has reached its initial equilibrium by time t_1. At time t_2, more A is injected, bringing its concentration up to 5M. This perturbs the system and it is no longer at equilibrium. The concentrations of the species adjust themselves to new values and settle into a new equilibrium at time t_3.

Taking advantage of this aspect of Le Châtelier's principle is a common way in industry to increase the yield of a useful product or drive a reaction to completion. If D were constantly removed from the above reaction, the net reaction would produce more D and concurrently more C. Likewise, using an excess of the least expensive reactant helps to drive the reaction forward.

Changes in Pressure or Volume

In a system at constant temperature, a change in pressure may cause a change in volume, and vice versa. Since liquids and solids are practically incompressible, a change in the pressure or volume of systems involving only these phases has little or no effect on their equilibrium. Reactions involving gases, however, may be greatly affected by changes in pressure or volume, since gases are highly compressible.

Pressure and volume are inversely related from the ideal gas law (or Boyle's law to be more precise). An increase in the pressure (or decrease in the volume) of a system will shift the equilibrium so as to decrease the number of moles of gas present. This reduces the total pressure of the system and relieves the stress. Consider the following reaction:

$$3 H_2 (g) + N_2 (g) \rightarrow 2 NH_3 (g)$$

The left side of the reaction has 4 moles of gas, whereas the right side has only 2 moles. When the pressure of this system is increased, the equilibrium will shift so that the side of the reaction with fewer moles is favored. Since there are fewer moles on the right, the equilibrium will shift toward the right. Conversely, if the volume of the same system is increased, its pressure immediately decreases, which, according to Le Châtelier's principle, leads to a shift in the equilibrium to the left.

Changes in Temperature

Changes in temperature also affect equilibrium. To predict this effect, heat may be considered as a product in reactions that release energy (exothermic reactions), and as a reactant in reactions that absorb energy (endothermic reactions). Consider the following exothermic reaction:

$$A \rightarrow B + heat$$

If this system were placed in an ice bath, its temperature would decrease, driving the reaction to the right to produce more heat. Conversely, if the system were placed in a boiling-water bath, the reaction equilibrium would shift to the left because of the increased "concentration" of heat.

Not only does a temperature change alter the position of the equilibrium, it also alters the numerical value of the equilibrium constant. In contrast, changes in the concentration of a species in the reaction will alter the position of the equilibrium without changing the numerical value of the equilibrium constant.

Basic Concept

Higher temperature: favors reactant if exothermic (heat as product)

Ionic Equilibria

In Chapter 9, we discussed the concepts of solubility and solvation. The notion of the equilibrium constant gives us a way of describing the process of dissolution more quantitatively. The process of solvation, like other reversible chemical and physical changes, tends toward an equilibrium. Immediately after solute has been introduced into a solvent, most of the change taking place is dissociation, because no dissolved solute is initially present. However, as solute dissociates, the reverse reaction (precipitation of the solute) also begins to occur. Eventually an equilibrium is reached, with the rate of solute dissociation equal to the rate of precipitation.

The Solubility Product Constant

An ionic solid introduced into a polar solvent (for example, water) dissociates into its component ions. The dissociation of such a solute in these solvents may be represented by:

$$A_mB_n\ (s) \rightarrow mA^{n+}\ (aq) + nB^{m-}\ (aq)$$

The equilibrium constant for this reaction is $[A^{n+}]^m[B^{m-}]^n$, where the concentrations are those of the ions in the *saturated* solution. (Recall that the solid does not appear in the mass action expression.) To specify that this constant is for a dissolution process, it is also referred to as the solubility product constant and given the symbol K_{sp}. Each salt has its own distinct K_{sp} at a given temperature.

Just as in the case of the more general law of mass action, it is important to realize that one can always write an expression of the form $[A^{n+}]^m[B^{m-}]^n$; it is only when the solution is saturated that the concentrations are related in such a way that $[A^{n+}]^m[B^{m-}]^n = K_{sp}$. An unsaturated solution will have $[A^{n+}]^m[B^{m-}]^n < K_{sp}$, while a supersaturated solution will have $[A^{n+}]^m[B^{m-}]^n > K_{sp}$. If the supersaturated solution is disturbed by adding more salt, other solid particles, or jarring the solution by a sudden decrease in temperature, the solid salt will precipitate until the equality holds.

Don't Mix These Up on Test Day

- Every slightly soluble salt of the general formula MX will have $K_{sp} = x^2$, where x is the molar solubility.

- Every slightly soluble salt of the general formula MX_2 will have a $K_{sp} = 4x^3$, where x is the molar solubility.

- Every slightly soluble salt of the general formula MX_3 will have a $K_{sp} = 27x^4$, where x is the molar solubility.

Solubility and the Solubility Product Constant

Since the solubility product constant contains information about the maximum concentration of ions in the solution of a particular salt, it should come as no surprise that K_{sp} is related to the salt's solubility, the maximum amount that can be dissolved in the solvent. For ionic compounds that dissociate into constituent ions, the way these two quantities relate to each other depends on the stoichiometry of the compound.

For the general dissolution and dissociation of the type:

$$A_mB_n\ (s) \rightarrow mA^{n+}\ (aq) + nB^{m-}\ (aq),$$

we have already determined the relationship $K_{sp} = [A^{n+}]^m[B^{m-}]^n$. Relating that to the solubility of A_mB_n, we need to keep in mind that for every x moles of A_mB_n that dissolve, mx moles of A^{n+} and nx moles of B^{m-} will result. Thus if x is the molar solubility of the salt, then:

$$K_{sp} = (mx)^m(nx)^n$$

$$= m^m n^n\ x^{(m+n)}$$

So for salts of the form MX, e.g., AgCl, the reaction is:

$$MX\ (s) \rightarrow M^+\ (aq) + X^-\ (aq),$$

and $K_{sp} = [M^+][X^-] = x^2$. Whereas for salts of the form MX_2, e.g., $CaCl_2$, the reaction is:

$$MX_2 \ (s) \rightarrow M^{2+} \ (aq) + 2X^- \ (aq),$$

and

$$K_{sp} = [M^{2+}][X^-]^2 = x \ (2x)^2 = 4x^3; \ \text{etcetera}$$

In general, note how the stoichiometry comes into play twice: once to determine the relationship between the amount of solid dissolved and the amount of ions present, and again as the exponent in the law of mass action.

Example: The solubility of $Fe(OH)_3$ in aqueous solution was determined to be 4.5×10^{-10} mol/L. What is the value of the K_{sp} for $Fe(OH)_3$?

Solution: The molar solubility (the solubility of the compound in mol/L) is given as 4.5×10^{-10} M. The equilibrium concentration of each ion can be determined from the molar solubility and the balanced dissociation reaction of $Fe(OH)_3$. The dissociation reaction is:

$$Fe(OH)_3 \ (s) \rightarrow Fe^{3+} \ (aq) + 3OH^- \ (aq)$$

Thus, for every mole of $Fe(OH)_3$ that dissociates, one mole of Fe^{3+} and three moles of OH^- are produced. Since the solubility is 4.5×10^{-10} M, the K_{sp} can be determined as follows:

$$K_{sp} = [Fe^{3+}][OH^-]^3$$

$$[OH^-] = 3[Fe^{3+}]; \qquad [Fe^{3+}] = 4.5 \times 10^{-10} \ M$$

$$K_{sp} = [Fe^{3+}](3[Fe^{3+}])^3 = 27[Fe^{3+}]^4$$

$$K_{sp} = (4.5 \times 10^{-10})[3(4.5 \times 10^{-10})]^3 = 27(4.5 \times 10^{-10})^4$$

$$K_{sp} = 1.1 \times 10^{-36}$$

Example: What are the concentrations of each of the ions in a saturated solution of $PbBr_2$, given that the K_{sp} of $PbBr_2$ is 2.1×10^{-6}? If 5 g of $PbBr_2$ are dissolved in water to make 1 L of solution at 25°C, would the solution be saturated, unsaturated, or supersaturated?

Solution: The first step is to write out the dissociation reaction:

$$PbBr_2 \ (s) \rightarrow Pb^{2+} \ (aq) + 2Br^- \ (aq)$$

$$K_{sp} = [Pb^{2+}][Br^-]^2$$

Test Strategy

Predict your answer *before* you go to the answer choices so you don't get persuaded by the wrong choices you'll find there. This helps protect you from persuasive or tricky wrong answer choices, which are often logical twists on the correct choice.

Let x be the concentration of Pb^{2+}. Note that this is also the maximum molarity of $PbBr_2$ that can dissolve (i.e., its solubility). Since every mole of $PbBr_2$ that dissolves yields one mole of Pb^{2+} and two moles of Br^-, $[Br^-]$ is two times $[Pb^{2+}]$, i.e., the concentration of Br^- in the saturated solution at equilibrium $= 2x$.

$$(x)(2x)^2 = 4x^3 = K_{sp} = 2.1 \times 10^{-6}$$

Solving for x, the concentration of Pb^{2+} in a saturated solution is 8.07×10^{-3} M and the concentration of Br^- is $2x = 1.61 \times 10^{-2}$ M.

For the second part of the problem, we convert 5 g of $PbBr_2$ into moles:

$$5 \text{ g} \times \tfrac{1 \text{ mol PbBr}_2}{367 \text{ g}} = 1.36 \times 10^{-2} \text{ mol}$$

1.36×10^{-2} mol of $PbBr_2$ is dissolved in 1 L of solution, so the concentration of the solution is 1.36×10^{-2} M. Since this is higher than the concentration of a saturated solution, this solution would be supersaturated. Any slight disturbance would cause the salt to precipitate out of solution, reducing the ion concentration until the solution is just saturated.

The Common-Ion Effect

The solubility of a salt is considerably reduced when it is dissolved in a solution that already contains one of its ions, rather than in a pure solvent. For example, if a salt such as CaF_2 is dissolved in a solution already containing Ca^{2+} ions, the dissociation equilibrium will shift toward the production of the solid salt. This reduction in solubility, called the common-ion effect, is another example of Le Châtelier's principle.

Example: The K_{sp} of AgI in aqueous solution is 1×10^{-16} mol/L. If a 1×10^{-5} M solution of $AgNO_3$ is saturated with AgI, what will be the final concentration of the iodide ion?

Solution: The concentration of Ag^+ in the original $AgNO_3$ solution will be 1×10^{-5} mol/L. After AgI is added to saturation, the iodide concentration can be found by the formula:

$$[Ag^+][I^-] = 1 \times 10^{-16}$$

$$(1 \times 10^{-5} + x)\, x = 1 \times 10^{-16}$$

where x = molar solubility of AgI. Since the K_{sp} of AgI is so small compared to the initial concentration of Ag^+ ions, we can assume that dissociation of AgI will not contribute much to the final concentration of Ag^+ ions; that is, that $(1 \times 10^{-5} + x)$ $\approx 1 \times 10^{-5}$. Solving for x:

$$x = [I^-] = \frac{1 \times 10^{-16}}{1 \times 10^{-5}} = 1 \times 10^{-11} \text{ mol/L}$$

If the AgI had been dissolved in pure water, the concentration of both Ag^+ and I^- would have been 1×10^{-8} mol/L. The presence of the common ion, silver, at a concentration one thousand times higher than what it would normally be in a silver iodide solution, has reduced the iodide concentration to one thousandth of what it would have been otherwise. An additional 1×10^{-11} mol/L of silver will, of course, dissolve in solution along with the iodide ion, but this will not significantly affect the final silver concentration, which is much higher, thus validating the approximation we made.

Chemical Equilibrium Review Problems

1. Given the following reaction:

$$2NO_2 (g) + 2H_2 (g) \rightarrow N_2 (g) + 2H_2O (g)$$

what is the law of mass action equation?

2. If $K \gg 1$,

   A. the equilibrium mixture will favor products over reactants

   B. the equilibrium mixture will favor reactants over products

   C. the equilibrium amounts of reactants and products are equal

   D. the reaction is irreversible

3. Answer the following questions using the reaction given below.

   $$CH_3OH (l) + H_2 (g) \rightarrow CH_4 (g) + H_2O (l)$$

   a. If the reaction releases energy, in which direction would the reaction be shifted if the temperature were increased?

   b. In which direction would the reaction be shifted if the volume were doubled?

   c. In which direction would the reaction be shifted if CH_4, methane, were removed from the reaction vessel?

4. What is the concentration of the Ag^+ ion in a saturated solution of AgCl? (K_{sp} for AgCl = 1.7×10^{-10}).

   A. 1.7×10^{-10} M

   B. 3.4×10^{-10} M

   C. 1.3×10^{-5} M

   D. 2.6×10^{-5} M

Turn the page
for answers and explanations
to the Review Problems.

Solutions to Review Problems

1. $K_c = \dfrac{[N_2][H_2O]_2}{[NO_2]_2[H_2]_2}$

It can also be expressed in terms of pressures.

2. **A**

The larger the value of K, the more heavily the products are favored over the reactants.

3.

a.

The reaction shifts to the left.

b.

The reaction will remain unchanged. Changing the volume constraints on a reaction that involves gases will affect the equilibrium only when one side of the reaction has a greater number of moles of gases than the other. For this reaction, there is one mole of gas on each side.

c.

The reaction would shift to the right. By removing methane gas from the reaction vessel, the reactant concentrations are effectively increased relative to the product concentrations, and the reaction will go to the right to correct for this.

4. **C**

1.3×10^{-5} M

$$K_{sp} = [Ag^+][Cl^-]$$

Let $x = [Ag^+]$

Since $[Ag^+] = [Cl^-]$, $1.7 \times 10^{-10} = x^2$

$$x = 1.3 \times 10^{-5} \text{ M}$$

CHAPTER 11

THERMOCHEMISTRY

In the last chapter, we introduced the equilibrium constant, which allows us to decribe in a quantitative way just how favorable a reaction is. Le Châtelier's principle, in addition, allows us to predict qualitatively how the equilibrium position of a system would react to changes in external conditions. A more fundamental question we did *not* try to answer, however, is: What makes a reaction favorable in the first place? Why does Mother Nature prefer some reactions to others? In this chapter we will introduce and examine the concepts of enthalpy, entropy and free energy, all of which will give us a fuller understanding of why reactions take place.

From HCl dissociating in water to give protons and chloride ions, to the synthesis of polypeptide chains (proteins) from amino acids in our cells, to the breakdown of ozone in the stratosphere, all chemical reactions have in common the fact that they are accompanied by energy changes. Whenever a reaction takes place, the atoms involved find themselves in a different environment than before. Old bonds have been broken while new ones have been formed; they may now be surrounded by solvent molecules rather than "one of their own," and so on. The amount of energy still residing in the system, and the way that this energy is being distributed among the species in it, may very well have changed. It is these changes that dictate the favorability of a reaction, and form the heart of the study of chemical thermodynamics, or thermochemistry.

Systems and Surroundings: the First Law of Thermodynamics

The term *system* is used to describe the particular part of the universe we are focusing our attention on: a beaker, a cell, Earth and its atmosphere, etcetera; everything outside the system is considered the surroundings or environment (i.e., system + surroundings = universe). A system may be classified as:

- Isolated—when it cannot exchange energy or matter with the surroundings, as with an insulated bomb reactor or a well-insulated thermos flask;

Basic Concept

Three Types of Systems

- Isolated
- Closed
- Open

- Closed—when it can exchange energy but not matter with the surroundings, as with a steam radiator or a stoppered test tube out of (and into) which heat can flow

- Open—when it can exchange both matter and energy with the surroundings, as with a pot of boiling water: Water molecules are escaping into the gas phase, bringing energy with them

You may have learned that energy, though interconvertible among all its different forms (kinetic, potential, etcetera), is conserved: The total amount of energy has to be constant. This is true of a particular system only if it is isolated. Since energy can neither go in nor go out, it has to be conserved. If the system is closed or open, the amount of energy in the system can certainly change. A system can exchange energy with its surroundings in two general ways: as heat or as work. The first law of thermodynamics states that the change in the internal energy of a system is equal to the heat added to the system, q, minus the work that a system does, w:

$$\Delta E \text{ (or } \Delta U) = q - w$$

If work is done on a system, w is negative. Note, however, that sometimes w is defined as the work done on, rather than by, the system, in which case the equation is written as $\Delta E = q + w$, and work done by the system is considered negative. Regardless of which convention is used, if work is done on a system, its energy would increase; if work is done by the system, its energy would decrease. Work is generally associated with movement against some force. For ideal gas systems, for example, expansion against some external pressure means that work is done by the system, while compression implies work being done on the system.

Heat

Heat is often considered a form of energy associated with temperature. Based on the equation above, we now see that a more accurate, though perhaps more abstract, way of looking at it is as a *means* by which energy is transferred. More specifically, heat is energy transfer that occurs as a result of a temperature difference between the system and its surroundings. This transfer will occur spontaneously from a warmer system to a cooler system. Heat, being an exchange of energy, is measured in the same units of energy, e.g., calories (cal) or joules (J), although kcal (kilocalorie, equals 1000 cal) or kJ (kilojoule, or 1000 J) is often more convenient. The conversion between calories and joules is done via the relation: 1 cal = 4.184 J; similarly, 1 kcal = 4.184 kJ. According to convention, heat absorbed by a system (from its surroundings) is considered positive, while heat lost by a system (to its surroundings) is considered negative. This is consistent with the first law: heat absorbed would lead to a positive ΔE, meaning that energy has increased.

Specific Heat and Heat Capacity

Heat is supplied to (or absorbed by) the system to raise its temperature; conversely, heat is released if it cools. The heat absorbed or released by an object as a result of a change in temperature is calculated from the equation:

$$q = mc\Delta T$$

where m is the mass of the object, ΔT is the change in temperature and is equal to the final temperature minus the initial temperature, and c is a quantity known as the specific heat of the substance, a notion that will be discussed further below.

It seems intuitively obvious that the more of the substance there is (the more massive it is), the more heat is required to bring about a particular change in temperature. Recall from our earlier discussion on the kinetic theory of gases that temperature is a measurement of the average kinetic energy of the particles. This applies, at least conceptually, to other states of matter as well: Even though motion is more restricted for particles in the condensed phases, they can still carry kinetic energy. In solids, for example, the atoms vibrate about their equilibrium positions; the stronger these vibrations, the higher the temperature of the solid. A certain amount of heat supplied to a large number of particles would not increase their average energy by much; however, if there were only a small number of particles in the system, that same amount of heat is now spread not as thinly, and thus each particle would gain a larger amount of energy, bringing up the temperature more.

Yet not every substance is responsive to heat to the same degree. Even though we expect from the last paragraph that to raise 2 kg of a substance by 1 °C requires more heat than raising 1 kg of the same substance by 1 °C (in fact, it requires twice the amount of heat), we would not expect that the same amount of heat is required to raise the temperature of 1 kg of steel versus 1 kg of plastic. The specific heat, c, is a proportionality constant that gives an indication of the ease with which one can raise the temperature of something: the larger it is, the larger the amount of heat required to raise its temperature a certain number of degrees, and also the more heat is released if it cools by a certain number of degrees. Its value is a property of the nature of the substance and does not change based on the amount of stuff we have (that has already been taken into account by the mass). The specific heat is often more formally defined as the heat necessary to raise the temperature of 1 kg or 1 g of a material by 1 °C or 1 K. Iron, for example, has a specific heat of about 0.1 kcal/kg°C, while water has a specific heat of 1.0 kcal/kg°C. It is therefore much easier to raise the temperature of 1 kg of iron by 10 °C than it is to do the same to 1 kg of water. In fact, you should be able to see that ten times the heat is needed.

The mass and the specific heat is sometimes lumped together to give a quantity known as the heat capacity. This quantity then describes the heat needed to raise the temperature of the object as a whole by 1 °C or 1 K.

Basic Concept

The more of a substance there is, the more heat is required to bring about a change in its temperature.

While heat is associated with exchange of thermal energy, and we have so far been talking about the mathematical relationship between heat and temperature changes, a system that is heated does not necessarily increase in temperature. Heat can also increase the potential (rather than kinetic) energy of the particles in a system; this occurs during a phase change. Heat is required to melt something (change its phase from solid to liquid) or to vaporize something (change its phase from liquid to gas). In both cases (and also in the case of sublimation, where a solid is converted into a gas directly), the molecules are overcoming the attractive forces that hold them together. This is where the energy supplied by heating is being "put to use." Conversely, heat is released as a substance crystallizes (or freezes) or condenses. During such phase changes the temperature remains constant, and the heat involved in these processes can be expressed as:

$$q = mL$$

where m is again the mass of the substance undergoing the phase change and L is the heat of transformation, the value of which depends on both the substance and the particular process we are talking about: vaporization, sublimation, fusion (melting), etcetera.

The heats of transformation for two reverse processes have the same magnitude; that is, the heat of fusion is the same as the heat of crystallization with opposite sign; the heat of vaporization is the same as the heat of condensation with opposite sign; and so forth.

To bring a cube of ice at –50 °C to water vapor at 130 °C, then, the heat required would be:

$$q = mc_{ice} \, (50 \, °C) + mL_{fus} + mc_{water} \, (100 \, °C) + mL_{vap} + mc_{steam} \, (30 \, °C)$$

where L_{fus} is the heat of fusion of water and L_{vap} the heat of vaporization of water. Only one m value is needed since mass is conserved. The "heating curve," a plot of the temperature versus the amount of heat added, would look like:

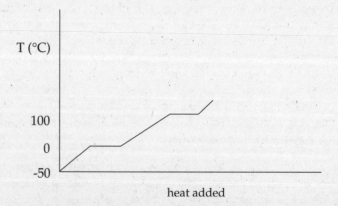

The flat portions imply that the temperature is not changing, and correspond to the processes of melting and boiling.

Reactions that absorb heat energy are said to be endothermic, while those that release heat are said to be exothermic. An adiabatic process is one in which no heat exchange occurs (no heat goes into or out of the system). Melting and vaporization are therefore endothermic, while freezing and condensation are exothermic processes. An isothermal process is one in which the temperature of the system remains constant.

Calorimetry

Calorimetry is an experimental technique that measures the heat change associated with a process. In constant-volume calorimetry, the volume of the container holding the reacting mixture does not change during the course of the reaction. The heat of reaction is measured using a device called a bomb calorimeter. This apparatus consists of a steel bomb into which the reactants are placed. The bomb is immersed in an insulated container containing a known amount of water. The reactants are electrically ignited and heat is absorbed or evolved as the reaction proceeds. The heat of the reaction, q_{rxn}, can be determined as follows. Since no heat enters or leaves the system, the net heat change for the system is zero; therefore, the heat change for the reaction is compensated for by the heat change for the water and the bomb, which is easy to measure.

$$q_{system} = q_{rxn} + q_{water} + q_{steel} = 0$$

Thus:
$$q_{rxn} = -(q_{water} + q_{steel})$$

$$= -(m_{water} c_{water} \Delta T + m_{steel} c_{steel} \Delta T)$$

States and State Functions

The state of a system is described by the macroscopic properties of the system. Examples of macroscopic properties include temperature (T), pressure (P), and volume (V). When the state of a system changes, the values of the properties also change. If the change in the value of a property depends only on the initial and final states of the system, and not on the path of the change (how the change was accomplished), that property is known as a state function. Pressure, temperature, and volume are important state functions. Other examples are enthalpy (H), entropy (S), free energy (G) (all discussed below), and internal energy (E or U).

A set of standard conditions (25°C and 1 atm) is normally used for measuring the enthalpy, entropy, and free energy of a reaction. A substance in its most stable form under standard conditions is said to be in its standard state. Examples of substances in their standard states include hydrogen as $H_2(g)$, water as $H_2O(l)$, and salt as $NaCl(s)$. The changes in enthalpy, entropy, and free energy that occur when a reaction takes place under standard conditions are symbolized by $\Delta H°$, $\Delta S°$, and $\Delta G°$, and are known as the standard change in enthalpy (or standard enthalpy change), etcetera.

Don't Mix These Up on Test Day

Do not confuse standard conditions with standard temperature and pressure (STP) used in gas law calculations.

Enthalpy

Most reactions in the lab occur under constant pressure (at 1 atm, in open containers). (Reactions carried out under a constant pressure are said to be isobaric.) To express heat changes at constant pressure, chemists use the term enthalpy (H), often thought of as the "heat content" of a system. The change in enthalpy (ΔH) of a process is equal to the heat absorbed or evolved by the system at constant pressure. Since enthalpy is a state function, the enthalpy change of a process depends only on the enthalpies of the initial and final states, not on the path. Thus, to find the enthalpy change of a reaction, ΔH_{rxn}, one may subtract the enthalpy of the reactants from the enthalpy of the products:

$$\Delta H_{rxn} = H_{products} - H_{reactants}$$

A positive ΔH corresponds to an endothermic process (absorbs heat), and a negative ΔH corresponds to an exothermic process (releases heat). The value of enthalpy change (or of enthalpies in general) is dependent on external conditions such as temperature and pressure. As we shall see, enthalpy plays an important role in determining the favorability of a reaction, but first we need to discuss ways in which one can calculate or obtain values for changes in enthalpy.

Hess's Law

Hess's law is simply the application of the concept of path-independence to enthalpy. It states that if a reaction can be broken down into a series of steps, the enthalpy change for the overall net reaction is just the sum of the enthalpy changes of each step. The steps need not even correspond to actual processes carried out in the real world or in the lab, but can be purely hypothetical. For example, consider the reaction:

$$Br_2 (l) \rightarrow Br_2 (g) \quad \Delta H = 31 \text{ kJ}$$

The enthalpy change of the above reaction will always be 31 kJ/mol provided that the same initial and final states $Br_2 (l)$ and $Br_2 (g)$ are operative. Instead of direct vaporization, $Br_2 (l)$ could first be decomposed to Br atoms and then recombined to form $Br_2 (g)$; since the net reaction is the same (the two possible sequences share the same initial state and the same final state), the change in enthalpy will be the same.

$Br_2 (l) \rightarrow$	2 Br (g)	ΔH_1
2 Br (g) $\rightarrow$	$Br_2 (g)$	ΔH_2

$Br_2 (l) \rightarrow$	$Br_2 (g)$	$\Delta H = \Delta H_1 + \Delta H_2 = 31$ kJ

Example: Given the following thermochemical equations:

Basic Concept

ΔH negative   = exothermic
                    = heat given off

ΔH positive   = endothermic
                    = heat absorbed

a) C_3H_8 (g) + 5 O_2 (g) → 3 CO_2 (g) + 4 H_2O (l)

ΔH_a = −2220.1 kJ

b) C (graphite) + O_2 (g) → CO_2 (g)

ΔH_b = −393.5 kJ

c) H_2 (g) + 1/2 O_2 (g) → H_2O (l)

ΔH_c = −285.8 kJ

Calculate ΔH for the reaction:

d) 3 C (graphite) + 4 H_2 (g) → C_3H_8 (g)

Solution: Equations a, b, and c must be combined to obtain equation d. Since equation d contains only C, H_2, and C_3H_8, we must eliminate O_2, CO_2, and H_2O from the first three equations. Equation a is reversed to get C_3H_8 on the product side.

e) 3 CO_2 + (g) + 4 H_2O + (l) → C_3H_8 (g) + 5 O_2 (g)

ΔH_e = 2220.1 kJ

(Note that when we reverse the reaction, the sign of the enthalpy change is reversed as well.)

Next, equation b is multiplied by 3 (this gives equation f) and c by 4 (this gives equation g). The following addition is done to obtain the required equation d: 3b + 4c + e.

3 CO_2 (g) + 4 H_2O (l) → C_3H_8 (g) + 5 O_2

ΔH_e = 2220.1 kJ

f) 3 × [C (graphite) + O_2 (g) → CO_2 (g)]

ΔH_f = 3 × −393.5 kJ

g) 4 × [H_2 (g) + O_2 (g) → H_2O (l)]

ΔH_g = 4 × −285.8 kJ

3 C (graphite) + 4 H_2 (g) → C_3H_8 (g)

ΔH_d = −103.6 kJ

where $\Delta H_d = \Delta H_e + \Delta H_f + \Delta H_g$.

Standard Enthalpy of Formation

Hess's law is useful because measuring the change in enthalpy for a process directly can be challenging experimentally, but by taking advantage of the fact that this quantity is independent of path, one can calculate ΔH for any process if the values for certain other reactions are known. The most common approach is to express the enthalpy change of a reaction in terms of the standard enthalpies of formation of the products and the reactants.

The standard enthalpy of formation of a compound, $\Delta H°_f$, is the enthalpy change that would occur if one mole of a compound were formed directly from its elements in their standard states. For example, the standard enthalpy of formation of H_2O is just the enthalpy change for the reaction:

$$H_2\,(g) + 1/2\,O_2\,(g) \rightarrow H_2O\,(l)$$

if the reaction were carried out under standard conditions. We have picked the gaseous forms of hydrogen and oxygen because that is the most stable form in which they exist under such conditions.

Note that $\Delta H°_f$ of an element in its standard state is zero. The $\Delta H°_f$'s of most known substances are tabulated. The enthalpy of formation is also often referred to as the heat of formation.

Standard Enthalpy of Reaction

The standard enthalpy (or heat) of a reaction, $\Delta H°_{rxn}$, is the hypothetical enthalpy change that would occur if the reaction were carried out under standard conditions; that is, when reactants in their standard states are converted to products in their standard states at 298K. It can be expressed as:

$$\Delta H°_{rxn} = (\text{sum of } \Delta H°_f \text{ of products}) - (\text{sum of } \Delta H°_f \text{ of reactants})$$

Earlier we have given a general definition $\Delta H_{rxn} = H_{products} - H_{reactants}$. This expression, however, is not very useful. The actual values of the enthalpies of the species cannot be measured (and are in fact more a matter of definition); what we can measure is only enthalpy changes, and those often only with difficulty. But from our discussion on Hess's law, we know we can concoct a scheme consisting of a series of steps that yield the same net reaction, and the sums of enthalpy changes would be the same as the net enthalpy change of the reaction. The equation above, $\Delta H°_{rxn} = (\text{sum of } \Delta H°_f \text{ of products}) - (\text{sum of } \Delta H°_f \text{ of reactants})$, establishes a common scheme for all reactions: We first break the reactants down to give elements in their standard states, then we combine these rudimentary "building blocks" in new ways to give the products of the reaction. For example, when we express the enthalpy change for the reaction

$$CaCO_3\,(s) \rightarrow CaO\,(s) + CO_2\,(g)$$

as $\Delta H°_{rxn} = \Delta H°_f$ of CaO $+ \Delta H°_f$ of $CO_2 - \Delta H°_f$ of $CaCO_3$, we are essentially reporting the reaction as follows: First we break down each mole of $CaCO_3$ into 1 mole of Ca, 1 of C and 3/2 of O_2, then we combine the Ca with half a mole of O_2 to form a mole of CaO, and use the remaining oxygen to form CO_2 with carbon. The enthalpy changes for the last two steps are the standard enthalpies of formation of CaO and CO_2, while the first step is the reverse of the the formation of $CaCO_3$, from which we get the minus sign in front of the enthalpy of formation of $CaCO_3$, and which explains why we subtract the enthalpies of formation of the reactants in the general equation.

It should be emphasized again that this scheme is not actually carried out, but merely used for "accounting conveniences": Instead of tabulating an enthalpy change for every reaction imaginable, we need only a list of enthalpies of formation of different compounds which, while still numerous, do not possess this almost infinite possible number of combinations.

Bond Dissociation Energy

Enthalpies or heats of reaction are related to changes in energy associated with the breakdown and formation of chemical bonds. The reason why bonds are formed in the first place is because it is energetically favorable for the atoms to come together—it corresponds to a state of lower energy. This implies, therefore, that energy needs to be supplied to break a bond and separate the atoms (i.e., bond-breaking is endothermic.) Bond energy, or bond dissociation energy, is an average of the energy required to break a particular type of bond in one mole of gaseous molecules. It is tabulated as the magnitude of the energy absorbed as the bonds are broken. For example:

$$H_2 \, (g) \rightarrow 2H \, (g) \quad \Delta H = 436 \text{ kJ}$$

A molecule of H_2 gas is cleaved to produce two gaseous, unassociated hydrogen atoms. For each mole of H_2 gas cleaved, 436 kJ of energy is absorbed by the system. This is the bond energy of the H-H bond. For other types of bonds, the energy requirements are averaged by measuring the enthalpy of cleaving many different compounds with that bond. The averaging is needed because the energy required to break a bond is not uniquely determined by what atoms are being separated; it also depends on what else may be bonded to those atoms. For example, the energy needed to break a C-H bond is different on going from CH_4 to CH_3Cl to CCl_3H, etcetera. The C-H bond dissociation energy one would find in a table (415 kJ/mol) was compiled from measurements on thousands of different organic compounds.

Bond energies can be used to estimate enthalpies of reactions. The enthalpy change of a reaction is given by:

ΔH_{rxn} =   (ΔH of bonds broken) – (ΔH of bonds formed)

=   total energy input – total energy released

Example: Calculate the enthalpy change for the following reaction:

$$C \, (s) + 2 \, H_2 \, (g) \rightarrow CH_4 \, (g)$$

Basic Concept

Bond-breaking is endothermic; bond formation is exothermic.

Bond dissociation energies of H-H and C-H bonds are 436 kJ/mol and 415 kJ/mol, respectively.

ΔH_f of C (g) = 715 kJ/mol

Solution: For each mole of CH_4 formed, 2 moles of H-H bonds are broken and 4 moles of C-H bonds are formed (each molecule of CH_4 contains 4 C-H bonds). The one additional thing we have to take into consideration is that bond dissociation energies are always in reference to the gas phase, yet here we have carbon in its solid form. This is why we need the enthalpy of formation of gaseous carbon: The first step in our hypothetical scheme is to convert carbon from its solid to its gaseous form. Thus the enthalpy change is:

$\Delta H°_{rxn}$ = 715 kJ/mol + 2 × H-H bond energy – 4 × C-H bond energy

= 715 kJ/mol + 2 × 436 kJ/mol – 4 × 415 kJ/mol

= –73 kJ/mol

Again, we subtract the bond dissociation energy to reflect the fact that the C-H bonds are being formed rather than broken.

Quick Quiz

At any given temperature, which will have a higher degree of entropy, a solid or a gas?

Answer: Gas

Heats of Combustion

One more type of standard enthalpy change which is often used is the standard heat of combustion, $\Delta H°_{comb}$. A combustion reaction is one in which the reactant reacts with (excess) oxygen to yield (in most cases) carbon dioxide and water, producing a flame during the reaction. (Excess oxygen is specified because inadequate oxygen leads to the generation of carbon monoxide rather than carbon dioxide.) The burning of a log, for example, is a combustion reaction. These reactions are exothermic (release energy, have a negative enthalpy change). The reactions used in the C_3H_8 (g) example earlier were combustion reactions, and the corresponding values ΔH_a , ΔH_b and ΔH_c in that example were thus heats of combustion.

Entropy

Entropy (S) is a measure of the disorder, or randomness, of a system. The units of entropy are energy/temperature, commonly J/K or cal/K. The greater the order in a system, the lower the entropy; the greater the disorder or randomness, the higher the entropy. At any given temperature, a solid will have lower entropy than a gas, because individual molecules in the gaseous state are moving randomly, while individual molecules in a solid are constrained in place. Entropy is a state function, so a change in entropy depends only on the initial and final states:

$$\Delta S = S_{final} - S_{initial}$$

A change in entropy is also given by:

$$\Delta S = \frac{q_{rev}}{T}$$

where q_{rev} is the heat added to the system undergoing a reversible process (a process that proceeds with infinitesimal changes in the system's conditions) and T is the absolute temperature.

A standard entropy change for a reaction, $\Delta S°$, is calculated using the standard entropies of reactants and products:

$$\Delta S°_{rxn} = (\text{sum of } S°_{products}) - (\text{sum of } S°_{reactants})$$

The Second Law of Thermodynamics

Entropy is an important concept because it determines whether a process will occur spontaneously. The second law of thermodynamics states that all spontaneous processes proceeding in an isolated system lead to an increase in entropy. Since the universe as a whole is one big isolated system, we can also rephrase this law in a way that is perhaps more stimulating to our imagination—the entropy of the universe either increases (spontaneous, irreversible processes) or stays the same (reversible processes). It can never decrease:

$$\Delta S_{universe} = \Delta S_{system} + \Delta S_{surroundings} \geq 0$$

Note that the entropy of a system can decrease, as long as it is compensated for by a larger increase in entropy in the surroundings. The mechanism of refrigeration decreases the entropy within the refrigerator by maintaining a low temperature that would not persist if left to nature; heat is however generated and dumped outside into the kitchen that increases the entropy. Our cells are constantly engaging in biochemical reactions that increase the order locally: synthesizing large biomolecules from disordered "building blocks," sequestering ions in different compartments when it would be more "natural" for them to diffuse over a larger volume, etcetera. All these processes come at the expense of entropy increases elsewhere; for example, disorder that has been generated when we digested our meal the previous evening. A system will spontaneously tend toward an equilibrium state (one of maximum entropy) if left alone.

The Third Law of Thermodynamics

Instead of just working with changes or relative magnitudes (as in the case of enthalpy), there is a standard with which one can assign the actual value of entropy of a substance. The third law of thermodynamics states that the entropy of a pure crystalline substance at absolute zero is zero. This corresponds to a state of "perfect order" because all the atoms in this hypothetical state possess no kinetic energy and do not vibrate at all; thus, there is absolutely no randomness and no disorder in the spatial arrangement of the atoms.

Gibbs Free Energy

What makes a reaction favorable? In the quantity of entropy, we have an unambiguous criterion of whether a reaction would occur spontaneously: The total entropy of the universe has to increase. The only problem with this is that it is not very practical—who knows how to keep track of the entropy of the entire universe? It would be nice to have a quantity that deals only with the system itself that we can examine to determine the favorability of a reaction. The thermodynamic state function, G (known as the Gibbs Free Energy), is just such a quantity. It combines the two factors that affect the spontaneity of a reaction-changes in enthalpy, ΔH, and changes in entropy, ΔS, of the system. The change in the free energy of a system, ΔG, represents the maximum amount of energy released by a process, occurring at constant temperature and pressure, that is available to perform useful work. ΔG is defined by the equation:

$$\Delta G = \Delta H - T\Delta S$$

where T is the absolute temperature.

Spontaneity of Reaction

In the equilibrium state, free energy is at a minimum. A process can occur spontaneously if the Gibbs function decreases, i.e., $\Delta G < 0$.

1. If ΔG is negative, the reaction is spontaneous.

2. If ΔG is positive, the reaction is not spontaneous.

3. If ΔG is zero, the system is in a state of equilibrium; thus, $\Delta G = 0$ and $\Delta H = T\Delta S$ at equilibrium.

Because the temperature is always positive, i.e., in Kelvins, the effects of the signs of ΔH and ΔS and the effect of temperature on spontaneity can be summarized as follows:

ΔH	ΔS	Outcome
–	+	spontaneous at all temperatures
+	–	nonspontaneous at all temperatures
+	+	spontaneous only at high temperatures
–	–	spontaneous only at low temperatures

Qualitatively, the equation $\Delta G = \Delta H - T\Delta S$, and more generally the whole notion of free energy tell us that there are two factors which favor a reaction: a decrease in energy in the form of enthalpy and an increase in disorder. If these two factors are working against each other in a reaction, then temperature will determine which is the more dominant factor—the higher the temperature, the easier entropic considerations override enthalpic ones. Note also the implication that a reaction that is favorable at one temperature may not be favorable at another. This, actually, should not be a surprise; after all, for example, the melting of ice is expected to occur at 15 °C, but not at –100 °C (assuming atmospheric pressure).

Standard Free Energy Change

Standard free energy change, $\Delta G°$, is defined as the ΔG of a process occurring under standard conditions, and for which the concentrations of any solutions involved are 1 M. The standard free energy of formation of a compound, $\Delta G°_f$, is the free-energy change that occurs when 1 mole of a compound in its standard state is formed from its elements in their standard states. The standard free energy of formation of any element in its most stable form (and, therefore, its standard state) is zero. The standard free energy of a reaction, $\Delta G°_{rxn}$, is the free energy change that occurs when that reaction is carried out under standard state conditions; i.e., when the reactants in their standard states are converted to the products in their standard states, at standard conditions of T and P. Bearing in mind what we did for enthalpy changes, we can write:

$$\Delta G°_{rxn} = \text{(sum of } \Delta G°_f \text{ of products)} - \text{(sum of } \Delta G°_f \text{ of reactants)}.$$

Relation Between Free Energy and the Equilibrium Constant

The value of the free energy change in general (under nonstandard conditions) is related to the standard free energy change by the following equation:

$$\Delta G = \Delta G° + RT \ln Q$$

where R is the gas constant, T is the temperature in Kelvins, ln stands for the natural logarithm function, and Q is the reaction quotient we mentioned briefly in passing in the last chapter. For the reaction $a\,A + b\,B \leftrightarrows c\,C + d\,D$, the reaction quotient is

$$Q = \frac{[C]^c[D]^d}{[A]^a[B]^b}$$

Unlike the mass action expression, the values for the concentrations are not necessarily those at equilibrium. Because the concentrations of the species change as the reaction progresses towards equilibrium, the value of Q, and hence also the value of ΔG, will change as the reaction progresses. As pointed out above, when the system is at equilibrium ΔG is zero, and in that case we would obtain from the equation

$$0 = \Delta G° + RT \ln Q_{eq}$$

Basic Concept

$\Delta G°$ positive: small K, reaction not favored

$\Delta G°$ negative: large K, reaction favored

where Q_{eq} is the value of the reaction quotient at equilibrium. But that is just K, the equilibrium constant! Hence we can rewrite and rearrange to get

$$\Delta G° = - RT \ln K$$

or equivalently,

$$K = e^{-\Delta G°/RT}$$

Thus, we have obtained the final missing link to put it all together: the more negative $\Delta G°$ is, the larger the equilibrium constant, and hence the more the products are favored at equilibrium.

Examples

a.   Vaporization of water at one atmosphere pressure

$$H_2O \ (l) + heat \rightarrow H_2O \ (g)$$

When water boils, hydrogen bonds (H-bonds) are broken. Energy is absorbed (the reaction is endothermic), and thus ΔH is positive. Entropy increases as the closely packed molecules of the liquid become the more randomly moving molecules of a gas; thus, $T\Delta S$ is also positive. Since ΔH and $T\Delta S$ are both positive, the reaction will proceed spontaneously only if $T\Delta S > \Delta H$. For the particular values of ΔS and ΔH applicable for this reaction, this condition is true only at temperatures above 100°C. Below 100°C, ΔG is positive and the water remains a liquid. At 100°C, $\Delta H = T\Delta S$ and $\Delta G = 0$: an equilibrium is established between water and water vapor. The opposite is true when water vapor condenses: H-bonds are formed, and energy is released; the reaction is exothermic (ΔH is negative) and entropy decreases, since a liquid is forming from a gas ($T\Delta S$ is negative). Condensation will be spontaneous only if $\Delta H < T\Delta S$. This is the case at temperatures below 100°C; above 100°C, $T\Delta S$ is more negative than H, ΔG is positive, and condensation is not spontaneous. Again, at 100°C, an equilibrium is established.

b.   The combustion of C_6H_6 (benzene)

$$2 \ C_6H_6 \ (l) + 15 \ O_2 \ (g) \rightarrow 12 \ CO_2 \ (g) + 6 \ H_2O \ (g) + heat$$

In this case, heat is released (ΔH is negative) as the benzene burns and the entropy is increased ($T\Delta S$ is positive), because two gases (18 moles total) have greater entropy than a gas and a liquid (15 moles gas and 2 liquid). ΔG is negative and the reaction is spontaneous.

Thermochemistry Review Problems

1. A process involving no heat exchange is known as

   A. an isothermal process.

   B. an isobaric process.

   C. an adiabatic process.

   D. an isometric process.

2. What is the heat capacity of a 10 g sample that has absorbed 100 cal over a temperature change of 30°C?

   A. 0.333 cal/g°C

   B. 0.666 cal/g°C

   C. 3 cal/g°C

   D. 300 cal/g°C

3. Calculate the enthalpy of formation of N (g) in the following reaction:

   N_2 (g) → 2N (g) $\Delta H°_{rxn} = 945.2$ kJ

   A. –945.2kJ/mol

   B. 0.0kJ/mol

   C. 472.6kJ/mol

   D. 945.2kJ/mol

4. Calculate the amount of heat needed to bring 10 g of ice from –15°C to 110°C. (Heat of fusion = 80 cal/g; heat of vaporization = 540 cal/g. The heat capacities of both ice and steam vary with temperature; for this problem, use the estimate of 0.5 cal/g • K for both. The heat capacity of water is 1.0 cal/g • K.)

   A. 7.325 cal

   B. 7.450 cal

   C. 7.325 kcal

   D. 7.450 kcal

5. Using the information given in the reaction equations below, calculate the heat of formation for 1 mole of carbon monoxide.

   $2 C$ (s) $+ 2 O_2 → 2 CO_2$ $\Delta H_{rxn} = -787$ kJ

   $2 CO + O_2 → 2 CO_2$ $\Delta H_{rxn} = -566$ kJ

   A. –221 kJ/mol

   B. –110 kJ/mol

   C. 110 kJ/mol

   D. 221 kJ/mol

6. If the free energy change accompanying a reaction is negative,

   A. the reaction can occur spontaneously.

   B. the reaction can be used to do work by driving other reactions.

   C. the entropy must always be negative.

   D. both A and B.

7. All of the following are correct statements concerning entropy EXCEPT

   A. all spontaneous processes tend towards an increase in entropy

   B. the more highly ordered the system, the higher the entropy

   C. the entropy of a pure crystalline solid at 0 K is 0

   D. the change in entropy of an equilibrium process is 0

8. A 50 g sample of metal was heated to 100°C and then dropped into a beaker containing 50 g of water at 25°C. If the specific heat capacity of the metal is 0.25 cal/g • °C, what is the final temperature of the water?

   A. 27°C

   B. 40°C

   C. 60°C

   D. 86°C

9. Calculate the maximum amount of work that can be done by the following reaction at 30°C ($\Delta H = -125$ kJ, $\Delta S = -200$ J/K).

   $FeCl_2 \ (aq) + 1/2 \ Cl_2 \ (g) \rightarrow FeCl_3 \ (aq)$

   A. 64.4 kJ

   B. 119 kJ

   C. 5875 kJ

   D. 60475 kJ

10. Calculate the bond energy of a BrF bond using the following reaction equation. (ΔH_f of $BrF_5 \ (g) = -429$ kJ/mol, ΔH_f of Br $(g) = 112$ kJ/mol, ΔH_f of F $(g) = 79$ kJ/mol)

    $Br \ (g) + 5 \ F \ (g) \rightarrow BrF_5 \ (g)$

    A. 936 kJ/mol

    B. 187 kJ/mol

    C. 86 kJ/mol

    D. 47 kJ/mol

11.

Bond	Average Bond Energy
$C\equiv O$	1075 kJ/mol
$C=O$	728 kJ/mol
$C-Cl$	326 kJ/mol
$Cl-Cl$	243 kJ/mol

Calculate the heat of reaction for the following equation using the information given above.

$CO + Cl_2 \rightarrow COCl_2$

A. 62 kJ

B. -62 kJ

C. -409 kJ

D. 706 kJ

Turn the page
for answers and explanations
to the Review Problems.

Solutions to Review Problems

1. C

An adiabatic process is one in which no heat flow takes place. An isothermal process is one in which the temperature remains the same. An isobaric process is one in which the pressure remains the same.

2. A

In calorimetry, the amount of heat absorbed in a given process is calculated using the following equation:

$$q = mc\Delta T$$

Knowing that the heat absorbed is 100 cal, the mass is 10 g, and the temperature change is 30°C, the specific heat capacity can be calculated:

$$100 \text{ cal} = 10 \text{ g (c)}(30°C)$$

$$c = 0.333 \text{ cal/g°C}$$

3. C

The reaction given is a formation reaction, since the reactant N_2 is in its stable elemental state (gaseous dimer). The enthalpy of the change of the reaction is the enthalpy of formation of N, except that since two moles of N are formed, we need to divide the value by two to get the value per mole.

4. C

To answer this question, the ice must be imagined as passing through the following stages: ice from –15°C to ice at 0°C; ice at 0°C to water at 0°C; water from 0°C to water at 100°C; water at 100°C to steam at 100°C; steam at 100°C to steam at 110°C. Once these steps have been outlined, the amount of heat needed to perform each of them must be calculated using the following equation:

$$\text{heat absorbed} = (10g) (0.5 \text{ cal/g} \times K) (15 \text{ K})$$

$$+(10g) (80 \text{ cal/g})$$

$$+(10g) (1 \text{ cal/g} \times K) (100 \text{ K})$$

$$+(10g) (540 \text{ cal/g})$$

$$+(10g) (0.5 \text{ cal/g} \times K) (10 \text{ K})$$

$$= 7325 \text{ cal}$$

$$= 7.325 \text{ kcal}$$

5. B

This problem uses Hess's Law, which states that heats of reaction may be added to determine the enthalpy of another reaction. To calculate the heat of formation for one mole of carbon monoxide, the reaction equations must be manipulated as follows:

$$2\,C\,(s) + 2\,O_2 \rightarrow \quad 2\,CO_2 \qquad \Delta H_{rxn} = -787\text{ kJ}$$

$$2\,CO + O_2 \rightarrow \quad 2\,CO_2 \qquad \Delta H_{rxn} = -566\text{ kJ}$$

The second equation must be reversed:

$$2\,C\,(s) + 2\,O_2 \rightarrow \quad 2\,CO_2 \qquad \Delta H_{rxn} = -787\text{ kJ}$$

$$2\,CO_2 \rightarrow \quad 2\,CO + O_2 \quad \Delta H_{rxn} = 566\text{ kJ}$$

If the two equations are added together, a third equation is obtained:

$$2\,C + O_2 \rightarrow \quad 2\,CO \qquad \Delta H_{rxn} = -221\text{ kJ}$$

This is the formation reaction of carbon monoxide. As in question 3, however, we need to divide the enthalpy change by two to obtain the value for 1 mole.

6. D

A negative free energy change signifies that the reaction is spontaneous and that work can be done. If the energy that is released is coupled to less favorable reactions, they can be driven to completion. This, for example, is how your body uses the breakdown of ATP to accomplish certain reactions that normally wouldn't occur.

7. B

The opposite of the statement in choice B is true: the more highly ordered a system, the *lower* its entropy. Choices A and C are statements of the second and third laws of thermodynamics respectively. Choice D is correct because if a system is at equilibrium, it is not getting more or less disordered and thus there should be no change in its entropy.

8. B

This problem uses the concept of conservation of energy. When the metal is put in the water, it will lose heat; that heat will be transferred to the water. Thus, the amount of heat released by the metal is the same as the amount of heat absorbed by the water. Use the following equation for heat transfer:

$$q = mc\Delta T$$

In addition, given a long enough time, the metal and the water will reach thermal equilibrium: they will eventually have the same temperature. The expressions for the heat released and absorbed by the metal and water respectively are:

$$q = 50\text{ g }(0.25\text{ cal/g} \cdot {}^\circ C)(100^\circ C - T_f)$$

$$q = 50\text{ g }(1.0\text{ cal/g} \cdot {}^\circ C)(T_f - 25^\circ C)$$

However, since q should be the same for both equations, the expression can be rewritten:

$$50\text{g }(0.25\text{ cal/g} \cdot {}^\circ C)\,(100^\circ C - T_f)$$

$$= 50\text{ g }(1.0\text{ cal/g} \cdot {}^\circ C)(T_f - 25^\circ C)$$

Canceling the 50 g from each side and solving for T_f, which is the final temperature,

$$(0.25)(100^\circ C - T_f) = (T_f - 25^\circ C)$$

$$25^\circ C - (0.25)T_f = (T_f - 25^\circ C)$$

$$50^\circ C = 1.25\,T_f$$

or, simply, $\quad T_f = 40^\circ C$

9. A

The maximum amount of work that can be done by a spontaneous reaction is the absolute value of its Gibbs free energy change, and the following equation can be used to solve for it:

$$\Delta G = \Delta H - T\Delta S$$

The free energy change due to the reaction can be calculated by substituting the information given for ΔH, T, and ΔS. Be sure that the units are consistent when solving the equation.

$$\Delta G = -125 \text{ kJ} - (30°C + 273)(-0.2 \text{ kJ/K})$$

$$\Delta G = -125 \text{ kJ} - (303 \text{ K})(-0.2 \text{ kJ/K})$$

$$\Delta G = -64.4 \text{ kJ}$$

Thus, 64.4 kJ can be done by the system.

10. B

The change in enthalpy, or heat of reaction, for any reaction is the sum of the enthalpies of formation of the products minus the sum of the enthalpies of formation of the reactants:

$$\Delta H_{rxn} = \Sigma \Delta H_f \text{ (products)} - \Sigma \Delta H_f \text{ (reactants)}$$

Substituting the heats of formation of BrF_5 (g), Br (g), and F (g) into the equation, the heat of reaction can be calculated:

$$\Delta H_{rxn} = [(-429 \text{ kJ/mol})(1 \text{ mol})] - [(112 \text{ kJ/mol})(1 \text{ mol}) + (79 \text{ kJ/mol}) (5 \text{ mol})]$$

$$= -936 \text{ kJ}$$

Since no bonds were broken in this reaction the (magnitude of the) heat of formation here is equal to the sum of the bond energies of the product. As there are five equivalent Br-F bonds each one would contribute to one fifth of the total bond energy. This means that for each bond formed, an average of 187 kJ/mol of energy was released (note the negative sign in front of the enthalpy). The bond energy is defined to be the energy required to break the bond, the reverse of the process we have just described. The Br-F bond energy is therefore 187 kJ/mol. As it always requires energy to break a bond, all bond enthalpies are positive.

11. B

Heats of reaction can be calculated from bond energies using the following equation:

$$\Delta H_{rxn} = \Delta H \text{ of bonds broken} - \Delta H \text{ of bonds formed}$$

$$= [(-728 \text{ kJ/mol})(1 \text{ mol})$$
$$+ (-326 \text{ kJ/mol})(2 \text{ mol})]$$
$$+ [(243 \text{ kJ/mol})(1 \text{ mol})$$
$$+ (1075 \text{ kJ/mol})(1 \text{ mol})]$$

$$= -62 \text{ kJ}$$

CO contains a triple bond between C and O that is being broken. $COCl_2$ contains a double bond between C and O and 2 C–Cl single bonds that are being formed.

CHAPTER 12

CHEMICAL KINETICS

Thermodynamics and the study of chemical equilibrium tell us whether the occurrence of a reaction is favorable and to what extent the reaction goes towards completion. However, a lot of reactions that, from a free energy perspective, are expected to favor products heavily, are not seen to proceed readily in our everyday experience. For example, graphite is the more thermodynamically stable state of carbon under standard conditions compared to diamond, but we probably have better things to worry about than our diamonds turning into pencils. Similarly, combustion reactions of hydrocarbons (a constituent of the human body) are exothermic and also tend to be entropically favored, but we probably shouldn't count on our enemies spontaneously erupting into flames. The reason behind these observations is that thermodynamics only reveals part of the story about chemical reactions. The inherent tendency of a reaction to occur does not necessarily have anything to do with how readily or quickly it does take place. Furthermore, thermodynamics does not give us a microscopic picture of how exactly a reaction is proceeding: How do the individual molecules interact with one another to lead to the end product? How many steps does the reaction have to go through? All these issues are investigated within the realm of chemical kinetics—the study of the rates of reactions, the effect of reaction conditions on these rates, and the mechanisms implied by such observations.

Reaction Rates

The rate of a reaction is an indication of how rapidly it is occurring. First of all, we need to have an exact, quantitative way of describing the rate. Then we shall explore what the rate depends on.

Definition Of Rate

Consider a reaction $2A + B \rightarrow C$, in which 1 mole of C is produced from every 2 moles of A and 1 mole of B. We want to come up with some quantitative way of describing just how fast the reaction has proceeded or is proceeding at any instant in time. The most natural way is to use either the disappearance of reactants over time, or the appearance of products over time. The faster either of these rates are, the faster the rate of reaction:

$$rate \sim \frac{decrease\ in\ reactant\ concentration}{time} \sim \frac{increase\ in\ product\ concentration}{time}$$

Because the concentration of a reactant decreases during the reaction while we want the rates to be positive numbers, a minus sign needs to be placed before a rate that is expressed in terms of reactants. For the reaction above, the rate of disappearance of A is $\frac{-\Delta[A]}{\Delta t}$, the rate of disappearance of B is $\frac{-\Delta[B]}{\Delta t}$, and the rate of appearance of C is $\frac{-\Delta[C]}{\Delta t}$. In this particular reaction, the three rates are not equal. According to the stoichiometry of the reaction, A is used up twice as fast as B ($\frac{-\Delta[A]}{\Delta t} = -2 \times \frac{-\Delta[B]}{\Delta t}$), and A is consumed twice as fast as C is produced ($\frac{-\Delta[A]}{\Delta t} = 2 \times \frac{-\Delta[C]}{\Delta t}$). To show a standard rate of reaction in which the rates with respect to all substances are equal, the rate for each substance should be divided by its stoichiometric coefficient. In this particular case, then:

$$rate\ of\ reaction = -\frac{1}{2}\frac{\Delta[A]}{\Delta t} = -\frac{\Delta[B]}{\Delta t} = \frac{\Delta[C]}{\Delta t}$$

In general, for the reaction

$$a\,A + b\,B \rightarrow c\,C + d\,D,$$

$$rate\ of\ reaction = -\frac{1}{a}\frac{\Delta[A]}{\Delta t} = -\frac{1}{b}\frac{\Delta[B]}{\Delta t} = \frac{1}{c}\frac{\Delta[C]}{\Delta t} = \frac{1}{d}\frac{\Delta[D]}{\Delta t}$$

Rate is expressed in units of concentration per unit time, most often moles per liter per second ($mol/L \times s$), which is the same as molarity per second ($molarity/s$).

Rate Law

For nearly all forward, irreversible reactions, the rate is proportional to the product of the concentrations of the reactants, each raised to some power. For the general reaction:

Don't Mix These Up on Test Day

Keep in mind the difference between the rate law and the equilibrium expression. The exponents in the rate law, unlike those in the equilibrium expression, do not necessarily have to do with the balanced chemical equation!

$$a A + b B \rightarrow c C + d D$$

the rate is proportional to $[A]^x [B]^y$, that is:

$$\text{rate} = k [A]^x [B]^y.$$

This expression is the rate law for the general reaction above, where k is known as the rate constant, and is different for different reactions and may also change depending on the reaction conditions (more on this below). Multiplying the units of k by the concentration factors raised to the appropriate powers gives the rate in units of concentration/time. (The unit of k, therefore, depends on the values of x and y.) The exponents x and y are called the orders of reaction; x is the order of the reaction with respect to A and y is the order with respect to B. These exponents may be integers, fractions, or zero, and must be determined experimentally. It is most important to realize that the exponents of the rate law are not necessarily equal to the stoichiometric coefficients in the overall reaction equation. It is generally *not* the case that $x = a$ and $y = b$, for example, unless if the reaction is a one-step process in which the stoichiometric equation is actually a microscopic description of how the molecules collide.

The overall order of a reaction (or the reaction order) is defined as the sum of the exponents, here equal to $x + y$.

Example: Given the data below, find the rate law for the following reaction at 300K.

$$A + B \rightarrow C + D$$

Trial	[A]initial(M)	[B]initial(M)	rinitial(M/sec)
1	1.00	1.00	2.0
2	1.00	2.00	8.1
3	2.00	2.00	15.9

Solution: a) In trials 1 and 2, the concentration of A is kept constant while the concentration of B is doubled. The rate increases by a factor of approximately 4. Write down the rate expression of the two trials.

Trial 1: $r_1 = k [A]^x [B]^y = k(1.00)^x (1.00)^y$

Trial 2: $r_2 = k [A]^x [B]^y = k(1.00)^x (2.00)^y$

Dividing the second equation by the first,

$$\frac{r_2}{r_1} = 4 = \frac{k(1.00)^x(2.00)^y}{k(1.00)^x(1.00)^y} = (2.00)^y$$

$$4 = (2.00)^y$$

$$y = 2$$

b)   In trials 2 and 3, the concentration of B is kept constant while the concentration of A is doubled; the rate is increased by a factor of 15.9/8.1, approximately 2. The rate expression of the two trials are:

Trial 2: $r_2 = k(1.00)^x (2.00)^y$

Trial 3: $r_3 = k(2.00)^x (2.00)^y$

Dividing the second equation by the first,

$$\frac{r_3}{r_2} = 2 = \frac{k(2.00)^x(2.00)^y}{k(1.00)^x(2.00)^y} = (2.00)^x$$

$$2 = (2.00)^x$$

$$x = 1$$

So $r = k[A] [B]^2$, i.e., the order of the reaction with respect to A is 1 and with respect to B is 2; the overall reaction order is $1 + 2 = 3$.

To calculate k, substitute the values from any one of the above trials into the rate law, e.g.:

$$2.0 \text{ M/sec} = k \times 1.00 \text{ M} \times (1.00 \text{ M})^2$$

$$k = 2.0 \text{ M}^{-2} \text{ sec}^{-1}$$

Therefore, the rate law is $r = 2.0 \text{ M}^{-2} \text{ sec}^{-1} [\dot{A}][B]^2$.

Note, however, that this result could have been obtained more easily if we had been able to see without cumbersome substitution that, for example, from trial 1 to trial 2 the rate has quadrupled when the concentration of B is doubled. Therefore, the rate has to be dependent on the square of [B].

Changes in Concentration Over Time

As a reaction proceeds, the concentrations of the species involved (reactants and products) will change. The value of the concentration of each reactant and product at different points in time can in general be calculated if the rate law is known, although the mathematics needed to perform the manipulations may be quite complicated, depending on the order of the reaction. The relationships derived are often also not general enough to be of interest. The simplest cases, however, are worth examining in more detail.

Zero-Order Reactions

A zero-order reaction has a constant rate, which is independent of the reactants' concentrations. Thus, the rate law is: rate = k, where k has units of M sec^{-1}. The concentration of the reactants decreases linearly over time, i.e., it decreases by the same amount in each period of time, until it is completely used up. A plot of reactant concentration ([A]) versus time for a zero-order reaction is a straight line with slope equal to negative k.

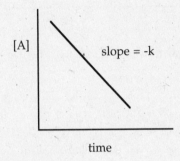

First-Order Reactions

A first-order reaction (order = 1) has a rate proportional to the concentration of one reactant:

$$\text{rate} = -\frac{\Delta[A]}{\Delta t} = k[A]$$

First-order rate constants have units of sec^{-1}.

A Closer Look

The most common example of a first order reaction is the process of radioactive decay.

The most common example of a first-order reaction is the process of radioactive decay. Each atom of a radioactive species has a certain probability of undergoing decay within a window of time. The more of these "reactants" there are, therefore, the more of these decay events we will see, which means that the process occurs faster the more radioactive atoms are around. From the rate law above, one can, with the use of calculus, derive the following relationship between the concentration of radioactive substance A and the time t:

$$[A]_t = [A]_0\, e^{-kt}$$

where

$[A]_0$ = initial concentration of A

$[A]_t$ = concentration of A at time t

k = rate constant

t = elapsed time

A graph of [A] versus time is an exponential function:

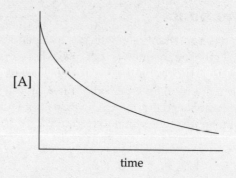

Plotting the natural logarithm of [A] versus time, however, would yield a straight line:

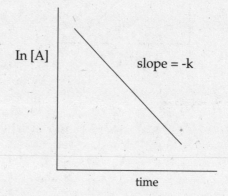

The half-life ($\tau_{1/2}$) of a reaction is the time needed for the concentration of the radioactive substance to decrease to one-half of its original value. In other words, the half-life is the value of t such that the following equality holds:

$$[A]_t \ (\text{at } t = \tau_{1/2}) = \frac{1}{2} \ [A]_0$$

We can substitute in the expression for $[A]_t$:

$$[A]_0 \ e^{-kt} = \frac{1}{2}[A]_0 \text{ when } t = \tau_{1/2}$$

$$e^{-kt} = \frac{1}{2} \text{ when } t = \tau_{1/2}$$

Manipulation of the above equation leads to the following formula for half-life:

$$\tau_{1/2} = \frac{\ln 2}{k} = \frac{0.693}{k}$$

where k is the first order rate constant.

After one half-life, half of the reactants will be left. After another half-life has elapsed, half of that half will remain; in other words, the concentration of the reactants will be one quarter the initial value. More generally, the concentration of A is $\frac{1}{2^n} \ [A]_0$ after the passage of n number of half-lives.

Collision Theory, Transition States, and Energy Profiles

In order for a reaction to occur, molecules must collide with each other. The collision theory of chemical kinetics states that the rate of a reaction is proportional to the number of collisions per second between the reacting molecules.

Not all collisions, however, result in a chemical reaction. An effective collision (one that leads to the formation of products) occurs only if the molecules collide with correct orientation and sufficient force to break the existing bonds and form new ones. The minimum energy of collision necessary for a reaction to take place is called the activation energy, or the energy barrier, designated E_a. Only a fraction of colliding particles have enough kinetic energy to exceed the activation energy. This means that only a fraction of all collisions are effective.

When molecules collide with sufficient energy, they go through what is known as a transition state (also called the activated complex), in which the old bonds are weakened and the new bonds are beginning to form. The transition state then dissociates into products and the new bonds are fully formed. For example, in a reaction $A_2 + B_2 \rightarrow 2AB$, the transition state or activated complex may look like the middle species in the following diagram, where the dashed lines represent partial bonds (bonds that are not quite as strong as a single bond):

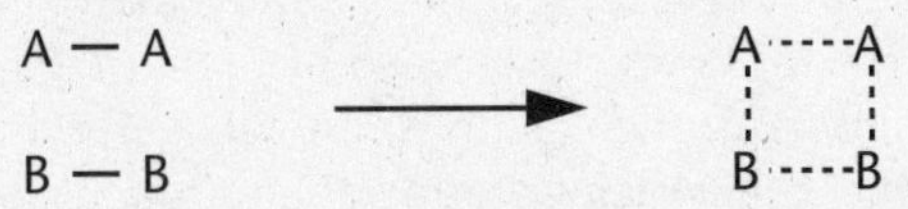

```
A — A                        A ----A
                             |      |
                             |      |
B — B                        B ----B
```

The activated complex exists only in one fleeting instant in time and is thus only a snapshot of how the molecules are arranged somewhere along the reaction; it is not a stable isolatable species by itself.

A potential energy diagram is very helpful in visualizing the progress of a reaction. The x-axis in these diagrams corresponds to the "reaction coordinate," which essentially measures how far along one is in a reaction by charting the progress from reactants to products: As one moves from left to right, the configuration of all the species participating in the reaction goes from being in the reactant stage to resembling more and more that of the products, until the reaction has gone to completion and the products are formed. The y-axis plots the potential energy of the varying configuration of the atoms. The most important thermodynamic factor in these diagrams is the relative energy of the products and reactants. The overall energy change of the reaction is the difference between the potential energy of the products and the potential energy of the reactants. If the energy level of the products is lower than that of the reactants, the products are more stable thermodynamically and there is a net release of energy for the reaction. If the energy of the reactants is lower than that of the products, the reverse is true: The reactants are more stable thermodynamically and energy is absorbed in the process.

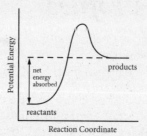 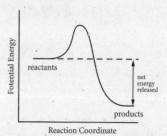

The activated complex exists at the top of the energy barrier. It has greater energy than either the reactants or the products, and is denoted by the symbol ‡. The energy required to bring the reactants up to this level is the activation energy for the reaction, E_a. Once an activated complex is formed, it can either go on to form the products or revert to reactants without any additional energy input. (The difference in energy between the activated complex and the *products* is the activation energy for the *reverse* reaction: that of products going to reactants.)

It is important to note that even if the reaction results in a net release of energy, i.e., even if the products are more stable thermodynamically than the reactants, the reactants need to have sufficient energy initially to overcome this activation barrier, since the transition state is always at a higher energy than either the reactants or the products. This is the reason why we do not observe diamond turning into graphite: Under standard conditions, the activation energy for these processes is so high that regardless of the thermodynamic favorability of the reactions, the reactants simply cannot overcome the energy barrier. In plainer terms, the ultimate payoff may be great for these reactions, but the effort called for is too much.

In terms of the rate law, the height of this barrier (the activation energy) is what determines the value of the rate constant k: The higher the barrier, the slower the rate, i.e., the smaller the value of k.

Reaction Mechanisms

The mechanism of a reaction is the actual series of steps through which a chemical reaction occurs. Consider the reaction below:

$$\text{Overall Reaction: } A_2 + 2\,B \rightarrow 2\,AB$$

This equation seems to imply some sort of encounter between two molecules or atoms of B and one molecule of A_2 to form two molecules of AB. But suppose instead that the reaction actually takes place in two steps.

Step 1: $A_2 + B \rightarrow A_2B$ (Slow)

Step 2: $A_2B + B \rightarrow 2\,AB$ (Fast)

Note that these two steps add up to the overall (net) reaction. A_2B, which does not appear in the overall reaction because it is neither a reactant nor a product, is called an intermediate. Reaction intermediates, unlike activated complexes, are real molecules that exist at least for a while. Nonetheless, they may still be difficult to detect.

The slowest step in a proposed mechanism is called the rate-determining step (or the rate-limiting step), so called because as the bottleneck in the progression of the reaction, it determines the rate by imposing an upper limit on how fast it goes. In the reaction mechanism given above, for example, the first step, which has been described as slow, is the rate-determining step; the overall reaction simply cannot occur any faster than this step, in the same way that in a family outing, no matter how fast the other members are, everyone will still have to wait for the slowest one to be ready before the family can set off.

In the discussion on the potential energy diagram above, we have limited ourselves to considering reactions occurring in a single step. For one that involves several steps, the graph will go through a series of "hilltops and valleys": Each step will involve a transition state. The low points between these maxima are intermediates. The activation energy of each step is the difference in energy between the transition state and the "valley" immediately before it (corresponding to either reactants or intermediates). The step with the highest activation energy is the rate-determining or rate-limiting step. The two plots below show the potential energy diagrams for a two-step reaction. In one, the first step is the rate-determining step while in the other, the second step is rate-determining.

A Closer Look

The factors discussed here change the rate of reaction mathematically by affecting the value of k, the rate constant, in the rate law.

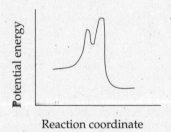

Reaction coordinate

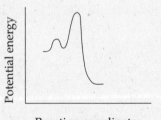

Reaction coordinate

Factors Affecting Reaction Rate

The rate of a chemical reaction as expressed in the applicable rate law, as we have seen, involves both a rate constant and, except for zero order reactions, the concentration of reactants. The rate of a reaction, then, could be increased by either increasing the concentration of the reactants (which increases the number of effective collisions between the reactant molecules), or by altering the value of the rate constant. We have pointed out above that the rate constant ultimately depends upon the energy difference between the reactants and the transition state. The smaller this activation energy is (the smaller the gap between the two energy levels), the larger the rate constant, and the faster the reaction will proceed. Two factors that most commonly affect the rate of a reaction are temperature and the presence of a catalyst.

For nearly all reactions, the reaction rate will increase as the temperature of the system increases. Since the temperature of a substance is generally a measure of the particles' average kinetic energy, increasing the temperature increases the average kinetic energy of the molecules. Consequently, the proportion of molecules having energies greater than E_a (thus capable of undergoing reaction) increases with higher temperature. Again, this is valid even for reactions that are exothermic because the activated complex is at a higher potential energy than the reactants. Raising the temperature of a system in which an exothermic reaction is occurring would shift the equilibrium in favor of the reactants (from Le Châtelier's principle), but the system would attain this equilibrium faster.

Catalysts are substances that increase the reaction rate without themselves being consumed; they do this by lowering the activation energy. Catalysts are important in biological systems and in industrial chemistry; enzymes are biological catalysts. Catalysts accomplish this lowering of activation energy by a variety of ways: They may, for example, increase the frequency of collision between the reactants, or change the relative orientation of the reactants making a higher percentage of collisions effective. The following figure compares the energy profiles of catalyzed and uncatalyzed reactions.

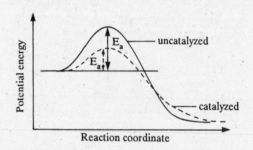

The energy barrier for the catalyzed reaction is much lower than the energy barrier for the uncatalyzed reaction. Note that the rates of both the forward and the reverse reactions are increased by catalysis, since E_a of the forward and reverse reactions are lowered by the same amount. Therefore, the presence of a catalyst causes the reaction to proceed more quickly toward equilibrium, without changing the position of the equilibrium, i.e., without changing the value of the equilibrium constant.

A Closer Look

The rate for most reactions approximately doubles for every 10°C increase in the temperature.

Dynamic Equilibrium

Although we have learned that the kinetics of a reaction may override predictions based purely on thermodynamics, the two aspects are not totally separate. This is because, as mentioned in the chapter on chemical equilibrium, one of the things that characterizes equilibrium is that the forward and reverse reactions are occurring at the same rate: The relative rates of reactions set up a thermodynamic equilibrium in the system.

The ratio of the forward rate constant to the reverse rate constant is the equilibrium constant for that one-step process, i.e.:

$$A + B \underset{k_r}{\overset{k_f}{\rightleftharpoons}} C$$

$$K_{eq} = \frac{k_f}{k_r}$$

where k_f is the rate constant of the forward reaction and k_r is the rate constant for the reverse reaction. The larger the value of k_f, the faster the forward reaction occurs (relative to the reverse reaction); that is, products are formed more rapidly than they revert back to reactants, and so there will tend to be more products around at equilibrium, which is reflected in a large equilibrium constant.

It may now be more obvious why a catalyst does not affect the position of equilibrium: Both k_f and k_r are increased by the same proportion, such that their ratio remains unchanged.

Chemical Kinetics Review Problems

1. All of the following are true statements concerning reaction orders EXCEPT:

   A. The rate of a zero-order reaction is constant.

   B. After three half-lives, a radioactive sample will have one-ninth of its original concentration.

   C. The units for the rate constant for first order reactions are $\sec^{-1}$.

   D. If doubling the concentration of a reactant doubles the rate of the reaction, then the reaction is first order in that reactant.

2. The half-life of radioactive sodium is 15.0 hours. How many hours would it take for a 64 g sample to decay to one-eighth of its original concentration?

   A. 3

   B. 15

   C. 30

   D. 45

3. Consider the following hypothetical reaction and experimental data:

   $$A + B \rightarrow C + D$$

   $$T = 273K$$

	$[A]_0$ (mol/L)	$[B]_0$ (mol/L)	rate
Exp 1	0.10	1	0.035
Exp 2	0.10	4	0.070
Exp 3	0.20	1	0.140
Exp 4	0.10	16	0.140

   a. What is the order with respect to A?

   b. What is the order with respect to B?

   c. What is the rate equation?

   d. What is the overall order of the reaction?

   e. Calculate the rate constant.

4. Consider the following chemical reaction and experimental data:

$$A\ (aq) \rightarrow B\ (aq) + C(g)$$

Trial 1

[A] (mol/L)	rate
0.10	0.6
0.20	0.6
0.30	0.6
0.40	0.6

Trial 2

[A] (mol/L)	rate
0.10	0.9
0.20	0.9
0.30	0.9
0.40	0.9

a. What is the rate expression for trial 1?

b. What is the rate constant for trial 1?

c. What is the most likely reason for the increased rate in trial 2?

5. Consider the following reaction and experimental data:

$$SO_3 + H_2O \rightarrow H_2SO_4$$

	[SO₃] (mol/L)	[H₂O] (mol/L)	Rate
Trial 1	0.1	0.01	0.013
Trial 2	0.2	0.01	0.052
Trial 3	X	0.02	0.234
Trial 4	0.1	0.03	0.039

a. What is the value of X?

b. What is the order of the reaction?

c. What is the rate constant?

d. What would be the rate if [SO₃] in trial 4 were raised to 0.2?

6. In the following diagram, which labeled arrow represents the activation energy for the reverse reaction?

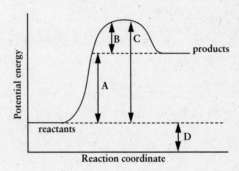

A. A

B. B

C. C

D. D

7. The activation energy for a reaction in the forward direction is 78 kJ. The activation energy for the same reaction in reverse is 300 kJ. If the energy of the products is 25 kJ, then:

a. What is the energy of the reactants?

b. Is the forward reaction endothermic or exothermic?

c. Is the reverse reaction endothermic or exothermic?

d. What is the enthalpy change for the forward reaction?

8. According to chemical kinetic theory, a reaction can occur

   A. if the reactants collide with the proper orientation.

   B. if the reactants possess sufficient energy of collision.

   C. if the reactants are able to form a correct transition state.

   D. All of the above.

9. The number of undecayed nuclei in a sample of bromine-87 decreased by a factor of 4 over a period of 112 sec. What is the rate constant for the decay of bromine-87?

   A. 56 sec

   B. 6.93×10^{-1} sec^{-1}

   C. 1.24×10^{-2} sec-1

   D. 6.19×10^{-3} sec^{-1}

10. Which of the following is most likely to increase the rate of a reaction?

    A. Decreasing the temperature

    B. Increasing the volume of the reaction vessel

    C. Reducing the activation energy

    D. Decreasing the concentration of the reactant in the reaction vessel

11. All of the following are true statements concerning catalysts EXCEPT

    A. A catalyst will speed the rate-determining step.

    B. A catalyst will be used up in a reaction.

    C. A catalyst may induce steric strain in a molecule to make it react more readily.

    D. A catalyst will lower the activation energy of a reaction.

12. The equilibrium constant, K_{eq}, of a certain single-reactant reaction is 0.16. Suppose an appropriate catalyst is added.

    a. What will be the equilibrium constant?

    b. Will the activation energy increase or decrease?

13. At equilibrium

    A. the forward reaction will continue.

    B. a change in reaction conditions may shift the equilibrium.

    C. the reverse reaction will not continue.

    D. Both A and B.

Solutions to Review Problems

1. B

Radioactivity is a first order process. After three half-lives, the concentration of the sample will be $\left(\frac{1}{2}\right)^3$, or $\frac{1}{8}$, of the original.

2. D

In order for a 64 g sample to decay to one-eighth of its original activity, or 8 g, the sample would have to go through three half-lives. Therefore, the amount of time needed for the decay is $3 \times 15 = 45$ hours.

3.

First, the general rate equation must be written out. It is:

rate = k[A]^x[B]^y

where:

k = the rate constant

x = the order with respect to reactant A

y = the order with respect to reactant B

a. x = 2

To solve for x, it is necessary to find two trials in which B is held constant; here, experiments 1 and 3. The data shows that if the concentration of A is doubled, the rate increases by a factor of 4. Thus, the rate varies as the square of the concentration of A. The order is therefore equal to 2.

b. y = 0.5

To solve for y, follow the steps as in a. In experiments 1 and 2 (and 2 and 4) [A] is held constant, the concentration of B quadruples, and the rate doubles. The rate, therefore, varies as the square root of the concentration of B. The order, y, is therefore equal to 0.5.

c. rate = $k[A]^2 [B]^{\frac{1}{2}}$

d. 2.5

e. 3.5

Given the rate expression, the rate constant can easily be calculated by substituting the rate and concentrations for any of the four trials into the rate expression; the rate constant will work out to 3.5 in each case.

Trial 1: $0.035 = k[0.10]^2 [1]^{0.5} : k = 3.5$

Trial 2: $0.070 = k[0.10]^2 [4]^{0.5} : k = 3.5$

Trial 3: $0.140 = k[0.20]^2 [1]^{0.5} : k = 3.5$

Trial 4: $0.140 = k[0.10]^2 [16]^{0.5} : k = 3.5$

4.

a. rate $= k[A]^0 = k$

This reaction has only one reactant. It is evident from the data that the rate of the reaction is not affected by reactant concentration. This is a zero-order reaction, and the rate is equal to its rate constant, k.

b. $k = 0.6$

c.

The most likely reason for the increased rate in trial 2 is a change in temperature or the addition of a catalyst. The rate is still independent of the reactant concentration, but is faster overall. This increase can only come from an increase in the value of the rate constant, which is affected by temperature and the activation energy, which in turn is lowered in the presence of a catalyst.

5.

a. $X = 0.3$

To calculate X, first write the rate expression for this reaction. From the data, the rate expression is calculated as:

$$rate = k[SO_3]^2 [H_2O]$$

The order with respect to SO_3 is 2, since the rate quadruples while the concentration of SO_3 doubles (with the concentration of H_2O remaining constant) between trials 1 and 2. The order with respect to H_2O is 1, as the rate triples as the concentration of H_2O triples (with the concentration of SO_3 remaining constant) between trials 1 and 4.

X can be calculated by plugging in the values from trial 3 into the rate expression. First, however, calculate the rate constant, k, by plugging in the known values from trial 1, 2, or 4. For instance:

Trial 4: $0.039 = k[0.1]^2 [0.03]$: $k = 130$

To calculate X, plug in the values of rate and $[H_2O]$ for trial 3, using k = 130:

$$0.234 = 130[X]^2 [0.02]$$

$$X = 0.3$$

One can also arrive at this answer without first calculating the rate constant by noting that the concentration of H_2O is doubled on going from trial 1 to trial 3. Since the reaction is first order in H_2O, we would expect the rate to have doubled from the change in $[H_2O]$ alone. The fact is, however, that the rate has been increased 18 times ($0.013 \times 18 = 0.234$), and so the remaining factor of 9 ($18 = 9 \times 2$) in the increase has to come from the change in $[SO_3]$. We know that the reaction is second order in SO_3, and so a threefold increase in $[SO_3]$ would give us the overall increase we are looking for. Therefore the concentration of SO_3 is $3 \times 0.1 = 0.3$.

b. 3

The order of the reaction is the sum of the exponents in the rate expression: in this case, $2 + 1 = 3$.

c. 130

For calculations see solution to part a.

d. 0.156 units

Substitute 0.2 instead of 0.1:

$$rate = (130)(0.2)^2(0.03) = 0.156$$

Again, even if we had not known what the rate constant is, we would not need to calculate it explicitly for this problem. If $[SO_3]$ were raised to 0.2 in trial 4, it would have been doubled. Therefore the rate would be quadrupled, and would equal $4 \times 0.039 = 0.156$.

6.   B

The activation energy is the minimum amount of energy needed for a reaction to proceed. The activation energy for the reverse reaction is the change in potential energy between the products and the transition state indicated by arrow B. A is the overall (thermodynamic) energy change of the reaction. C is the activation energy for the forward, not the reverse, reaction.

7.

a.

The best way to visualize the solution to this set of problems is to draw a diagram.

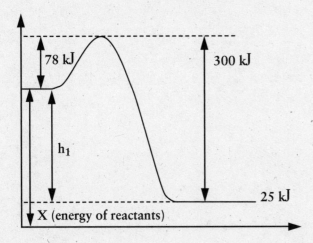

ΔE = Activation Energy (forward) – Activation Energy (reverse)

$$= 78 \ kJ - 300 \ kJ = -222 \ kJ.$$

And since

ΔE = Energy (products) – Energy (reactants)

$-222 \ kJ = 25 \ kJ - X \ kJ,$

$X = 247 \ kJ$ = energy of reactants.

b.

The forward reaction is exothermic, since ΔE is negative.

c.

The reverse reaction is endothermic.

d.

The enthalpy change, ΔE, of the reaction is $-222 \ kJ$.

8.   D

In order for products to form, the reactant atoms or molecules need to collide at an orientation that allows them to react, and with sufficient energy to surmount the activation barrier as it goes through the transition state.

9.   C

If the number of nuclei decaying in a sample has decreased by a factor of 4, the sample has been through 2 half-lives, and the half-life will be

$$\frac{112 \ sec}{2} = 56 \ sec = \tau_{1/2}$$

The equation to determine the decay constant for the first-order reaction is

$$\tau_{1/2} = \frac{\ln 2}{k} = \frac{0.693}{k}$$

Thus, given that the half life is 56 sec, the decay constant will be

$$56 \ sec = \frac{0.693}{k}$$

$$k = 0.0124 \ sec^{-1}$$

10. C

Reducing the activation energy makes it easier for molecules to overcome the barrier to reaction: At any temperature, there will now be more molecules that have sufficient energy to react. The rate of reaction will therefore be faster. All the other choices tend to decrease the reaction rate: Lowering the temperature would mean decreasing the energy of the molecules; it will be more difficult for them to overcome the energy barrier. Increasing the volume of the reaction vessel, with all else remaining the same, would decrease the density (and hence concentration) of the reactant molecules, causing collisions among them to be less frequent.

11. B

The definition and properties of a catalyst are discussed in the section on factors affecting the reaction rate.

12.

a.

K_{eq} remains constant at 0.16. Catalysts do not affect equilibrium position.

b.

Addition of a catalyst decreases the activation energy.

13. D

At equilibrium, both the forward and reverse reactions are proceeding. Any change in the equilibrium conditions will shift the equilibrium in order to alleviate the stress on the reaction.

CHAPTER 13

ACIDS AND BASES

Many important reactions in chemical and biological systems involve two classes of compounds called acids and bases. The presence of acids and bases can often be easily detected because they lead to color changes in certain compounds called indicators, which may be in solution or on paper. A particular common indicator is litmus paper, which turns red in acidic solution and blue in basic solution. A more extensive discussion of the chemical properties of acids and bases is outlined below.

Definitions

Three different definitions of acids and bases exist. The Brønsted-Lowry definition is the most common, although you should be aware of the other two as well.

Arrhenius Definition

The first definitions of acids and bases were formulated by Svante Arrhenius toward the end of the 19th century. Arrhenius defined an acid as a species that produces H^+ (protons) in an aqueous solution, and a base as a species that produces OH^- (hydroxide ions) in an aqueous solution. These definitions, though useful, fail to describe acidic and basic behavior in non-aqueous media.

Brønsted-Lowry Definition

A more general definition of acids and bases was proposed independently by Johannes Brønsted and Thomas Lowry in 1923. A Brønsted-Lowry acid is a species that donates protons, while a Brønsted-Lowry base is a species that accepts protons. For example, NH_3 and Cl^- are both Brønsted-Lowry bases because they accept protons. However, they cannot be called Arrhenius bases since in aqueous solution they do not dissociate to form OH^-. Another advantage of the Brønsted-Lowry concept of acids and bases is that it is not limited to aqueous solutions.

Basic Concept

Arrhenius defined an acid as a species that produces H^+ in an aqueous solution, and a base as a species that produces OH^- in aqueous solution.

Basic Concept

A Brønsted-Lawry acid is a species that donates protons, while a Brønsted-Lawry base is a species that accepts protons.

Brønsted-Lowry acids and bases always occur in pairs, called conjugate acid-base pairs. The two members of a conjugate pair are related by the transfer of a proton. For example, H_3O^+ is the conjugate acid of H_2O, and NO_2^- is the conjugate base of HNO_2:

$$H_3O^+ \ (aq) \Leftrightarrow H_2O \ (aq) + H^+ \ (aq)$$

$$HNO_2 \ (aq) \Leftrightarrow NO_2^- \ (aq) + H^+ \ (aq)$$

Conversely, one can also say that H_2O is the conjugate base of H_3O^+, and that HNO_2 is the conjugate acid of NO_2^-.

Lewis Definition

Basic Concept

A Lewis acid is an electron-pair acceptor, and a Lewis base is an electron-pair donor.

At approximately the same time as Brønsted and Lowry, Gilbert Lewis also proposed definitions for acids and bases. Lewis defined an acid as an electron-pair acceptor, and a base as an electron-pair donor. Lewis's are the most inclusive definitions. Just as every Arrhenius acid is a Brønsted-Lowry acid, every Brønsted-Lowry acid is also a Lewis acid (and likewise for bases). However, the Lewis definition encompasses some species not included within the Brønsted-Lowry definition. For example, BCl_3 and $AlCl_3$ can each accept an electron pair and are therefore Lewis acids, despite their inability to donate protons. We shall, however, focus our attention on Brønsted-Lowry acids and bases.

Nomenclature of Acids

The name of an acid is related to the name of the parent anion (the anion that combines with H^+ to form the acid). Acids formed from anions whose names end in *-ide* have the prefix *hydro-* and the ending *-ic*.

F^-	Fluoride	HF	Hydrofluoric acid
Br^-	Bromide	HBr	Hydrobromic acid

Acids formed from oxyanions are called oxyacids. If the anion ends in *-ite* (less oxygen), then the acid will end with *-ous* acid. If the anion ends in *-ate* (more oxygen), then the acid will end with *-ic* acid. Prefixes in the names of the anions are retained. Some examples:

ClO^-	Hypochlorite	HClO	Hypochlorous acid
ClO_2^-	Chlorite	$HClO_2$	Chlorous acid
ClO_3^-	Chlorate	$HClO_3$	Chloric acid
ClO_4^-	Perchlorate	$HClO_4$	Perchloric acid

NO_2^-	Nitrite	HNO_2	Nitrous acid
NO_3^-	Nitrate	HNO_3	Nitric acid

Properties of Acids and Bases

The behavior of acids and bases in solution is governed by equilibrium considerations. Some concepts you already may be familiar with will appear with new names in this context.

Hydrogen Ion Equilibria (pH and pOH)

Hydrogen ion or proton concentration, $[H^+]$, like concentrations of other particles, can of course be measured in the familiar units like molarity. However, it is more generally measured as pH, where:

$$pH = -\log [H^+] = \log \left(\frac{1}{[H^+]}\right)$$

where $[H^+]$ is its molarity and the logarithm is of base 10. (Log x) is the power to which 10 would be raised to obtain the number x, i.e.:

$$\log x = p \Leftrightarrow 10^p = x$$

Likewise, hydroxide ion concentration, $[OH^-]$, can be measured as pOH where:

$$pOH = -\log [OH^-] = \log \left(\frac{1}{[OH^-]}\right)$$

It turns out, however, that pH and pOH are not totally independent of each other: Knowing one would allow us to calculate the other. This is because in any aqueous solution, the H_2O solvent dissociates slightly:

$$H_2O(l) \Leftrightarrow H^+ (aq) + OH^- (aq)$$

This dissociation is an equilibrium reaction and is therefore described by an equilibrium constant, K_w, known as the water dissociation constant:

$$K_w = [H^+][OH^-] = 10^{-14} \text{ (at 25 °C)}$$

One can take the logarithm of both sides and manipulate the equation by using the properties of logarithms:

$\log ([H^+][OH^-]) = \log (10^{-14})$

$\log [H^+] + \log [OH^-] \qquad = \log (10^{-14}), \quad$ since $\log (xy) = \log x + \log y$

$\qquad\qquad\qquad\qquad\quad = -14, \qquad\quad$ since $\log (10p) = p$

$$-\log [H^+] - \log [OH^-] = 14,$$ where we have taken the negative of both sides

$$\therefore pH + pOH = 14$$

In pure H_2O, $[H^+]$ is equal to $[OH^-]$, since equimolar amounts of H^+ and of OH^- are formed from the dissociation process. The pH and pOH would therefore also be equal, both having a value of 7. A solution with equal concentrations of H^+ and OH^- is neutral. A pH below 7 indicates a relative excess of H^+ ions, and therefore an acidic solution; a pH above 7 indicates a relative excess of OH^- ions, and therefore a basic solution.

It is important to realize that even when the pH deviates from the value of 7, the water dissociation equilibrium still holds. If an acid, for example HCl, dissociates in water at 25 °C and causes an increase in proton concentration such that the pH falls below 7, the hydroxide ion concentration will have to decrease so as to maintain the relation $K_w = [H^+][OH^-] = 10^{-14}$. Despite the higher concentration of H^+ relative to OH^-, the solution does not acquire a net positive charge because the conjugate base of the dissociated acid will be negatively charged (for example, Cl^-) and thus will maintain charge neutrality.

It should also be pointed out that even though we have written acid-base reactions so far as involving protons, in aqueous solution the protons will interact with other water molecules, forming H_3O^+, known as the hydronium ion. The water dissociation reaction can therefore be written as

$$H_2O(l) + H_2O(l) \rightleftharpoons H_3O^+ (aq) + OH^- (aq)$$

We shall be using H^+ and H_3O^+ interchangeably, unless otherwise stated.

Strong Acids and Bases

Basic Concept

Strong acids and bases completely dissociate into their component ions in aqueous solution.

Strong acids and bases are those that completely dissociate into their component ions in aqueous solution. For example, when NaOH is added to water, it dissociates completely:

$$NaOH(s) + (\text{excess}) \ H_2O(l) \rightarrow Na^+ (aq) + OH^- (aq)$$

Hence, in a 1 M solution of NaOH, complete dissociation gives 1 mole of OH^- ions per liter of solution.

$$pH = 14 - pOH = 14 - (-\log [OH^-]) = 14 + \log [1] = 14$$

Virtually no undissociated NaOH remains. Note that the $[OH^-]$ contributed by the dissociation of H_2O is considered to be negligible in this case. The contribution of OH^- and H^+ ions from the dissociation of H_2O can be neglected only if the concentration of the acid or base is much greater than 10^{-7} M.

Strong acids commonly encountered in the laboratory include $HClO_4$ (perchloric acid), HNO_3 (nitric acid), H_2SO_4 (sulfuric acid), and HCl (hydrochloric acid). Commonly encountered strong bases include NaOH (sodium hydroxide), KOH (potassium hydroxide), and other soluble hydroxides of Group IA and IIA metals. Calculation of the pH and pOH of strong acids and bases assumes complete dissociation of the acid or base in solution: $[H^+]$ = normality of strong acid and $[OH^-]$ = normality of strong base.

Weak Acids and Bases

Weak acids and bases are those that only partially dissociate in aqueous solution. A weak monoprotic acid, HA, in aqueous solution will achieve the following equilibrium after dissociation:

$$HA \ (aq) + H_2O \ (l) \rightleftharpoons H_3O^+ \ (aq) + A^- \ (aq)$$

The acid dissociation constant, K_a, is a measure of the degree to which an acid dissociates.

$$K_a = \frac{[H_3O^+][A^-]}{[HA]}$$

The weaker the acid, the smaller the K_a. Weak acids have values of K_a that are much smaller than 1. Note that K_a, like other equilibrium constants, does not contain an expression for the pure liquid, water. Many weak acids are organic compounds.

A weak base, BOH, undergoes dissociation to give B^+ and OH^-. The base dissociation constant, K_b, is a measure of the degree to which a base dissociates. The weaker the base, the smaller its K_b. For a monovalent base, K_b is defined as follows:

$$K_b = \frac{[B^+][OH^-]}{[BOH]}$$

Conjugate Acids and Bases Revisited

As mentioned above, a conjugate acid is defined as the acid formed when a base gains a proton. Similarly, a conjugate base is formed when an acid loses a proton. For example, in the acetic acid/acetate conjugate acid/base pair CH_3COOH/CH_3COO^- (also written as $HC_2H_3O_2/C_2H_3O_2^-$ although the former gives more structural information about the molecule), CH_3COO^- is the conjugate base and CH_3COOH is the conjugate acid:

$$CH_3COOH \ (aq) \rightleftharpoons H^+ \ (aq) + CH_3COO^-(aq)$$

or:

$$CH_3COOH \ (aq) + H_2O \ (l) \rightleftharpoons H_3O^+ \ (aq) + CH_3COO^-(aq)$$

Basic Concept

Weak acids and gases only partially dissociate in aqueous solution.

Likewise, for the K_b of CH_3COO^-:

$$CH_3COO^- (aq) + H_2O (l) \rightleftharpoons CH_3COOH (aq) + OH^- (aq)$$

The equilibrium constants for these reactions are as follows.

$$K_a = \frac{[H_3O^+][CH_3COO^-]}{[CH_3COOH]}$$

and

$$K_b = \frac{[CH_3COOH][OH^-]}{[CH_3COO^-]}$$

Adding the two reactions shows that the net reaction is simply the dissociation of water:

$$H_2O (l) \rightleftharpoons H^+ (aq) + OH^- (aq)$$

The equilibrium constant for this net reaction is $K_w = [H^+][OH^-]$, which is the product of K_a and K_b. Thus, if the dissociation constant either for an acid or for its conjugate base is known, then the dissociation constant for the other can be determined, using the equation:

$$K_a \times K_b = K_w = 1 \times 10^{-14}$$

Thus K_a and K_b are inversely related. In other words, if K_a is large (the acid is strong), then K_b will be small (the conjugate base will be weak), and vice versa. This is a more mathematical way of describing a general rule: The stronger an acid is, the weaker its conjugate base is (as a base); the weaker the acid, the stronger its conjugate base. For example, HCl ($K_a \sim 10^7$) is a much stronger acid than acetic acid ($K_a = 1.8 \times 10^{-5}$); CH_3COO^- is therefore expected to be a much stronger base than Cl^-. The mathematics should not be allowed to obscure how much sense this makes: after all, a weak acid means that it undergoes dissociation reluctantly; the equilibrium lies to the left, favoring undissociated HA. A^- in solution is therefore likely to grab a proton to reconstitute HA; that is, A^- is reactive as a base.

Applications of K_a and K_b

To calculate $[H^+]$ in a 2.0 M aqueous solution of acetic acid, first write the equilibrium reaction:

$$CH_3COOH (aq) \rightleftharpoons H^+ (aq) + CH_3COO^- (aq)$$

Next, write the expression for the acid dissociation constant:

$$K_a = 1.8 \times 10^{-5} = \frac{[H^+][CH_3COO^-]}{[CH_3COOH]}$$

The concentration of CH_3COOH at equilibrium is equal to its initial concentration, 2.0 M, less the amount dissociated, x. Likewise $[H^+]$ = $[CH_3COO^-]$ = x, since each molecule of CH_3COOH dissociates into one H^+ ion and one CH_3COO^- ion. Thus, the equation can be rewritten as follows:

$$K_a = \frac{x \cdot x}{2.0 - x} = 1.8 \times 10^{-5}$$

We can approximate that $2.0 - x \approx 2.0$ since acetic acid is a weak acid, and only slightly dissociates in water. This simplifies the calculation of x:

$$K_a = \frac{x^2}{2.0} = 1.8 \times 10^{-5}$$

$$x = 6.0 \times 10^{-3} \text{ M}$$

The fact that x is so much less than the initial concentration of acetic acid (2.0 M) validates the approximation; otherwise, it would have been necessary to solve for x using the quadratic formula.

Polyvalence and Normality

The relative acidity or basicity of an aqueous solution is determined by the relative concentrations of acid and base equivalents. An acid equivalent is equal to one mole of H^+ (or H_3O^+) ions; a base equivalent is equal to one mole of OH^- ions. Some acids and bases are polyvalent, that is, each mole of the acid or base liberates more than one acid or base equivalent. For example, the diprotic acid H_2SO_4 undergoes the following dissociation in water:

$$H_2SO_4(aq) \rightarrow H^+(aq) + HSO_4^-(aq)$$

$$HSO_4^-(aq) \Leftrightarrow H^+(aq) + SO_4^{2-}(aq)$$

One mole of H_2SO_4 can thus produce 2 acid equivalents (2 moles of H^+) if dissociation is complete. The acidity or basicity of a solution depends upon the concentration of acidic or basic equivalents that can be liberated. The quantity of acidic or basic capacity is directly indicated by the solution's normality. Since each mole of H_3PO_4 can liberate 3 moles (equivalents) of H^+ a 2 M H_3PO_4 solution would be 6 N (6 normal).

Another useful measurement is equivalent weight. For example, the gram molecular weight of H_2SO_4 is 98 g/mol. Since each mole liberates 2 acid equivalents, the gram equivalent weight of H_2SO_4 would be $\frac{98}{2}$ = 49 g; that is, the dissociation of 49 g of H_2SO_4 would release one acid equivalent. Common polyvalent acids include H_2SO_4, H_3PO_4, and H_2CO_3.

Salt Formation

Acids and bases may react with each other, forming a salt and (often, but not always) water, in what is termed a neutralization reaction. For example, a generic acid and a generic base react as follows:

$$HA + BOH \rightarrow BA + H_2O$$

The salt, BA, may precipitate out or remain ionized in solution, depending on its solubility and the amount produced. Neutralization reactions generally go to completion. The reverse reaction, in which the salt ions react with water to give back the acid or base, is known as hydrolysis.

Four combinations of strong and weak acids and bases are possible:

1. strong acid + strong base: e.g., $HCl + NaOH \rightarrow NaCl + H_2O$
2. strong acid + weak base: e.g., $HCl + NH_3 \rightarrow NH_4Cl$
3. weak acid + strong base: e.g., $HClO + NaOH \rightarrow NaClO + H_2O$
4. weak acid + weak base: e.g., $HClO + NH_3 \rightarrow NH_4ClO$

The products of a reaction between equal concentrations of a strong acid and a strong base are a salt and water. The acid and base neutralize each other, so the resulting solution is neutral (pH = 7), and the ions formed in the reaction do not react with water. The product of a reaction between a strong acid and a weak base is also a salt but usually no water is formed since weak bases are usually not hydroxides; however, in this case, the cation of the salt will react with the water solvent, reforming the weak base. This reaction constitutes hydrolysis. For example:

$$HCl \, (aq) + NH_3 \, (aq) \rightarrow NH_4^+ \, (aq) + Cl^- \, (aq) \text{ Reaction I}$$

$$NH_4^+ \, (aq) + H_2O \, (aq) \rightarrow NH_3 \, (aq) + H_3O^+ \, (aq) \text{ Reaction II}$$

NH_4^+ is the conjugate acid of a weak base (NH_3), and is therefore stronger than the conjugate base (Cl^-) of the strong acid HCl. NH_4^+ will thus react with OH^-, reducing the concentration of OH^-. There will thus be an excess of H^+, which will lower the pH of the solution.

On the other hand, when a weak acid reacts with a strong base the solution is basic, due to the hydrolysis of the salt to reform the acid, with the concurrent formation of hydroxide ion from the hydrolyzed water molecules. The pH of a solution containing a weak acid and a weak base depends on the relative strengths of the reactants.

Titration and Buffers

Neutralization is an important concept in the performance of titrations. Titration (or more specifically acid-base titration) is a procedure used to determine the molarity of an acid or base. This is accomplished by reacting a known volume of a solution of unknown concentration with a known volume of a solution of known concentration. When the number of acid equivalents equals the number of base equivalents added, or vice versa, the equivalence point is reached. It is important to emphasize that, while a strong acid/strong base titration will have an equivalence point at pH 7, the equivalence point need not always occur at pH 7.

Strong Acid and Strong Base

Consider the titration of 10 mL of a 0.1 N solution of HCl with a 0.1 N solution of NaOH. Plotting the pH of the reaction solution versus the quantity of NaOH added gives the following curve:

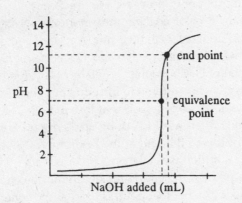

Because HCl is a strong acid and NaOH is a strong base, the equivalence point of the titration will be at pH 7 and the solution will be neutral. Note that the endpoint shown is close to, but not exactly equal to, the equivalence point; selection of a better indicator, such as one that changes colors at pH 8, would have given a better approximation.

In the early part of the curve (when little base has been added), the acidic species predominates, and so the addition of small amounts of base will not appreciably change either the [OH⁻] or the pH. Similarly, in the last part of the titration curve (when an excess of base has been added), the addition of small amounts of base will not change the [OH⁻] significantly, and the pH remains relatively constant. The addition of base most alters the concentrations of H⁺ and OH⁻ near the equivalence point, and thus the pH changes most drastically in that region.

Basic Concept

An indicator is a weak organic acid or base that has different colors in its undissociated and dissociated states. The indicator used most frequently in intro chemistry labs is phenolphthalein, which is colorless at low pH but turns red around a pH of 8.

As described above, titration is used to determine the concentration of an acid or a base. Imagine that we have an acidic solution of volume V_A and unknown normality N_A. We add to this acidic solution a basic solution of known normality N_B a little bit at a time (a drop at a time), keeping track of the amount of base we have added. The equivalence point, as defined above, is the point at which the amount of acid equals the amount of base, or:

$$V_A N_A = V_B N_B$$

where V_B is the volume of the base we have added so far when the equivalence point is reached. Using this equation we can determine the normality or concentration of the original acid:

$$N_A = \frac{V_B N_B}{V_A}$$

The question is, of course, how do we know we have reached the equivalence point? We can use a pH meter and monitor the pH as a function of base added. More commonly, however, we use a couple of drops of an indicator and watch for a color change. Indicators are weak organic acids or bases that have different colors in their undissociated and dissociated states. If the solution in which it finds itself is below a certain pH, it will be of one color; if the solution pH is above that, it will be of a different color. The indicator most commonly encountered in introductory chemistry labs is probably phenolphthalein, which is colorless at low pH but becomes red around pH = 8, and is most often used in strong acid-strong base titrations. If we add a few drops of this to the acid at the beginning, then at a certain point in the titration the solution will take on a pale reddish hue (pale because it is so dilute). This point is the endpoint, signifying the end of the titration. The volume of base added at that point is used in the equation above as V_B. You may be thinking, this is not the same as the equivalence point! Indeed: The equivalence point is expected to occur at a pH of 7 for this kind of titration, yet phenolphthalein does not change color until pH ~ 8. The reason we can get away with this is because the only piece of information we need from the equivalence point is the volume of base added by the time it is reached. Since the pH is rising so sharply near the equivalence point, a slight difference in pH between it and the endpoint translates to a minuscule difference in V_B.

Weak Acid and Strong Base

Titration of a weak acid, HA, with a strong base produces the following titration curve:

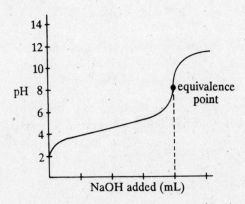

Comparing this figure with the previous one shows that the initial pH of the weak acid solution is greater than the initial pH of the strong acid solution. The equivalence point is in the basic range. This is consistent with what we discussed earlier about the salt of such a titration being a base.

Buffers

A buffer solution consists of a mixture, in roughly equal molar quantities, of a weak acid and its salt (which consists of its conjugate base and a cation), or a mixture of a weak base and its salt (which consists of its conjugate acid and an anion). Two examples of buffers are: a solution of acetic acid (CH_3COOH) and its salt, sodium acetate ($CH_3COO^-Na^+$); and a solution of ammonia (NH_3) and its salt, ammonium chloride ($NH_4^+Cl^-$). Buffer solutions have the useful property of resisting changes in pH when small amounts of acid or base are added.

Consider a buffer solution of acetic acid and sodium acetate:

$$CH_3COOH \rightleftharpoons H^+ + CH_3COO^-$$

When a small amount of NaOH is added to the buffer, the OH^- ions from the NaOH react with the H^+ ions present in the solution; subsequently, more acetic acid dissociates (equilibrium shifts to the right), restoring the $[H^+]$. Thus, an increase in $[OH^-]$ does not appreciably change pH. Likewise, when a small amount of HCl is added to the buffer, H^+ ions from the HCl react with the acetate ions to form acetic acid. Thus $[H^+]$ is kept relatively constant and the pH of the solution is relatively unchanged.

Amphoteric Species

An amphoteric, or amphiprotic, species is one that can act either as an acid or a base, depending on its chemical environment. In the Brønsted-Lowry sense, an amphoteric species can either gain or lose a proton. Water is the most common example. When water reacts with a base, it behaves as an acid:

A Closer Look

The pH of blood is maintained within a small range (slightly above 7) by a bicarbonate buffer system.

$$H_2O + B^- \rightleftharpoons HB + OH^-$$

When water reacts with an acid, it behaves as a base:

$$HA + H_2O \rightleftharpoons H_3O^+ + A^-$$

The partially dissociated conjugate base of a polyprotic acid is usually amphoteric (e.g., HSO_4^- can either gain an H^+ to form H_2SO_4, or lose an H^+ to form SO_4^{2-}). The hydroxides of certain metals, such as Al, Zn, Pb, and Cr, are also amphoteric.

Acids and Bases
Review Problems

1. A certain aqueous solution at 25° C has $[OH^-]$ $= 6.2 \times 10^{-5}$ M.

   a. Calculate $[H^+]$

   b. Calculate the pH of the solution.

   c. Is the solution acidic or basic?

2. What is the ratio of $[H^+]$ of a solution of pH = 4 to the $[H^+]$ of a solution of pH = 7?

3. Write equations expressing what happens to each of the following bases in aqueous solutions.

   a. LiOH

   b. $Ba(OH)_2$

   c. NH_3

   d. NO_2^-

4. What volume of a 3 M solution of NaOH is required to titrate 0.05 L of a 4 M solution of HCl to the equivalence point?

5. If 10 mL of 1 M NaOH is titrated with 1 M HCl to a pH of 2, what volume of HCl was added?

6. Identify the conjugate acids and bases in the following equation:

   $$NH_3 + H_2O \rightleftharpoons NH_4^+ + OH^-$$

7. Identify each of the following as an Arrhenius acid or base, Brønsted-Lowry acid or base, or Lewis acid or base:

   a. NaOH, in $NaOH \rightarrow Na^+ + OH^-$

   b. HCl, in $HCl \rightarrow H^+ + Cl^-$

   c. NH_3, in $NH_3 + H^+ \rightarrow NH_4^+$

   d. NH_4^+, in $NH_4^+ \rightarrow NH_3 + H^+$

   e. $(CH_3)_3N:$, in $(CH_3)_3N: + BF_3 \rightarrow (CH_3)_3N:BF_3$

   f. BF_3 in the above equation

8. At equilibrium, a certain acid, HA, in solution yields 0.94 M [HA] and 0.060 M [A⁻].

   a. Calculate K_a.

   b. Is this acid stronger or weaker than sulfurous acid ($K_a = 1.7 \times 10^{-2}$)?

   c. Calculate K_b.

   d. Calculate pH.

9. For each of the following choices, choose that which describes the weaker acid:

   a. $K_a = X$, $K_a = 3X$

   b. $[H^+] = X$, $[H^+] = 3X$

   c. $pH = X$, $pH = 3X$

10. For a certain acid, HA, $K_b(A^-) = 2.22 \times 10^{-11}$. Calculate the pH of a 0.5 M solution of HA.

11. Which of the following sets of materials would make the best buffer solution?

    A. H_2O, 1 M NaOH, 1 M H_2SO_4

    B. H_2O, 1 M CH_3COOH, 1 M $Na^+CH_3COO^-$

    C. H_2O, 1 M CH_3COOH, 6 M $Na^+CH_3COO^-$

    D. H_2O, 1 M CH_3COOH, 1 M NaOH

Turn the page
for answers and explanations
to the Review Problems.

Solutions to Review Problems

1.

a.

The concentration of H^+ is 1.6×10^{-10} M. K_w = $[H^+][OH^-] = 1.0 \times 10^{-14}$ M. If $[OH^-] = 6.2 \times 10^{-5}$, then $[H^+] = K_w /[OH^-] = 1.6 \times 10^{-10}$ M.

b.

$pH = -\log (1.6 \times 10^{-10}) = 9.79$.

c.

A pH of 9.79 indicates a basic solution.

2. 1000:1.

This problem can be solved by calculating the $[H^+]$ of the pH = 4 solution and the $[H^+]$ of the pH = 7 solution. Then divide the former by the latter: Since pH = $-\log[H^+]$, $[H^+]$ = antilog (–pH). For pH = 4, antilog (–4) = 10^{-4}. For pH = 7, antilog (–7) = 10^{-7}. 10^{-4}: 1×10^{-7} = 1000:1. Alternatively, we could subtract the pH's first, and then take the antilog:

$$7 - 4 = 3 \text{ implies } 10^3, \text{ or } 1000:1.$$

3.

a.

$$LiOH \rightarrow Li^+ + OH^-$$

b.

$$Ba(OH)_2 \rightarrow Ba^{2+} + 2OH^-$$

c.

$$NH_3 + H_2O \rightarrow NH_4^+ + OH^-$$

d.

$$NO_2^- + H_2O \rightarrow HNO_2 + OH^-$$

4.

At the equivalence point,

$$(\text{Normality})_{acid}(\text{Volume})_{acid} = (\text{Normality})_{base}(\text{Volume})_{base}$$

$$4 \text{ M HCl} = 4 \text{ N HCl}$$

$$3 \text{ M NaOH} = 3 \text{ N NaOH}$$

Plugging into the formula,

$$(4)(0.05) = (3)(V)$$

$$V_B = 0.067 \text{ L}$$

5.   10.2 mL

First, add enough HCl to neutralize the solution. Since both the acid and the base are 1 M, 10 mL of HCl will neutralize 10 mL of NaOH. This produces 20 mL of 0.5 M NaCl solution.

Next, calculate how much more HCl must be added to produce a $[H^+]$ of 1×10^{-2}. Let x be the amount of HCl to be added. The total volume of the solution will be $(20 + x)$ mL. Since this is now a dilution problem, the amount of HCl to be added can be found by using the formula:

$$M_1V_1 = M_2V_2$$

$$(1M)(x \text{ mL}) = (0.01 \text{ M})[(20 + x)\text{mL}]$$

When this equation is solved, x is found to have the value of 0.2. The final volume is 20.2 mL, so 10.2 mL of HCl was added to the original NaOH solution.

6.   (H_2O, OH^-) and (NH_4^+, NH_3)

NH_4^+ is the conjugate acid of the weak base, NH_3; OH^- is the conjugate base of the weak acid H_2O. The reaction in question is:

$$NH_3 + H_2O \rightarrow NH_4^+ + OH^-$$

According to the Brønsted-Lowry theory of acids and bases, an acid releases a proton while a base accepts a proton. In the case of weak acids and bases, an equilibrium is established whereby a weak acid, in this case H_2O, dissociates partially, donating a proton to a weak base, which is NH_3. The weak acid, H_2O, loses a proton and becomes a relatively stronger conjugate base, OH^-. This is one conjugate acid-base pair (H_2O, OH^-). Meanwhile, the weak base, NH_3, picks up a proton to become a relatively stronger conjugate acid, NH_4^+. This is the second conjugate acid-base pair (NH_4^+, NH_3).

7.

a.

NaOH is an Arrhenius base.

b.

HCl is an Arrhenius acid and a Brønsted-Lowry acid.

c.

NH_3 is a Brønsted-Lowry base and a Lewis base.

d.

NH_4^+ is an Arrhenius acid and a Brønsted-Lowry acid.

e.

$(CH_3)_3N$: acts only as a Lewis base.

f.

BF_3 acts only as a Lewis acid.

8.

a.

The dissociation of HA can be written as follows:

$$HA \rightarrow H^+ + A^-$$

The molar ratio of A^- to H^+ is 1:1. We are told that at equilibrium [HA] is 0.94 M while $[A^-]$ is 0.060 M. So $[H^+]$ must also be 0.060 M at equilibrium. It follows, then, that:

$$K_a = [A^-][H^+]/[HA] = (0.060)(0.060)/(0.94)$$
$$= 3.8 \times 10^{-3}$$

b.

An acid with a high K_a is a strong acid because its equilibrium position lies further to the right, meaning that dissociation is more complete. The K_a of sulfurous acid is 1.7×10^{-2} and the K_a of HA is 3.8×10^{-3}. The K_a of HA is less than that of sulfurous acid; therefore, HA is a weaker acid.

c.

$$K_b = \frac{1.0 \times 10^{-14}}{3.8 \times 10^{-3}} = 2.6 \times 10^{-12}$$

d.

$$pH = -\log [H^+] = -\log (0.060) = 1.22$$

9.

a.

A higher K_a indicates a stronger acid, a lower K_a indicates a weaker acid. X is one third the value of 3X and therefore a weaker acid.

b.

$[H^+]$ is a direct measure of the strength of an acid. The greater the concentration of H^+ in solution, the stronger the acid. An acid which liberates X moles of H^+ per liter is weaker, therefore, than an acid which liberates 3X moles of H^+ per liter.

c.

The lower the pH, the higher the concentration of H^+. The acid with a pH of 3X is thus the weaker acid.

10.

If $K_b = 2.22 \times 10^{-11}$, then

$$K_a = \frac{1.0 \times 10^{-14}}{2.22 \times 10^{-11}} = 4.5 \times 10^{-4}$$

The equilibrium expression for this dissociation is:

$$K_a = \frac{[H^+][A^-]}{[HA]}$$

We can let $[H^+]$ = x at equilibrium and, since $[H^+]:[A^-]$ = 1:1, $[A-]$ = x.

If the original $[HA]$ was 0.5M, and x mol/L are dissociated, then at equilibrium, $[HA] = 0.5 - x$.

Thus the equilibrium expression becomes:

$$4.5 \times 10^{-4} = \frac{[x][x]}{0.5 - x}$$

We can approximate that $0.5 - x \approx 0.5$ since HA has a small K_a, which indicates it is a weak acid.

$$4.5 \times 10^{-4} = x^2/0.5$$

$$x^2 = 2.25 \times 10^{-4}$$

$$x = 0.015 = [H^+]$$

$$pH = -\log [H^+] = -\log [0.015] = 1.82$$

11.   B

A buffer solution is prepared from a weak acid and its conjugate base, preferably in near-equal quantities. Choices A and D are wrong because they do not show conjugate acid/base pairs. Choice C is wrong because it shows a weak acid and its conjugate base, where the concentrations of the acid and the base are quite different. Thus, the best buffer solution would be that prepared from choice B, which shows a conjugate acid/base pair both present in 1 M concentrations.

CHAPTER 14

REDOX CHEMISTRY AND ELECTROCHEMISTRY

Redox chemistry involves the study of *red*uction and *ox*idation reactions: Reduction refers to reactions in which a species gains electrons, while oxidation refers to those in which a species gives up or loses electrons. Since electrons can neither be created nor destroyed in normal chemical reactions (as opposed to nuclear reactions, which will be discussed in the next chapter), an isolated loss or gain of electrons cannot occur; in other words, neither oxidation nor reduction can occur all by itself. Each occurs simultaneously in a redox reaction, resulting in net electron transfer between the species. The electrons released during oxidation are taken up in the reduction process. The species undergoing reduction is said to be reduced when it gains electrons; a reduced species is also called an oxidizing agent because it causes something else (the species giving up the electrons) to be oxidized. Similarly, a reducing agent causes another species to be reduced, and is itself oxidized. This is summarized below:

oxidizing agent	reducing agent
reduced	oxidized
gains electrons	loses electrons

Oxidation States and Assigning Oxidation Numbers

It is important to know which atom is oxidized and which is reduced. Oxidation states or oxidation numbers are assigned to atoms in order to keep track of the redistribution of electrons during a redox reaction. In a redox reaction, the oxidation numbers of some atoms have to change to reflect the gain or loss of electrons. By keeping track of and comparing the oxidation numbers of the atoms on the reactant and the product side, it is possible to determine how many electrons are gained or lost by each atom. The oxidation number of an atom in a compound is assigned according to the following rules:

1. The oxidation number of free elements is zero. For example, the atoms in N_2, P_4, S_8, and He all have oxidation numbers of zero.

2. The oxidation number for a monatomic ion is equal to the charge of the ion. For example, the oxidation numbers for Na^+, Cu^{2+}, Fe^{3+}, Cl^-, and N^{3-} are +1, +2, +3, –1, and –3 respectively.

3. The oxidation number of each Group IA element in a compound is +1. The oxidation number of each Group IIA element in a compound is +2.

4. The oxidation number of each Group VIIA element (halogens) in a compound is –1, except when combined with an element of higher electronegativity. For example, in HCl, the oxidation number of Cl is –1; in HOCl, however, the oxidation number of Cl is +1 because of the oxygen (see rule 6 below).

5. The oxidation number of hydrogen is –1 in compounds with less electronegative elements than hydrogen (Groups IA and IIA). Examples include NaH and CaH_2. The more common oxidation number of hydrogen is +1.

6. In most compounds, the oxidation number of oxygen is –2. This is not the case, however, in molecules such as OF_2. Here, because F is more electronegative than O, the oxidation number of oxygen is +2. Also, in peroxides such as BaO_2, the oxidation number of O is –1 instead of –2 because of the structure of the peroxide ion, $[O–O]^{2-}$. (Note that Ba, a group IIA element, cannot be a +4 cation.) e/H_2O_2

7. The sum of the oxidation numbers of all the atoms present in a neutral compound is zero. The sum of the oxidation numbers of the atoms present in a polyatomic ion is equal to the charge of the ion. Thus, for SO_4^{2-}, the sum of the oxidation numbers must be –2.

Example: Assign oxidation numbers to the atoms in the following reaction in order to determine the oxidized and reduced species and the oxidizing and reducing agents.

$$SnCl_2 + PbCl_4 \rightarrow SnCl_4 + PbCl_2$$

Solution: All these species are neutral, so the oxidation numbers of all the atoms in each compound must add up to zero. In $SnCl_2$, since there are two chlorines present, and chlorine has an oxidation number of –1, Sn must have an oxidation number of +2. Similarly, the oxidation number of Sn in $SnCl_4$ is +4; the oxidation number of Pb is +4 in $PbCl_4$ and +2 in $PbCl_2$. Notice that the oxidation number of Sn goes from +2 to +4; i.e., it loses electrons and thus is oxidized, making it the reducing agent. Since the oxidation number of Pb has decreased from +4 to +2, it has gained electrons and been reduced. Pb is the oxidizing agent. The sum of the charges on both sides of the reaction is equal to zero, so charge has been conserved.

Quick Quiz

Match the substance with its correct oxidation number below.

1. Na^+
2. Cu^{2+}
3. Cl^-
4. Fe^{3+}

(A) –1
(B) +3
(C) +2
(D) +1

Answers: 1. D
2. C
3. A
4. B

Note that even though we have been making the connection between oxidation states and charge distribution, the oxidation number is not in general the charge (nor even the formal charge) of the atom in a compound. It merely reflects a way of accounting for how electrons are transferred between species.

Balancing Redox Reactions

By assigning oxidation numbers to the reactants and products, one can determine how many moles of each species are required for conservation of charge and mass, which is necessary to balance the equation. In general, to balance a redox reaction, both the net charge and the number of atoms must be equal on both sides of the equation. The most common method for balancing redox equations is the half-reaction method, also known as the ion-electron method, in which the equation is separated into two half-reactions–the oxidation part and the reduction part. Each half-reaction is balanced separately, and they are then added to give a balanced overall reaction, in which electrons do not appear explicitly by convention. Consider a redox reaction between $KMnO_4$ and HI in an acidic solution:

$$MnO_4^- + I^- \rightarrow I_2 + Mn^{2+}$$

Step 1:   Separate the two half-reactions.

oxidation half-reaction: $MnO_4^- \rightarrow Mn^{2+}$

reduction half-reaction: $I^- \rightarrow I_2$

Step 2:   Balance the atoms of each half-reaction. First, balance all atoms except H and O. Next, in an acidic solution, add H_2O to balance the O atoms and then add H^+ to balance the H atoms. (In a basic solution, use OH^- and H_2O to balance the O's and H's.)

To balance the iodine atoms, place a coefficient of two before the I^- ion.

$2 I^- \rightarrow I_2$

For the permanganate half-reaction, Mn is already balanced. Next, balance the oxygens by adding $4H_2O$ to the right side.

$MnO_4^- \rightarrow Mn^{2+} + 4H_2O$

Finally, add H^+ to the left side to balance the 4 H_2Os. These two half-reactions are now balanced in mass (but not in charge).

$MnO_4^- + 8 H^+ \rightarrow Mn^{2+} + 4H_2O$

Step 3:   Balance the charges of each half-reaction. The reduction half-reaction must consume the same number of electrons as are supplied by the oxidation half. For the oxidation reaction, add 2 electrons to the right side of the reaction:

$$2 \, I^- \rightarrow I_2 + 2e^-$$

For the reduction reaction, a charge of +2 must exist on both sides. Add 5 electrons to the left side of the reaction to accomplish this:

$$5 \, e^- + 8 \, H^+ + MnO_4^- \rightarrow Mn^{2+} + 4H_2O$$

Step 4:   Both half-reactions must have the same number of electrons so that they will cancel. Multiply the oxidation half by 5 and the reduction half by 2 and add the two:

$$5(2 \, I^- \rightarrow I_2 + 2e^-)$$

$$2(5 \, e^- + 8 \, H^+ + MnO_4^- \rightarrow Mn^{2+} + 4H_2O)$$

The final equation is:

$$10 \, I^- + 10 \, e^- + 16 \, H^+ + 2 \, MnO_4^- \rightarrow 5 \, I_2 + 2 \, Mn^{2+} + 10 \, e^- + 8 \, H_2O$$

To get the overall equation, cancel out the electrons and any H_2Os, H^+s, OH^-s or e^-s that appear on both sides of the equation.

$$10 \, I^- + 16 \, H^+ + 2 \, MnO_4^- \rightarrow 5 \, I_2 + 2 \, Mn^{2+} + 8 \, H_2O$$

Step 5:   Finally, confirm that mass and charge are balanced. There is a +4 net charge on each side of the reaction equation, and the atoms are stoichiometrically balanced.

As you may have noticed, balancing redox equations can be trickier and more involved than other types of reactions because often one needs to supply additional chemical species like water and protons to the equation, rather than simply playing with stoichiometric coefficients. The above scheme is the most general one that can be applied; it may be that many intermediate steps can be omitted if the equation is simpler. For example, for the equation:

$$Zn \, (s) + HCl \, (aq) \rightarrow ZnCl_2 \, (aq) + H_2 \, (g)$$

one may balance it by just supplying coefficients.

Oxygen as an Oxidizing Agent and the Chemistry of Oxides

Oxygen is a powerful oxidizing agent; after all, one can say that the process of oxidation is named after it! The reason why this is so is because of its high electronegativity. Upon reacting with other species, it will undergo reduction and take on a negative oxidation state, oxidizing its partner in the process. The only exception is in its reaction with fluorine, which is more electronegative than oxygen is (see rules for assigning oxidation numbers above). Because of the relative abundance of oxygen relative to fluorine, however, the "electron-grabbing" effect of oxygen is much more evident in everyday life. Rusting, for example, occurs when iron is oxidized to ferric oxide (Fe_2O_3) and complexes with water molecules to form a hydrate($Fe_2O_3 \bullet xH_2O$ where x is the number of water molecules to which it is complexed and may vary).

Oxide is the general name usually given to binary compounds in which the oxygen is in the –2 oxidation state (as distinguished from peroxides and superoxides, for example). Certain oxides dissolve in water to give acidic solutions; such oxides are called acidic anhydrides. Other oxides may dissolve in water to give basic solutions; such oxides are called basic anhydrides. Acidic anhydrides are mostly oxides of nonmetals, such as SO_3, which dissolves in water to give sulfuric acid, H_2SO_4. Oxides of Groups IA and IIA metals, on the other hand, tend to be basic anhydrides, such as BaO and CaO.

Instead of thinking of anhydrides in terms of what they would do in water, one can also think in the opposite direction: Anhydrides are obtained by the removal of water from acidic and basic compounds. CaO, for example, can be obtained by removing water from calcium hydroxide, $Ca(OH)_2$:

$$Ca(OH)_2 \ (s) \rightarrow CaO \ (s) + H_2O \ (l)$$

Acidic and basic anhydrides are Lewis acids and bases.

Electrochemical Cells

The concept of separating a redox reaction into separate reduction and oxidation parts is not purely a theoretical mechanism to help in this paper-and-pencil task. In real life, one can often actually carry out the two half-reactions in separate compartments or beakers, and couple the two so that the electrons are forced to flow through an external circuit. Such a configuration occurs in the galvanic cell, one class of electrochemical cells.

An electrochemical cell is a contained system in which a redox reaction occurs in conjunction with the passage of electric current. There are two types of electrochemical cells, galvanic cells (also known as voltaic cells), and electrolytic cells. (We shall investigate the differences between the two types shortly.) Both kinds of electrochemical cells contain two electrodes, which are essentially two pieces of metal, that serve as the sites for the oxidation and reduction half-reactions separately. The electrode at which oxidation occurs is called the anode, and the electrode at which reduction occurs is called the cathode. This is true for both galvanic (voltaic) and electrolytic cells.

Galvanic Cells

A redox reaction occurring in a galvanic cell has a negative ΔG and is therefore a spontaneous reaction. Galvanic cell reactions supply energy and are used to do work. This energy can be harnessed by placing the oxidation and reduction half-reactions in separate containers called half-cells. The half-cells are then connected by an apparatus that allows for the flow of electrons. The spontaneous flow of electrons is forced to go through external circuitry in which their potential energy is extracted.

A common example of a galvanic cell is the Daniell cell, shown below:

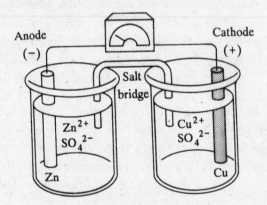

In the Daniell cell, a zinc bar is placed in an aqueous $ZnSO_4$ solution, and a copper bar is placed in an aqueous $CuSO_4$ solution. The anode of this cell is the zinc bar where Zn (s) is oxidized to $Zn^{2+}(aq)$. The cathode is the copper bar, and it is the site of the reduction of $Cu^{2+}(aq)$ to Cu (s). The half-cell reactions are written as follows:

$$Zn\ (s) \rightarrow Zn^{2+}\ (aq) + 2e^- \qquad \text{(anode)}$$

$$Cu^{2+}\ (aq) + 2e^- \rightarrow Cu\ (s) \qquad \text{(cathode)}$$

If the two half-cells were not separated, the Cu^{2+} ions would react directly with the zinc bar and no useful electrical work would be obtained. To complete the circuit, the two solutions must be connected. Without connection, the electrons from the zinc oxidation half reaction would not be able to get to the copper ions, so a wire (or other conductor) is necessary. If only a wire were provided for this electron flow, the reaction would soon cease anyway because an excess negative charge would build up in the solution surrounding the cathode and an excess positive charge would build up in the solution surrounding the anode. This charge gradient is dissipated by the presence of a salt bridge, which permits the exchange of cations and anions. The salt bridge contains an inert electrolyte, usually KCl or NH_4NO_3, whose ions will not react with the electrodes or with the ions in solution. At the same time the anions from the salt bridge (such as Cl^-) diffuse from the salt bridge of the Daniell cell into the $ZnSO_4$ solution to balance out the charge of the newly created Zn^{2+} ions, the cations of the salt bridge (such as K^+) flow into the $CuSO_4$ solution to balance out the charge of the SO_4^{2-} ions left in solution when the Cu^{2+} ions deposit as copper metal.

During the course of the reaction, electrons flow from the zinc bar (anode) through the wire and the ammeter, toward the copper bar (cathode). The anions (Cl^-) flow externally (via the salt bridge) into the $ZnSO_4$, and the cations (K^+) flow into the $CuSO_4$. This flow depletes the salt bridge and, along with the finite quantity of Cu^{2+} in the solution, accounts for the relatively short lifetime of the cell.

Instead of an ammeter that simply measures the current, one can place a device that is powered by electric current so as to extract the potential energy of the electrons. That is, after all, why galvanic cells are useful. The common dry cell battery and the lead-acid storage battery found in cars are examples of galvanic cells.

Electrolytic Cells

A redox reaction occurring in an electrolytic cell has a positive ΔG and is therefore nonspontaneous. In electrolysis, electrical energy is required to induce reaction; i.e., instead of extracting work from a spontaneous redox reaction, we supply energy to force a nonspontaneous redox reaction to occur. The oxidation and reduction half-reactions are usually placed in one container. Where the ammeter or electrical device used to be for the galvanic cell, we need to place a source of electrical power, like a battery, instead (see figure below).

Michael Faraday was the first to define certain quantitative principles governing the behavior of electrolytic cells. He theorized that the amount of chemical change induced in an electrolytic cell is directly proportional to the number of moles of electrons that are exchanged during a redox reaction. The number of moles exchanged can be determined from the balanced half-reaction. In general, for a reaction which involves the transfer of n electrons per atom:

Basic Concept

One Faraday is equivalent to the amount of charge contained in one mole of electrons.

$$M^{n+} + n\,e^- \rightarrow M(s)$$

one mole of $M(s)$ will be produced if n moles of electrons are supplied.

The number of moles of electrons needed to produce a certain amount of $M(s)$ can now be related to a measurable electrical property. One electron carries a charge of 1.6×10^{-19} coulombs (C). The charge carried by one mole of electrons can be calculated by multiplying this number by Avogadro's number, as follows:

$$(1.6 \times 10^{-19})(6.022 \times 10^{23}) = 96{,}487 \text{ C/mol } e^-$$

This number is called Faraday's constant, and one Faraday (F) is equivalent to the amount of charge contained in one mole of electrons (1 F = 96,487 coulombs, or J/V).

An example of an electrolytic cell, in which molten NaCl is electrolyzed to form Cl_2 (g) and Na (l), is given below:

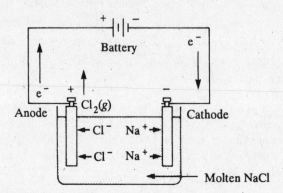

In this cell, Na^+ ions migrate towards the cathode, where they are reduced to Na (l). Similarly, Cl^- ions migrate towards the anode, where they are oxidized to Cl_2 (g).

Electrode Charge Designations

The anode of an electrolytic cell is considered positive, since it is attached to the positive pole of the battery and so attracts anions from the solution. The anode of a galvanic cell, on the other hand, is considered negative because the spontaneous oxidation reaction that takes place at the galvanic cell's anode is the original source of that cell's negative charge, i.e., is the source of electrons. In spite of this difference in designating charge, oxidation takes place at the anode in both types of cells, and electrons always flow through the wire from the anode to the cathode.

Reduction Potentials and the Electromotive Force

Sometimes when electrolysis is carried out in an aqueous solution, water rather than the solute is oxidized or reduced. For example, if an aqueous solution of NaCl is electrolyzed, water may be reduced at the cathode to produce H_2 (g) and OH^- ions, instead of Na^+ being reduced to Na (l), as occurs in the absence of water. The species in a reaction that will be oxidized or reduced can be determined from the reduction potential of each species, defined as the tendency of a species to acquire electrons and be reduced. Each species has its own intrinsic reduction potential; the more positive the potential, the greater the species' tendency to be reduced.

A reduction potential is measured in volts (V) and is defined relative to the standard hydrogen electrode (SHE), which is arbitrarily given a potential of 0.00 volts. Standard reduction potential, ($E°_{red}$), is measured under standard conditions: a 1 M concentration for each ion participating in the reaction, a partial pressure of 1 atm for each gas that is part of the reaction, and metals in their pure state. The relative reactivities of different half-cells can be compared to predict the direction of electron flow. A higher $E°_{red}$ means a greater tendency for reduction to occur, while a lower $E°_{red}$ means a greater tendency for oxidation to occur.

Example: Given the following half-reactions and $E°_{red}$ values, determine which species would be oxidized and which would be reduced.

$$Ag^+ + e^- \rightarrow Ag\ (s) \quad E°_{red} = +0.80\ V$$

$$Tl^+ + e^- \rightarrow Tl(s) \qquad E°_{red} = -0.34\ V$$

Solution: Ag^+ would be reduced to Ag (s) and Tl (s) would be oxidized to Tl^+, since Ag^+ has the higher $E°_{red}$. Therefore, the reaction equation would be:

$$Ag^+ + Tl\ (s) \rightarrow Tl^+ + Ag\ (s)$$

which is the sum of the two spontaneous half-reactions.

Note that reduction and oxidation are opposite processes. Therefore, in order to obtain the oxidation potential of a given half-reaction, the reduction half reaction and the sign of the reduction potential are both reversed. For instance, from the example above, the oxidation half reaction and oxidation potential of Tl(s) are:

$$Tl(s) \rightarrow Tl^+ + e^- \qquad\qquad E°_{ox} = +0.34\ V$$

Basic Concept

Galvanic/voltaic cell: positive EMF, spontaneous reaction, ΔG negative

Electrolytic cell: negative EMF, nonspontaneous reaction, ΔG positive

Standard reduction potentials are also used to calculate the standard electromotive force (EMF or $E°_{cell}$) of a reaction, the difference in potential between two half-cells. The EMF of a reaction is determined by adding the standard reduction potential of the reduced species and the standard oxidation potential of the oxidized species. *When adding standard potentials, it is very important to note that we do not multiply by the number of moles oxidized or reduced.*

$$EMF = E_{red} + E_{ox}$$

The standard EMF of a galvanic cell is positive, while the standard EMF of an electrolytic cell is negative. A spontaneous redox equation, therefore, will have a positive EMF, but a negative free energy change, and vice versa for a nonspontaneous reaction. We shall specify further the relation between EMF and ΔG below, but for now you should keep this reversal of sign in mind.

Example: Given that the standard reduction potentials for Sm^{3+} and $[RhCl_6]^{3-}$ are –2.41 V and +0.44 V respectively, calculate the EMF of the following reaction:

$$Sm^{3+} + Rh + 6\ Cl^- \rightarrow [RhCl_6]^{3-} + Sm$$

Solution: First, determine the oxidation and reduction half-reactions. As written, the Rh is oxidized and the Sm^{3+} is reduced. Thus the Sm^{3+} reduction potential is used as is, while the reverse reaction for Rh, $[RhCl_6]^{3-} \rightarrow Rh + 6\ Cl^-$, applies and the oxidation potential of $[RhCl_6]^{3-}$ must be used. Then the EMF can be calculated to be (–2.41 V) + (–0.44 V) = –2.85 V. Note that we have switched the sign in front of the potential for $[RhCl_6]^{3-}$. The cell is thus electrolytic as written.

Thermodynamics of Redox Reactions

The thermodynamic criterion for determining the spontaneity of a reaction is ΔG, Gibbs free energy, the maximum amount of useful work produced by a chemical reaction. In an electrochemical cell, the work done is dependent on the number of coulombs and the energy available. Thus, ΔG and EMF are related as follows:

$$\Delta G = -nFE_{cell}$$

where n is the number of moles of electrons exchanged, F is Faraday's constant, and E_{cell} is the EMF of the cell. Keep in mind that if Faraday's constant is expressed in coulombs (J/V), then ΔG must be expressed in J, not kJ.

If the reaction takes place under standard conditions, then the ΔG is the standard Gibbs free energy and E_{cell} is the standard cell potential. The above equation then becomes:

$$\Delta G^\circ = -nFE^\circ_{cell}$$

Recall that in the chapter on thermochemistry we have derived the following equation:

$$\Delta G^\circ = -RT \ln K_{eq}$$

where R is the gas constant 8.314 J/(K•mol), T is the temperature in K, and K_{eq} is the equilibrium constant for the reaction. Combining this with the equation above, we get:

$$\Delta G^\circ = -nFE^\circ_{cell} = -RT \ln K_{eq}$$

or simply:

$$nFE^\circ_{cell} = RT \ln K_{eq}$$

If the values for n, T, and K_{eq} are known, then the E°_{cell} for the redox reaction can be readily calculated.

Redox Chemistry and Electrochemistry Review Problems

1. If one F is equivalent to 96,487 C/mol e^-, what is the charge on an individual electron?

   A. 5.76×10^{28} C/e^-

   B. 6.022×10^{23} C/e^-

   C. 1.6×10^{19} C/e^-

   D. 1.6×10^{-19} C/e^-

2. How many F are required for the reduction of 1 mole of Ni^{2+} to Ni (s)?

   A. 1 F

   B. 2 F

   C. 96,487 F

   D. 6.022×10^{23} F

3. The gold-plating process involves the following reaction:

   Au^{3+} (aq) + 3e^- → Au (s).

   If 0.600 g of Au is plated onto a metal, how many coulombs are used?

   A. 299 C

   B. 868 C

   C. 2,990 C

   D. 8,680 C

4. The standard reduction potential of Cr^{3+} (aq) + 3 e^- → Cr (s) is –0.74 V. The standard reduction potential of Cl_2 (g) + 2 e^- → 2 Cl^- (aq) is 1.36 V.

   Based on the information given, it must be true that:

   A. Cl_2 is more easily oxidized than Cr^{3+}, and Cl_2 is thus a better oxidizing agent than Cr^{3+}.

   B. Cl_2 is more easily oxidized than Cr^{3+}, and Cl_2 is thus a better reducing agent than Cr^{3+}.

   C. Cl_2 is more easily reduced than Cr^{3+}, and Cl_2 is thus a better reducing agent than Cr^{3+}.

   D. Cl_2 is more easily reduced than Cr^{3+}, and Cl_2 is thus a better oxidizing agent than Cr^{3+}.

5. The standard reduction potential of Cu^{2+} (aq) is +0.34 V. What is the oxidation potential of Cu(s)?

   A. +0.68 V

   B. +0.34 V

   C. –0.34 V

   D. –0.68 V

6. (Equation 1) $F_2 + 2 e^- \rightarrow 2 F^- (aq)$
    $E° = +2.87$ V

   (Equation 2) $Ca^{2+} + 2 e^- \rightarrow Ca (s)$
    $E° = -2.76$ V

   When the above half-reactions are combined in a galvanic cell, which species will be reduced and which will be oxidized?

   A. F^- will be oxidized and Ca^{2+} will be reduced.

   B. Ca^{2+} will be oxidized and F_2 will be reduced.

   C. Ca (s) will be oxidized and F_2 will be reduced.

   D. F_2 will be oxidized and Ca (s) will be reduced.

7. What is the standard EMF if the following half-reactions are combined in a galvanic cell?

   (Equation 1) $Co^{3+}(aq) + e^- \rightarrow Co^{2+} (aq)$
    $E° = 1.82$ V

   (Equation 2) $Na^+ (aq) + e^- \rightarrow Na (s)$
    $E° = -2.71$

   A. 4.53 V

   B. 0.89 V

   C. −0.89 V

   D. −4.53V

8. What is $E°_{cell}$ for a reaction where the $\Delta G° = -553.91$ kJ and 2 electrons are transferred?

   A. −2.87 V

   B. −0.00287 V

   C. 0.00287 V

   D. 2.87 V

9. What is the $\Delta G°$ for the following reaction equation?

   $Ti^{2+} (aq) + Mg (s) \rightarrow Ti (s) + Mg^{2+} (aq)$

   $Ti^{2+} (aq) + 2 e^- \rightarrow Ti (s) \; E° = -1.63$ V

   $Mg^{2+} (aq) + 2 e^- \rightarrow Mg (s) \; E° = -2.38$ V

   A. 144.73 kJ

   B. 144.73 J

   C. −144.73 J

   D. −144.73 kJ

10. Assign oxidation numbers to each atom of the following reaction equation:

    $2 Fe (s) + O_2 (g) + 2 H_2O (l) \rightarrow 2 Fe(OH)_2 (s)$

11. Using the ion-electron method, balance the following equation of a reaction taking place in an acidic solution:

$$ClO_3^- + AsO_2^- \rightarrow AsO_4^{3-} + Cl^-$$

12. Given the standard reduction potentials for the following half-reactions,

$$ClO_4^- (aq) + 2H^+ (aq) + 2e^- \rightarrow ClO_3^- (aq) + H_2O (l)$$
$$E° = +1.19 \text{ V}$$

$$Ag^+ (aq) + e^- \rightarrow Ag (s)$$
$$E° = +0.799 \text{ V}$$

predict which half-reaction would occur at the anode and which would occur at the cathode in a galvanic cell.

Solutions to Review Problems

1. D

Avogadro's number (6.022×10^{23}) defines the number of particles present in 1 mole of anything. So there are 6.022×10^{23} electrons in 1 mole of electrons.

$$1 \text{ F} = 96{,}487 \, \frac{C}{\text{mol e}^-} \times \frac{1 \text{ mol e}^-}{6.022 \times 10^{23} \text{ e}^-}$$
$$= 1.602 \times 10^{-19} \text{ C/e}^-.$$

2. B

The reduction of one mole of Ni^{2+} to 1 mole of Ni (s) requires 2 moles of electrons. The transfer of 1 mole of electrons is equivalent to the transfer of 1 F of charge. Therefore, since 2 moles of electrons are required to reduce 1 mole of Ni^{2+}, 2 F are required.

3. B

In order to solve this problem, first determine the number of moles of Au present in 0.600 g.

$$0.600 \text{ g} \times \frac{1 \text{ mol Au}}{197.0 \text{g}} = 0.00300 \text{ mol Au}$$

Next, determine the number of moles of electrons used to reduce $Au^{3+}(aq)$ to 0.00300 mol of Au (s). From the given reduction half-reaction of Au, it is evident that for every mole of Au^{3+}, 3 moles of electrons are transferred in order to produce 1 mole of Au (s); therefore,

$$0.00300 \text{ mol Au} \times \frac{3 \text{ mol e}^-}{1 \text{ mol Au}} = 0.00900 \text{ mol e}^-$$

Finally, convert 0.00900 mol e^- to its equivalent in C:

$$0.00900 \text{ mol e}^- \times \frac{96487 \text{ C}}{1 \text{ mol e}^-} = 868 \text{ C}$$

4. D

Remember that in opposition to free energy, the more positive the E is for a reaction, the more favorable it is. Since the reduction of Cl_2 has a more positive E than Cr^{3+}, it will be more easily reduced, meaning that it is a stronger oxidizing agent.

5. C

The two processes are reverses of each other, and thus their potentials are the negative of each other.

6. C

The half-reaction with the greater reduction potential will proceed forward as written, while the half-reaction with the smaller reduction potential will proceed in the opposite direction (i.e., as oxidation). F_2 has a greater tendency to be reduced than Ca^{2+}, because it has the greater reduction potential. Therefore, Equation 1 will proceed as written, while Equation 2 will proceed in the opposite direction. As a result, F_2 is reduced to 2 F^- and Ca (s) is oxidized to Ca^{2+}.

7.   A

From the values of the reduction potentials, it is evident that Equation 1 will be the reduction half-reaction, since it has the larger reduction potential, and Equation 2 will be reversed for the oxidation half-reaction. The EMF of a reaction is determined by adding the standard potentials of the reduced and oxidized species. The standard reduction potential for of Co^{3+} is +1.82 V. The standard oxidation potential of Na(s) is +2.71 V, which is equivalent, but opposite in sign, to the standard reduction potential of Na^+ (aq).

$$EMF = 1.82 \text{ V} + 2.71 \text{ V} = 4.53 \text{ V}$$

8.   D

Use the relationship $\Delta G° = -nFE°$, which can be rearranged to $E° = -\Delta G°/(nF)$. $\Delta G° = -553.91$ kJ, n = 2, and F = 96,487 C/mol e^-. The next step is to convert the value of $\Delta G°$ from kJ to J. When this is done, $\Delta G°$ is equal to –553,910 J. Now, the final step is to substitute the values into the formula, to get the final answer of $E° = 2.87$ V.

9.   D

First, determine the EMF ($E°$) by adding the potentials of the reduced and oxidized species.

$$-1.63 \text{ V} + (+2.38 \text{ V}) = 0.75 \text{ V} = 0.75$$

By inspection, 2 moles of electrons are transferred to produce 1 mole of product; therefore n = 2. Now determine $\Delta G°$:

$$\Delta G° = -nFE°$$

$$\Delta G° = -(2 \text{ mol } e^-)\,(96487)\,(0.75\,)$$

$$\Delta G° = -144{,}730 \text{ J}$$

$$= -144.73 \text{ kJ}$$

10.   2 Fe (s) + O₂ (g) + H₂O (l) → Fe(OH)₂ (s)

To assign oxidation numbers, use the rules given in this chapter. Fe (s) and O_2 (g) have oxidation numbers of zero, because they are free elements. Hydrogen in H_2O (l) has an oxidation number of +1, because oxygen is more electronegative than hydrogen; likewise, oxygen in H_2O (l) has an oxidation number of –2. Oxygen and hydrogen in Fe(OH)$_2$ (s) have the same oxidation numbers as in H_2O (l). Each OH group contributes a charge of –1 to Fe(OH)$_2$ and since there are 2 OH groups, their overall contribution to the compound is –2. Since Fe(OH)$_2$ is a neutral compound and thus has no overall charge, the sum of all the oxidation numbers of the atoms in this compound is zero. Consequently, the Fe in Fe(OH)$_2$ must possess a charge of +2 in order to make the overall charge on the compound zero.

11.

The balanced equation is:

$$ClO_3^- + 3H_2O + 3AsO_2^- \rightarrow 3AsO_4^{3-} + Cl^- + 6H^+$$

12.

The rule is that the half-reaction with the greater reduction potential will be the reduction reaction and the reverse of the other will be the oxidation reaction. Since the first reaction (reduction of ClO_4^-) has the greater reduction potential, this will proceed as a forward reaction and ClO_4^- will be reduced. The second reaction (reduction of Ag^+), which has the lower reduction potential, will proceed as a reverse reaction, and Ag (s) will be oxidized to Ag^+. Oxidation occurs at the anode and reduction at the cathode. Thus, the reduction of ClO_4^- to ClO_3^- will occur at the cathode and the oxidation of Ag (s) to Ag^+ will occur at the anode.

CHAPTER 15

NUCLEAR CHEMISTRY

The Nucleus

At the center of an atom lies its nucleus, consisting of one or more nucleons (protons or neutrons) held together with considerably more energy than the energy needed to hold electrons in orbit around the nucleus. The radius of the nucleus is about 100,000 times smaller than the radius of the atom. Before we go on, let's revisit some concepts introduced in chapter 3.

Atomic Number (Z)

Z is always an integer, and is equal to the number of protons in the nucleus. As stated in chapter 3, the number of protons is what defines an element: An atom or an ion or a nucleus is identified as carbon, for example, if and only if it has 6 protons. Each element has a unique number of protons. Z is used as a presubscript to the chemical symbol in isotopic notation, that is, it appears as a subscript before the chemical symbol. The chemical symbols and the atomic numbers of all the elements are given in the Periodic Table.

Mass Number (A)

A is an integer equal to the total number of nucleons (neutrons and protons) in a nucleus. Let N represent the number of neutrons in a nucleus. The equation relating A, N, and Z is simply:

$$A = N + Z$$

In isotopic notation, A appears as a presuperscript to the chemical symbol: It appears as a superscript that comes before the chemical symbol. In general, then, a nucleus can be represented as

$$^A_Z X$$

where X is the chemical symbol for the element. This representation is sometimes referred to as a nuclide symbol. Note that since an element is defined by the atomic number, Z is technically redundant information if one has access to the Periodic Table. Often, then, this quantity is omitted.

Examples: $_1^1$H: a single proton; the nucleus of ordinary hydrogen. Number of neutrons = presuperscript – presubscript = 1 – 1 = 0.

$_2^4$He: the nucleus of ordinary helium, consisting of 2 protons and 2 neutrons. It is also known as an alpha particle (α-particle, see below).

$_{92}^{235}$U: a fissionable form of uranium, consisting of 92 protons and 235 – 92 = 143 neutrons.

Isotopes

Basic Concept

All isotopes of a given element have the same value of Z, but different values of N and A.

Different nuclei of the same element will by definition all have the same number of protons. The number of neutrons, however, can be different. Nuclei of the same element can therefore have different mass numbers. For a nucleus of a given element with a given number of protons (atomic number Z), the various nuclei with different numbers of neutrons are called **isotopes** of that element.

Example: The three isotopes of hydrogen are:

$_1^1$H: a single proton; the nucleus of ordinary hydrogen.

$_1^2$H: a proton and a neutron together; the nucleus of one type of heavy hydrogen called deuterium.

$_1^3$H: a proton and two neutrons together; the nucleus of a heavier type of heavy hydrogen called tritium.

Note that despite the existence of names like deuterium and tritium, they are all considered hydrogen bacause they have the same number of protons (one). The example shown here is a little bit of an anomaly because in general isotopes do not have specific names of their own.

Atomic Mass And Atomic Mass Unit

Atomic mass is most commonly measured in atomic mass units (abbreviated amu). By definition, 1 amu is exactly one-twelfth the mass of the neutral carbon-12 atom. In terms of more familiar mass units:

$$1 \text{ amu} = 1.66 \times 10^{-27} \text{ kg} = 1.66 \times 10^{-24} \text{ g}$$

Atomic Weight

Because isotopes exist, atoms of a given element can have different masses. The atomic weight refers to a weighted average of the masses of an element. The average is weighted according to the natural abundances of the various isotopic species of an element. The atomic weight can be measured in amu.

Example: 99.985499% of hydrogen occurs in the common ^{1}H isotope with a mass of 1.00782504 amu. About 0.0142972% occurs as deuterium with a mass (including the electron) of 2.01410 amu, and about 0.0003027% occurs as tritium with a mass of 3.01605 amu. The atomic weight of hydrogen is the sum of the mass of each isotope multiplied by its natural abundance (x):

atomic weight of H = $m_{1H}x_{1H} + m_{2H}x_{2H} + m_{3H}x_{3H}$

= (1.00782504)(0.99985499) + (2.01410)(0.000142972) + (3.01605)(0.000003027)

= 1.00797 amu

Nuclear Binding Energy and Mass Defect

Basic Concept

Every nucleus (other than $^1_1$H) has a smaller mass than the combined mass of its constituent protons and neutrons.

The mass of a nucleus is always less than the combined masses of its constituent protons and neutrons.

The difference is called the mass defect. Scientists had difficulty explaining why this mass defect occurred until Einstein discovered the equivalence of matter and energy, embodied by the equation $E = mc^2$, where c is the speed of light in vacuum: 3×10^8 m/s. Because its value is so big, even a small amount of mass will, upon conversion, release a large amount of energy. The mass defect is a result of matter that has been converted to energy. This energy is called the binding energy. This same amount of energy is needed to break apart the nucleus and separate the protons and neutrons: the larger the binding energy, the more stable the nucleus. The binding energy per nucleon peaks at iron, which implies that iron is the most stable atomic nucleus. In general, intermediate-sized nuclei are more stable than large and small nuclei.

Nuclear Reactions and Decay

Nuclear reactions such as fusion, fission, and radioactive decay involve either combining or splitting the nuclei of atoms. Since the binding energy per nucleon is greatest for intermediate-sized atoms, when small atoms combine or large atoms split a great amount of energy is released.

Fusion

Fusion occurs when small nuclei combine into a larger nucleus. As an example, many stars including the sun power themselves by fusing four hydrogen nuclei to make one helium nucleus. By this method, the sun produces 4×10^{26} J every second. Here on earth, researchers are trying to find ways to use fusion as an alternative energy source.

Fission

Fission is a process in which a large nucleus splits into smaller nuclei. Spontaneous fission rarely occurs. However, by the absorption of a low energy neutron, fission can be induced in certain nuclei. Of special interest are those fission reactions that release more neutrons, since these other neutrons will cause other atoms to undergo fission. This in turn releases more neutrons, creating a chain reaction. Such induced fission reactions power commercial nuclear electric-generating plants.

Example: A fission reaction occurs when uranium-235 (U-235) absorbs a low energy neutron, briefly forming an excited state of U-236 which then splits into xenon-140, strontium-94, and x more neutrons. In isotopic notation form the reactions are:

$$\,^{235}_{92}U + \,^{1}_{0}n \rightarrow \,^{236}_{92}U \rightarrow \,^{140}_{54}Xe + \,^{94}_{38}Sr + x\,^{1}_{0}n$$

How many neutrons are produced in the last reaction?

Solution: The question is asking "What is x in the equation above?" By treating each arrow as an equal sign, the problem is simply asking to balance the last "equation." The mass numbers (A) on either side of each arrow must be equal. Since $235 + 1 = 236$, the first arrow is indeed balanced. To find the number of neutrons solve for x in the last equation (arrow):

$$236 = 140 + 94 + x$$

$$x = 236 - 140 - 94$$

$$= 2$$

So there are two neutrons produced in this reaction. These neutrons are free to go on and be absorbed by more ^{235}U and cause more fissioning, and the process continues in a chain reaction. Note that it really was not necessary to know that the intermediate state was formed.

Some radioactive nuclei may be induced to fission via more than one decay channel or decay mode. For example, a different fission reaction may occur when uranium-235 absorbs a slow neutron and then immediately splits into barium-139, krypton-94, and three more neutrons with no intermediate state:

$$_{92}^{235}U + _{0}^{1}n \rightarrow \, _{92}^{236}U \rightarrow \, _{56}^{139}Ba + _{36}^{94}Kr + 3 \, _{0}^{1}n$$

Radioactive Decay

Radioactive decay is a naturally occurring spontaneous decay of certain nuclei accompanied by the emission of specific particles. It could be classified as a certain type of fission. Radioactive decay problems are of the following general types:

- The integer arithmetic of particle and isotope species

- Radioactive half-life problems

- The use of exponential decay curves and decay constants

Basic Concept

The parent nucleus undergoes radioactive decay to produce the daughter nucleus.

Isotope Decay Arithmetic and Nucleon Conservation

The "reactant" in a radioactive decay is known as the parent isotope while the "product" is the daughter isotope. Let X and Y be the parent and daughter isotopes respectively; a generic radioactive decay can be written as:

$$_{Z}^{A}X \rightarrow \, _{Z'}^{A'}Y + \text{emitted decay particle}$$

A' and Z' are the mass and atomic numbers of Y respectively. In the laboratory, radioactivity is monitored by a Geiger counter or a scintillation counter, which record the number of decay particles emitted.

Alpha Decay. Alpha decay is the emission of an α-particle, which is a ^{4}He nucleus that consists of two protons and two neutrons. The alpha particle is very massive (compared to a beta particle, see below) and doubly charged. Alpha particles interact with matter very easily; hence they do not penetrate shielding (such as lead sheets) very far.

The emission of an α-particle means that the daughter's atomic number Z will be 2 less than the parent's atomic number and the daughter's mass number will be 4 less than the parent's mass number. This can be expressed in two simple equations:

α decay: $\qquad\qquad Z_{daughter} = Z_{parent} - 2$

$$A_{daughter} = A_{parent} - 4$$

The generic alpha decay reaction is then:

$$^A_Z X \rightarrow \, ^{A-4}_{Z-2} Y + \alpha$$

where $\alpha = \, ^4_2 He$.

Example: Suppose a parent X alpha decays into a daughter Y such that:

$$^{238}_{92} X \rightarrow \, ^{A'}_{Z'} Y + \alpha$$

What are the mass number (A′) and atomic number (Z′) of the daughter isotope Y?

Solution: 238 = A′ + 4

A′ = 234

92 = Z′ + 2

Z′ = 90

So A′ = 234 and Z′ = 90. Note that it was not necessary to know the chemical species of the isotopes to do this problem. However, it would have been possible to look at the periodic table and see that Z = 92 means X is uranium-238 ($^{235}_{92} U$) and that Z = 90 means Y is thorium-234 ($^{234}_{90} Th$).

Beta Decay. Beta decay is the emission of a β-particle, which is an electron given the symbol e⁻ or β⁻. Despite the equivalence between electrons and β⁻particles, it is important to realize that these particles are not electrons that would normally be found around the nucleus in a neutral atom, but are products of decay emitted by the nucleus: in particular, when a neutron in the nucleus decays into a proton and an electron. Since an electron is singly charged, and about 1,836 times lighter than a proton, the beta radiation from radioactive decay is more penetrating than alpha radiation.

β decay means that a neutron disappears and a proton takes its place, ejecting a newly-formed electron as a β-particle in the process. Hence, the parent's mass number is unchanged and the parent's atomic number is increased by 1. In other words, the daughter's A is the same as the parent's, and the daughter's Z is one more than the parent's.

β decay: $Z_{daughter} = Z_{parent} + 1$

$A_{daughter} = A_{parent}$

The generic beta decay reaction is:

$$^A_Z X \rightarrow \, ^A_{Z+1} Y + \beta^-$$

Example: Suppose a cobalt-60 nucleus beta-decays:

$$^{60}\text{Co} \rightarrow {}^{A'}_{Z'}\text{Y} + e^-$$

What is the element Y and what are A' and Z'?

Solution: $60 = A' + 0$

$A' = 60$

Now balance the atomic numbers, taking into account that cobalt has 27 protons (you learn this by consulting the periodic table) and that there is one more proton on the right-hand side:

$$27 = Z' - 1$$

$$Z' = 28$$

By looking at the periodic table, one finds that Z' = 28 is nickel:

$$Y = {}^{60}_{28}\text{Ni}$$

Gamma Decay. Gamma decay is the emission of γ-rays, which are high energy photons. They carry no charge and simply lower the energy of the emitting (parent) nucleus without changing the mass number or the atomic number. In other words, the daughter's A is the same as the parent's and the daughter's Z is the same as the parent's.

γ decay: $$Z_{parent} = Z_{daughter}$$

$$A_{parent} = A_{daughter}$$

The generic gamma decay reaction is thus:

$$^A_Z\text{X}^* \rightarrow {}^A_Z\text{X} + \gamma$$

where the asterisk on the parent nucleus designates that it is in an unstable, high-energy state.

Positron Emission. Positron emission occurs when a positively-charged particle known as a positron is emitted. A positron is most conveniently thought of (and often referred to) as an antielectron: it has the same (negligible) mass but has a positive rather than a negative charge. The positron is given the symbol e^+ or β^+. Positron emission is sometimes referred to as β^+ decay, as distinguished from the kind of beta-decay (or β^- decay) discussed above.

Basic Concept

γ particles are high energy photons. γ-decay releases energy but does not change A or Z.

In positron decay, a proton splits into a positron and a neutron. Therefore, a β^+ decay means that the parent's mass number is unchanged and the parent's atomic number is decreased by 1. In other words, the daughter's A is the same as the parent's, and the daughter's Z is one less than the parent's. In equation form:

β^+ decay:

$$Z_{daughter} = Z_{parent} - 1$$

$$A_{daughter} = A_{parent}$$

The generic positive beta decay or positron emission is:

$$^A_Z X \rightarrow ^{\ \ A}_{Z+1} Y + \beta^+$$

Example: Suppose a parent isotope $^A_Z X$ emits a β^+ and turns into an excited state of the isotope $^{A'}_{Z'} Y^*$, which then γ-decays to $^{A''}_{Z''} Y$, which in turn α-decays to $^{A'''}_{Z'''} W$. If W is ^{60}Fe, what is $^A_Z X$?

Solution: Since the final daughter in this chain of decay is given, it will be necessary to work backward through the reactions. By looking at the periodic table one finds that W = Fe means $Z''' = 26$; hence the last reaction is the following α decay:

$$^{A''}_{Z''} Y \rightarrow ^{60}_{26} Fe + ^4_2 He$$

By balancing the atomic and mass numbers you find:

$$Z'' = 26 + 2 = 28$$

$$A'' = 60 + 4 = 64$$

The second-to-last reaction is a γ decay that simply releases energy from the nucleus but does not alter the atomic number or the mass number of the parent. That is: $Z' = Z'' = 28$ and $A' = A'' = 64$. So the second reaction is:

$$^{64}_{28} Y^* \rightarrow ^{64}_{28} Y + \gamma$$

The first reaction was a β^+ decay (positron emission) that must have looked like:

$$^A_Z X \rightarrow ^{64}_{28} Y^* + \beta^+$$

Again, balance the atomic numbers:

$$Z = 28 + 1 = 29$$

You carry out a balancing of mass numbers by taking into account that a proton has disappeared on the left and reappeared as a neutron on the right, leaving mass number unchanged:

$$A = 64 + 0 = 64$$

By looking at the periodic table you find that $Z = 29$ means that X is Cu. Since $A = 64$, that means that the solution is:

$$^A_Z X = ^{64}_{29} Cu$$

While the problem did not ask for it, it is possible again to look at the periodic table to find that $Z' = Z'' = 28$ means $Y^* = Y = Ni$.

Electron Capture. Certain unstable radionuclides are capable of capturing a core electron that combines with a proton to form a neutron. The atomic number is now one less than the original, but the mass number remains the same. Electron capture is a rare process that is perhaps best thought of as an inverse β^- decay.

Radioactive Decay Half-Life ($\tau_{1/2}$)

In the chapter on kinetics, we discussed radioactive decay as an example of first-order kinetics, and introduced the concept of the half-life ($\tau_{1/2}$): In a collection of a great many identical radioactive isotopes, the half-life of the sample is the time it takes for half of the sample to decay.

Example: If the half-life of a certain isotope is 4 years, what fraction of a sample of that isotope will remain after 12 years?

Solution: If 4 years is one half-life, then 12 years is three half-lives. During the first half-life—the first 4 years—half of the sample will have decayed. During the second half-life (years 5 to 8), half of the remaining half will decay, leaving one-fourth of the original. During the third and final period (years 9 to 12), half of the remaining fourth will decay, leaving one-eighth of the original sample. Thus the fraction remaining after 3 half-lives is $(1/2)^3$ or $(1/8)$.

Basic Concept

Fraction of original nuclei remaining after n half-lives $= (\frac{1}{2})^n$

Fraction of nuclei that has decayed away after n half-lives $= 1 - (\frac{1}{2})^n$

The fact that different radioactive species have different characteristic half-lives is what enables scientists to determine the age of organic materials. ^{14}C, for example, is generated from nuclear reactions induced by high-energy cosmic rays from outer space. There is therefore always a certain fraction of this isotope in the carbon found on Earth. Living things, like trees and animals, are constantly exchanging carbon with the environment, and thus will have the same ratio of carbon-14 to carbon-12 within them as the atmosphere. Once they die, however, they stop incorporating carbon from the environment, and start to lose carbon-14 because of its radioactivity. It undergoes a β-decay mechanism:

$$^{14}_{6}C \rightarrow {}^{14}_{7}N + \beta^-$$

The longer the species has been dead, the less carbon-14 it will still have: for example, if the ratio of ^{14}C to ^{12}C is half of that of the atmosphere, then we would conclude that the species existed about one half-life of ^{14}C ago.

Exponential Decay

Let n be the number of radioactive nuclei that have not yet decayed in a sample. It turns out that the rate at which the nuclei decay ($\Delta n/\Delta t$) is proportional to the number that remain (n). This suggests the equation:

$$\frac{\Delta n}{\Delta t} = -\lambda n$$

where λ is known as the decay constant which is just the rate constant for the decay reaction. The solution of this equation tells us how the number of radioactive nuclei changes with time. The solution is known as an exponential decay:

$$n = n_0 e^{-\lambda t}$$

where n_0 is the number of undecayed nuclei at time t = 0. The decay constant is related to the half-life by $\lambda = \dfrac{\ln 2}{\tau_{1/2}} = \dfrac{0.693}{\tau_{1/2}}$.

Example: If at time t = 0 there is a 2 mole sample of radioactive isotopes of decay constant 2 (hour)$^{-1}$, how many nuclei remain after 45 minutes?

Solution: Since 45 minutes is 3/4 of an hour the exponent is:

$$\lambda t = 2 \left(\frac{3}{4}\right) = \frac{3}{2}$$

The exponential factor will be a number smaller than 1:

$$e^{-\lambda t} = e^{-3/2} = 0.22$$

So only 0.22 or 22% of the original two-mole sample will remain. To find n_0 we can multiply the number of moles we have by the number of particles per mole (Avogadro's number):

$$n_0 = 2(6.02 \times 10^{23}) = 1.2 \times 10^{24}$$

From the equation that describes exponential decay, you can calculate the number that remain after 45 minutes:

$$n = n_0 e^{-\lambda t}$$

$$= (1.2 \times 10^{24})(0.22)$$

$$= 2.6 \times 10^{23} \text{ particles}$$

Nuclear Chemistry Review Problems

1. Element $^{102}_{20}\Omega$ is formed as a result of 3 α and 2 β^- decays. Which of the following is the parent element?

   A. $^{90}_{16}\Gamma$

   B. $^{114}_{24}\Phi$

   C. $^{114}_{28}\Theta$

   D. $^{12}_{8}\Delta + ^{90}_{12}\Xi$

2. Element X is radioactive and decays via α decay with a half-life of 4 days. If 12.5% of an original sample of element X remains after N days, determine N.

3. A patient undergoing treatment for thyroid cancer receives a dose of radioactive iodine (^{131}I) which has a half-life of 8.05 days. If the original dose contained 12 mg of ^{131}I, what mass of ^{131}I remains after 16.1 days?

4. In an exponential decay, if the natural logarithm of the ratio of intact nuclei (n) at time t to the intact nuclei at time t = 0 (n_0) is plotted against time, what does the slope of the graph correspond to?

Turn the page
for answers and explanations
to the Review Problems.

Solutions to Review Problems

1. B

Emission of three alpha particles by the (as yet unknown) parent results in the following changes:

> Mass number: decreases by 3×4 or 12 units
>
> Atomic number: decreases by 3×2 or 6 units

Emission of two negative betas results in the following changes:

> Mass number: no change
>
> Atomic number: increases by 2×1 or 2 units

So the net change is: Mass number decreases by 12 units; atomic number decreases by 4 units. Therefore, the mass number of the parent is 12 greater than 102, or 114; the atomic number of the parent is 4 greater than 20, or 24. The only choice given with these numbers is B.

2. N = 12 days

Since the half-life of element X is 4 days, then 50% of an original sample remains after 4 days, 25% of an original sample remains after 8 days, and 12.5% of an original sample remains after 12 days. Thus N = 12 days. A different approach is to set $(1/2)^n = 0.125$ where n is the number of half-lives that have elapsed. Solving for n, gives n = 3. Thus 3 half-lives have elapsed, so given the half-life is 4 days, we have that N = 12 days.

3. 3 mg

Given that the half-life of ^{131}I is 8.05 days, we have that 2 half-lives have elapsed after 16.1 days, which means that 25% of the original amount of ^{131}I is still present. Thus, only 25% of the original number of ^{131}I nuclei remain, which also means that only 25% of the original mass of ^{131}I remains. Since the original dose contained 12 mg of ^{131}I, only 3 mg remain after 16.1 days.

4. $-\lambda$

The expression $n = n_0 e^{-\lambda t}$ is equivalent to $n/n_0 = e^{-\lambda t}$. Taking the natural logarithm of both sides of the latter expression you find:

$$\ln(n/n_0) = -\lambda t$$

From this expression it is clear that plotting $\ln(n/n_0)$ versus t will give a straight line of slope $-\lambda$.

CHAPTER 16

ORGANIC CHEMISTRY

Organic chemistry is the study of compounds containing the element carbon. This covers a wide range of compounds including proteins, alcohols, steroids, sugars, and compounds found in petroleum, just to name a few. The reason we can study them as facets of one subject is because of the unifying way we can look at them through the bonding properties of carbon.

Carbon has four valence electrons and thus would like to have four more to complete its octet. Because of its moderate electronegativity, it tends to form covalent rather than ionic bonds. See chapter 5, which discusses Bonding and Molecular Structure, for discussions of hybridization, σ (sigma) and π (pi) orbitals, etcetera.

Basic Concept

Organic Chemistry is the study of compounds containing carbon.

sp³ Hybridized

The carbon atom forms four single (σ) bonds. According to VSEPR theory, the four bonds would be directed towards the corners of a tetrahedron, with a bond angle of 109.5° between any two. Despite this geometry, it is often drawn simply as a Lewis structure with the four bonds directed towards the corners of a two-dimensional diamond. It is important to realize that this Lewis structure does not give the right three-dimensional structure but only a conventional shorthand.

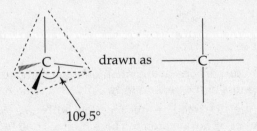

sp² Hybridized

The carbon atom forms a double bond and two single bonds, arranged in the same plane about 120° apart. The two single bonds are σ bonds while the double bond consists of one σ and one π bond. The π bond is formed from the interactions of unhybridized p orbitals which in the case below would be pointing in a direction perpendicular to the page.

$$\underset{120°}{\diagup C =}$$

sp Hybridized

The carbon atom forms a triple bond and a single bond, arranged linearly (180° apart). The triple bond consists of one σ and two π bonds.

$$— C \equiv$$

Different classes of organic compounds are named based on the nature of the bonds and the elements (besides carbon) that are present. Just to give a few examples, a carbon atom triple-bonded to a nitrogen results in a nitrile, a carbon atom double-bonded to an oxygen atom gives a carbonyl. Sometimes more than one such *functional groups* are present in a molecule. There are rules for systematically assigning names to organic compounds, known as the IUPAC system. Below are some of the more common functional groups or compounds:

Hydrocarbons

Hydrocarbons are compounds that contain only carbon and hydrogen atoms. Depending on the kinds of bonds found between the carbon atoms (only single bonds can exist between carbon and hydrogen), hydrocarbons can be classified into one of four classes: alkanes, alkenes, alkynes, and aromatics.

Alkanes

Alkanes are hydrocarbons that contain only single bonds. They all have a molecular formula of the general form $CnH_{2n + 2}$, where n is some positive integer. They are all named by attaching the suffix *-ane* to a prefix that indicates the number of carbon atoms. These prefixes will be used again in the naming of other hydrocarbons and it is therefore worth knowing at least a few:

A Closer Look

Alkanes are nonpolar and therefore tend to have low melting and boiling points in general.

# of C Atoms	Prefix	Name of Alkane	Molecular Formula
1	*meth-*	methane	CH_4
2	*eth-*	ethane	C_2H_6
3	*prop-*	propane	C_3H_8
4	*but-*	butane	C_4H_{10}
5	*pent-*	pentane	C_5H_{12}
6	*hex-*	hexane	C_6H_{14}

The simplest alkane is methane, CH_4. Ethane, the next in the series, has the molecular formula C_2H_6, but this does not convey its structure:

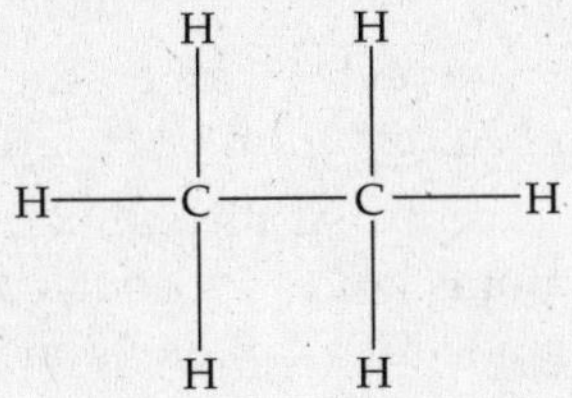

(Keep in mind the convention about drawing the Lewis structures of sp^3 hybridized carbon atoms discussed above.) The formula for ethane is more informatively (and commonly) written as CH_3CH_3. This tells us unambiguously that each carbon atom is attached to three hydrogen atoms. This kind of notation is known as a condensed structural formula. Similarly, the condensed structural formula for the next alkane, propane, is $CH_3CH_2CH_3$.

For alkanes with four or more carbons, there are different ways that the carbon atoms can be connected to each other, which makes the condensed structural formula all that more useful. Butane, for example, can have either one of the following structures:

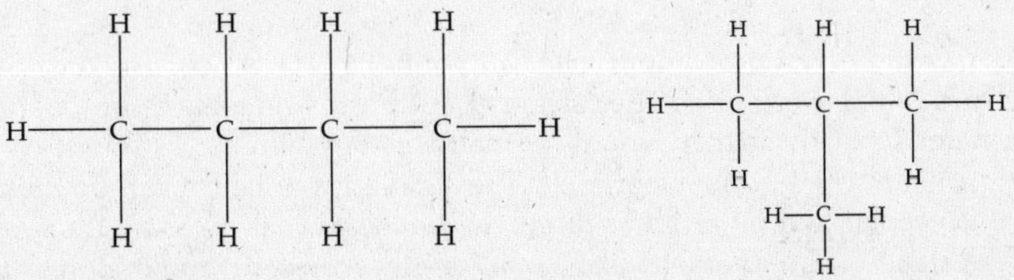

These two structures have the same molecular formula: C_4H_{10} (and hence the same molecular weight), but have different physical properties, such as boiling point and melting point. They are known as isomers (more specifically structural isomers) of each other. One can also say that the compound butane has two isomers: The top one is known as a straight-chain alkane for obvious reasons and is given the name *n*-butane, and the bottom one is a branched alkane and is known as isobutane.

The number of isomers increase for each alkane as the number of carbon atoms increases. Pentane, for example, has three isomers, while hexane has five and decane, with 10 carbon atoms, has 75.

Alkanes, especially straight-chain or *n*-alkanes, are the major constituents of petroleum. Since different alkanes, not to mention their respective isomers, have different boiling and melting points, the different alkane components in petroleum can be separated by distillation. Those compounds with the lowest boiling points would vaporize first; one can trap these vapors and condense them and thus achieve a separation of the more volatile from the less volatile components.

Alkenes

Alkenes are hydrocarbons involving carbon-carbon double bonds. They possess a molecular formula of the form CnH_2n. They are named using the same scheme as alkanes, except that the suffix used is *-ene*. Also, since it takes at least two carbon atoms to form a double bond, the smallest alkene is ethene, C_2H_4, which contains two carbon atoms.

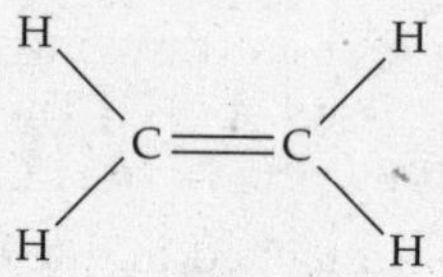

$$\begin{array}{ccc} H & & H \\ & \diagdown \quad \diagup & \\ & C = C & \\ & \diagup \quad \diagdown & \\ H & & H \end{array}$$

ethene

Alkynes

Alkynes are hydrocarbons involving carbon-carbon triple bonds. They follow the same naming scheme as alkanes and alkenes, but use the suffix *-yne*. Alkenes and alkynes are said to be unsaturated, while alkanes are said to be saturated.

We have only considered noncyclic compounds so far. Alkanes, alkenes, or alkynes can also be cyclic: The carbon atoms form a ring. Such compounds are named exactly as they would normally be, but with the additional prefix *cyclo-* attached at the beginning. The smallest number of carbon atoms that is needed to form a ring is three; the smallest cyclic alkane is therefore cyclopropane. The structures of cyclohexane, cyclohexene, and cyclohexyne are shown below.

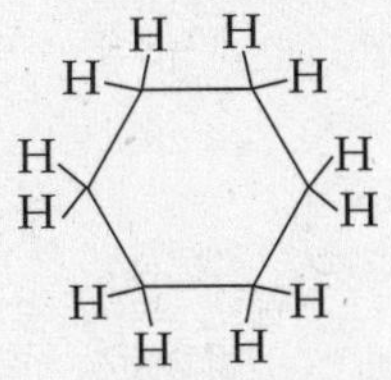

cyclohexane
C_6H_{12}

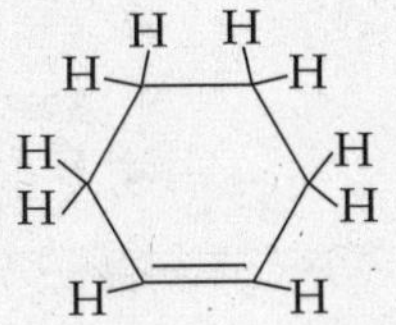

cyclohexane
C_6H_{10}

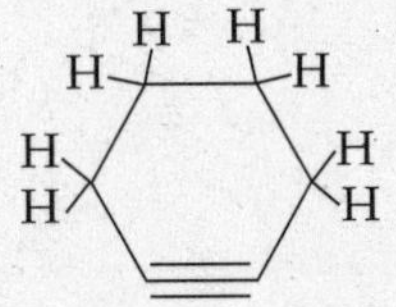

cyclohexane
C_6H_8

The carbon atoms are not explicitly drawn but occupy the positions where the bonds join together. This is a common convention in organic chemistry. Notice how the molecular formulas for cyclic compounds do not follow the generic formulas given above for alkanes, alkenes and alkynes: The extra carbon-carbon bond formed in making a ring upsets the ratio of carbon to hydrogen atoms. One final warning is that as Lewis structures, the drawings do not accurately reflect the three-dimensional appearance of the compounds; cyclohexane, for example, is not a planar hexagon but instead adopts a "chairlike" conformation in its most stable state.

Aromatics

Certain unsaturated cyclic hydrocarbons are known as aromatics. We need not concern ourselves with exactly what makes a compound aromatic, but all such compounds have in common a cyclic, planar structure and possess a higher degree of stability (a lower enthalpy of formation) than expected. This extra stability comes from the effects of resonance. The prime example of an aromatic compound is benzene, C_6H_6, which in a Lewis structure is represented as having alternating double and single bonds that can switch their positions to give an equivalent resonance structure:

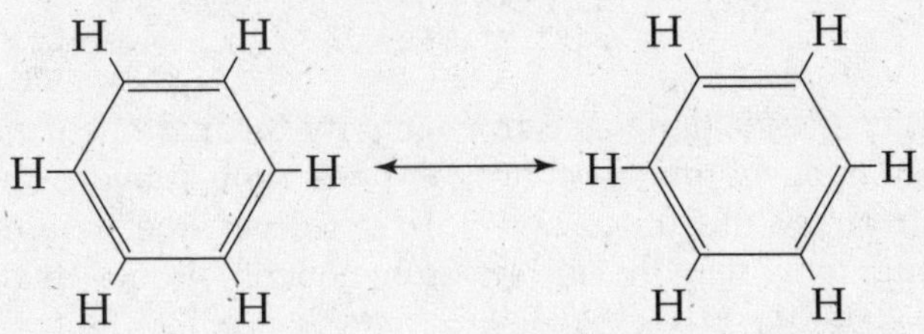

It is important, however, to keep in mind what resonance structures such as these really mean: Benzene does *not* exist as an equilibrium mixture of the two Lewis structures, nor does it flip back and forth between the two structures as time passes. Instead, a benzene molecule is always in a state that is intermediate between the two structures that cannot be accurately captured by a normal Lewis structure: Every carbon-carbon bond in the molecule has characteristics intermediate between those of a single and those of a double bond. If we have to try to depict this using a Lewis structure, we would draw the following:

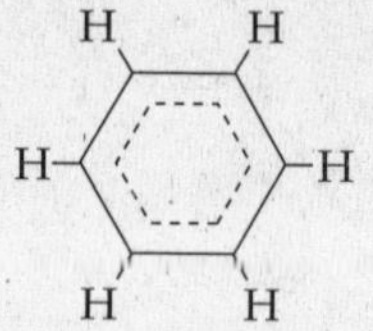

where the dashed lines indicate a partial π bond. Each of the six carbon atoms is sp^2 hybridized, and therefore each has an unhybridized p orbital that is coming out of (and going into) the plane of the paper. These orbitals interact through π bonding and form the partial π bonds: partial because there are six electrons total that are part of this "π cloud" (one from each carbon atom) and therefore there are enough electrons to form only three such bonds that need to be shared among six atoms.

Oxygen-Containing Compounds

Organic compounds that include oxygen in addition to carbon and hydrogen include alcohols, ethers, carbohydrates, and carbonyl compounds such as aldehydes, ketones, esters, and carboxylic acids.

Alcohols

Alcohols contain the functional group –OH, sometimes called the hydroxyl group. Ethanol, for example, can be considered a derivative of ethane, with the hydroxyl group in the place of a hydrogen atom:

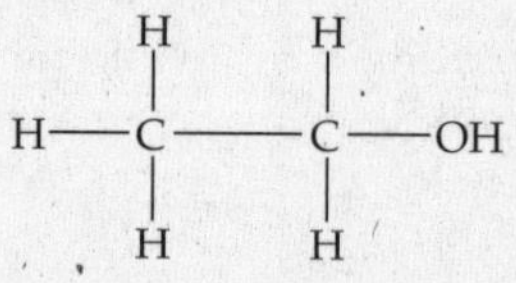

Because oxygen has such high electronegativity, the hydrogen attached to it (the hydrogen of the hydroxyl group) can participate in hydrogen bonding. The diagram below illustrates how this can occur; the R represents the organic group to which the alcohol functionality is attached (e.g., for ethanol, $R = CH_3CH_2-$):

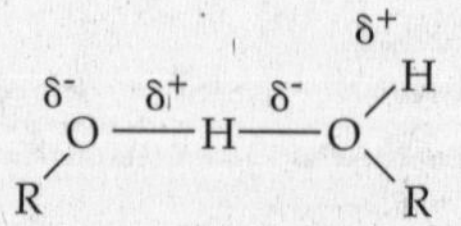

Hydrogen bonding is a strong intermolecular attractive force, and this causes alcohols to have boiling points that are significantly higher than those of the analogous hydrocarbons. The boiling point of propane ($CH_3CH_2CH_3$), for example, is –42.1 °C, while that of propanol ($CH_3CH_2CH_2OH$) is 97.4 °C.

Ethylene glycol, which is ethane with two hydroxyl groups attached (one to each carbon atom) is used as antifreeze because its high boiling point makes it a good nonvolatile solute when exploiting the colligative property of freezing point depression.

Ethers

Ethers are compounds containing a C-O-C bond. Following the same convention above where R is used to designate some organic group, the generic formula for an ether is ROR'. Examples of ethers include:

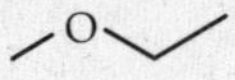

methoxyethane

(ethyl methyl ether)

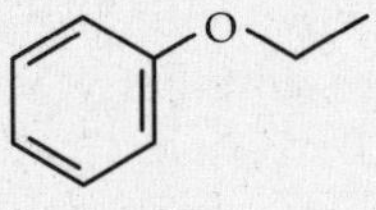

ethoxybenzene

(ethyl phenyl ether)

In the diagrams above, we have extended the common practice in organic chemistry where the carbon atoms are not explicitly depicted but are assumed to occupy the positions where there is a "kink" in the structure drawn. (Previously we have only done this in cyclic compounds.)

Ethers are not capable of hydrogen bonding and therefore have low boiling points. They are also relatively inert and are frequently used as solvents in organic chemistry reactions.

Carbonyl Compounds

Carbonyl compounds are those containing a carbon-oxygen double bond, the carbonyl bond. The generic structure for such a compound is:

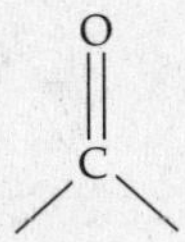

Depending on what exactly the two other groups attached to the carbon atom are, one can be more specific in naming the class of compounds to which the molecule belongs. Below are some examples:

Aldehydes

The carbon atom is attached to a hydrogen atom on one side and an R group (which may be another H atom in the case of formaldehyde) on the other.

Ketones

The carbon atom is attached to two R groups that are not hydrogen atoms.

Esters

The carbonyl carbon atom in an ester is bonded to an R group on one side and an OR′ group (that is not OH) on the other.

Carboxylic acids

The carbon atom is attached to an R group and a hydroxyl (–OH) group. Like alcohols that also contain the hydroxyl group, carboxylic acids can participate in hydrogen bonding. These compounds are weak acids (weak compared to inorganic acids like HCl) because the hydroxyl hydrogen can be donated as a proton. Fatty acids, for example, are carboxylic acids with a long hydrocarbon chain (the R group attached to the carbonyl carbon is a long hydrocarbon). After the donation of a proton, the carboxylic group left behind has a negative charge and is thus attracted to polar medium. The hydrocarbon chain, on the other hand, is nonpolar. In aqueous solution, therefore, these long carboxylate molecules (conjugate bases of carboxylic acids) arrange themselves into spherical structures known as micelles, in which the charged "heads" (the –COO⁻ groups) are exposed to the water while the organic chains are inside the sphere. Nonpolar molecules such as grease can dissolve in the hydrocarbon interior of the spherical micelle. This is why these molecules which are salts of long-chain carboxylic acids are called soaps.

Carbohydrates

Carbohydrates are so named because they are "hydrates of carbon": They have the general formula $Cn(H_2O)m$, where n is not necessarily the same as m. They serve as chemical sources of energy for most organisms.

Simple sugars or monosaccharides are carbohydrates and can be classified according to the number of carbons they possess, and have the general formula $Cn(H_2O)n$ or CnH_2nOn. Trioses, tetroses, pentoses, and hexoses have three, four, five, and six carbon atoms respectively. Glucose and fructose are the two most common examples of hexoses:

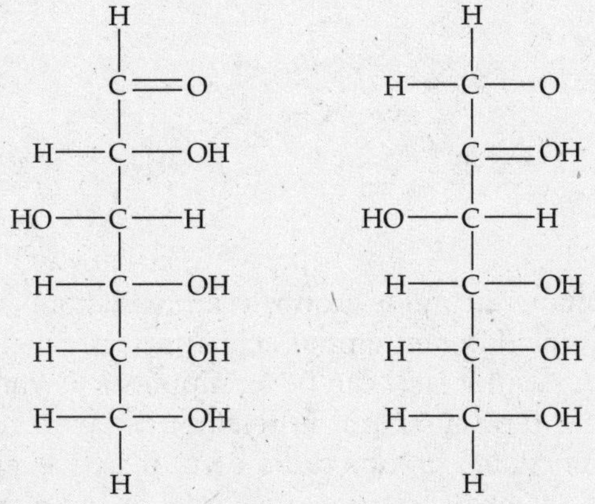

glucose   fructose

As you can see, then, these compounds also contain the carbonyl group. In particular, glucose has an aldehyde functionality while fructose contains a ketone functionality. In solution, however, the straight-chain forms of these sugars (structures shown above) exist in equilibrium with a cyclic form. In fact, it is in this cyclic form that most of the interesting chemistry occurs. Linking of monosaccharides to form disaccharides and polysaccharides, for example, takes place between cyclic sugars. An example of such a reaction is shown below:

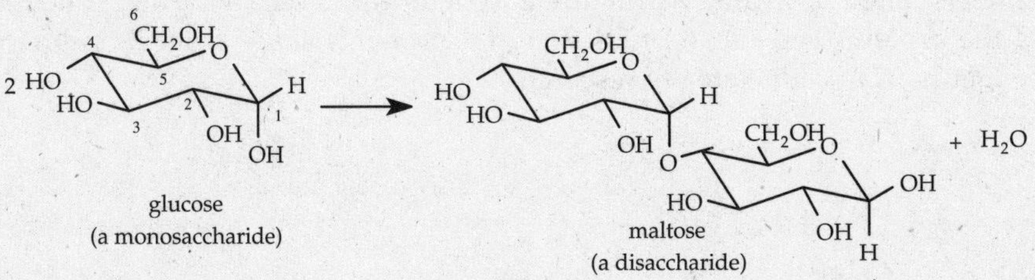

glucose
(a monosaccharide)

maltose
(a disaccharide)

$+ H_2O$

The cyclic form of the five-carbon sugar ribose is a component of nucleotides, which are the building blocks of nucleic acids in DNA and RNA.

Amino Acids

Nitrogen-containing compounds are another large class of organic compounds. The most important nitrogen-containing functional group is the amine group, $-NH_2$, which is found in amino acids, the basic building blocks of proteins.

Basic Concept

Proteins are made up of amino acids.

An amino acid contains an amine group, a carboxyl group, and a side group R that is different for different amino acids. In fact, what R is essentially defines what amino acid it is. It can be as simple as just another hydrogen atom (in which case the amino acid is glycine), or it can be more complex, with distinctive functional groups of its own. It may even contain atoms other than carbon, oxygen and nitrogen: Cysteine and methionine are two amino acids that contain sulfur atoms in their side chains.

There are 20 naturally occurring amino acids, and these amino acids can be joined together by bonds called peptide bonds to form small chains of amino acids known as peptides. Two amino acids joined together form a dipeptide, three form a tripeptide, and many amino acids linked together form a polypeptide. At some point (the exact boundary is not well defined), the polypeptide becomes long enough and we call it a protein. Proteins serve many diverse functions in biological systems, acting as enzymes, hormones, elements of cell structure, etcetera. The protein's amino acid sequence—the precise ordering and identity of each amino acid in the protein—is called its primary structure and determines the shape and function of the protein. The actual prediction of a protein's shape from its primary sequence is an active area of research.

Organic Chemistry Review Problems

For questions 1–5, use the following diagram:

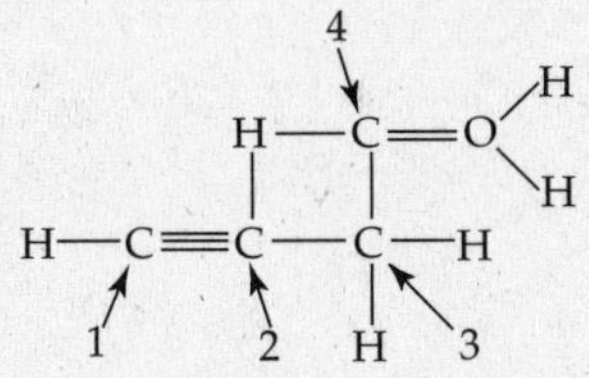

1. What is the hybridization of the carbon atom labeled 1?

2. What is the hybridization of the carbon atom labeled 2?

3. What is the hybridization of the carbon atom labeled 3?

4. What is the hybridization of the carbon atom labeled 4?

5. How many σ and π bonds are there total in the molecule?

6. Which of the following statements is true of ethene?

   A. Both carbon atoms are sp^2 hybridized and the molecule is planar.

   B. Both carbon atoms are sp^2 hybridized and all bond angles are approximately 109.5°.

   C. One carbon atom is sp hybridized while the other is sp^2.

   D. Both carbon atoms are sp^3 hybridized and all bond angles are approximately 109.5°.

   E. Both carbon atoms are sp hybridized and the molecule is planar.

7. Which of the following is the formula for a noncyclic, saturated hydrocarbon?

   A. C_7H_{12}

   B. C_7H_{14}

   C. C_7H_{16}

   D. C_7H_{18}

   E. C_7H_{12}

8.   What functional groups are present in the compound below?

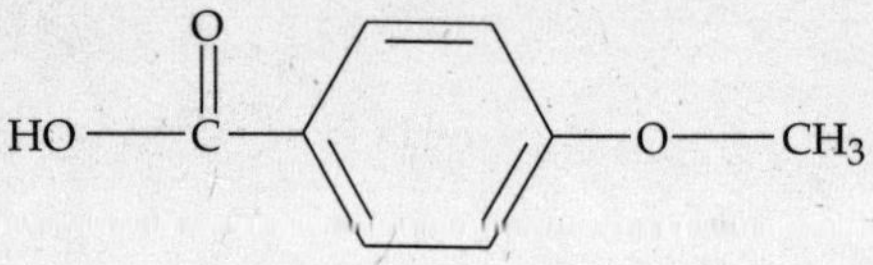

A.   Ester and ether

B.   Ester and amine

C.   Ester and carboxylic acid

D.   Ether and carboxylic acid

E.   Ether and ketone

Turn the page
for answers and explanations
to the Review Problems.

Solutions to Review Problems

1.

sp: the carbon forms a single bond and a triple bond.

2.

sp: the carbon forms a single bond and a triple bond.

3.

sp3: the carbon forms four single bonds.

4.

sp2: the carbon forms a double and two single bonds.

5.

There are 8 single bonds, one double bond and one triple bond. Each single bond gives a σ bond; a double bond gives one σ and one π bond; a triple bond gives one σ and two π bonds. So there is a total of $8 + 1 + 1 = 10$ σ bonds, and 1 (from double bond) + 2 (from triple bond) = 3 π bonds in the molecule.

6.   A

The two carbon atoms in ethene are bonded to each other via a double bond. They are thus both sp^2 hybridized and the three attached groups each has will be arranged in a planar configuration roughly 120° apart, since that will minimize the electron-pair repulsion. Choice B is wrong because of the bond angle. The other choices are wrong because both carbons are sp^2 hybridized; sp hybridization is found on carbon atoms with two adjacent double bonds (allenes) or, more commonly, a triple bond and a single bond (alkynes), while sp^3 hybridization is found on saturated carbon atoms.

7.   C

A saturated hydrocarbon is one that contains only single bonds. If it is noncyclic, its formula will conform to that of alkanes: CnH_{2n+2}. Only choice C satisfies this.

8.   D

The –COOH group is the carboxylic acid group. The ether functionality is represented by the formula R–O–R', and in this case R is the ring and R' is the methyl (CH_3) group.

READY, SET, GO!

CHAPTER 17

STRESS MANAGEMENT

The countdown has begun. Your date with THE TEST is looming on the horizon. Anxiety is on the rise. The butterflies in your stomach have gone ballistic. Perhaps you feel as if the last thing you ate has turned into a lead ball. Your thinking is getting cloudy. Maybe you think you won't be ready. Maybe you already know your stuff, but you're going into panic mode anyway. Worst of all, you're not sure of what to do about it.

Don't freak! It is possible to tame that anxiety and stress—before and during the test. We'll show you how. You won't believe how quickly and easily you can deal with that killer anxiety.

Make the Most of Your Prep Time

Lack of control is one of the prime causes of stress. A ton of research shows that if you don't have a sense of control over what's happening in your life, you can easily end up feeling helpless and hopeless. So, just having concrete things to do and to think about—taking control—will help reduce your stress. This section shows you how to take control during the days leading up to taking the test.

Identify the Sources of Stress

In the space provided, jot down anything you identify as a source of your test-related stress. The idea is to pin down that free-floating anxiety so that you can take control of it. Here are some common examples to get you started:

- I always freeze up on tests.

- I'm nervous about the subject matter.

- I need a good/great score to go to Acme College.

- My older brother/sister/best friend/girl- or boyfriend did really well. I must match their scores or do better.

- My parents, who are paying for school, will be really disappointed if I don't test well.

Avoid Must-y Thinking

Let go of "must-y" thoughts, those notions that you must do something in a certain way—for example, "I must get a great score, or else!" "I must meet Mom and Dad's expectations."

Don't Do It in Bed

Don't study on your bed, especially if you have problems with insomnia. Your mind might start to associate the bed with work, and make it even harder for you to fall asleep.

Think Good Thoughts

Create a set of positive but brief affirmations and mentally repeat them to yourself just before you fall asleep at night. (That's when your mind is very open to suggestion.) You'll find yourself feeling a lot more positive in the morning. Periodically repeating your affirmations during the day makes them more effective.

- I'm afraid of losing my focus and concentration.

- I'm afraid I'm not spending enough time preparing.

- I study like crazy, but nothing seems to stick in my mind.

- I always run out of time and get panicky.

- I feel as though thinking is becoming like wading through thick mud.

Sources of Stress

_____   _____

_____   _____

_____   _____

_____   _____

Take a few minutes to think about the things you've just written down. Then rewrite them in some sort of order. List the statements you most associate with your stress and anxiety first, and put the least disturbing items last. Chances are, the top of the list is a fairly accurate description of exactly how you react to test anxiety, both physically and mentally. The later items usually describe your fears (disappointing Mom and Dad, looking bad, etcetera). As you write the list, you're forming a hierarchy of items so you can deal first with the anxiety provokers that bug you most. Very often, taking care of the major items from the top of the list goes a long way toward relieving overall testing anxiety. You probably won't have to bother with the stuff you placed last.

Strengths and Weaknesses

Take one minute to list the areas of the test that you are good at. They can be general or specific. Put down as many as you can think of, and if possible, time yourself. Write for the entire time; don't stop writing until you've reached the one-minute stopping point.

Strong Test Subjects

_____   _____

_____   _____

_____   _____

_____   _____

Next, take one minute to list areas of the test you're not so good at, just plain bad at, have failed at, or keep failing at. Again, keep it to one minute, and continue writing until you reach the cutoff. Don't be afraid to identify and write down your weak spots! In all probability, as you do both lists, you'll find you are strong in some areas and not so strong in others. Taking stock of your assets and liabilities lets you know the areas you don't have to worry about, and the ones that will demand extra attention and effort.

Weak Test Subjects

_____   _____

_____   _____

_____   _____

_____   _____

Facing your weak spots gives you some distinct advantages. It helps a lot to find out where you need to spend extra effort. Increased exposure to tough material makes it more familiar and less intimidating. (After all, we mostly fear what we don't know and are probably afraid to face.) You'll feel better about yourself because you're dealing directly with areas of the test that bring on your anxiety. You can't help feeling more confident when you know you're actively strengthening your chances of earning a higher over-all test score.

Now, go back to the "good" list, and expand it for two minutes. Take the general items on that first list and make them more specific; take the specific items and expand them into more general conclusions. Naturally, if anything new comes to mind, jot it down. Focus all of your attention and effort on your strengths. Don't underestimate yourself or your abilities. Give yourself full credit. At the same time, don't list strengths you don't really have; you'll only be fooling yourself.

Very Superstitious

Stress expert Stephen Sideroff, Ph.D., tells of a client who always stressed out before, during, and even after taking tests. Yet, she always got outstanding scores. It became obvious that she was thinking superstitiously—sub-consciously believing that the great scores were a result of her worrying. She didn't trust herself, and believed that if she didn't worry she wouldn't study hard enough. Sideroff convinced her to take a risk and work on relaxing before her next test. She did, and her test results were still as good as ever—which broke her cycle of superstitious thinking.

Stress Tip

Don't work in a messy or cramped area. Before you sit down to study, clear yourself a nice, open space. And, make sure you have books, paper, pencils—whatever tools you will need—within easy reach before you sit down to study.

Expanding from general to specific might go as follows. If you listed "grammar" as a broad topic you feel strong in, you would then narrow your focus to include areas of this subject about which you are particularly knowledgeable. Your areas of strength might include a good ear for idiomatic English, a strong sense of logical construction, ability to spot faulty verb forms, etc.

Whatever you know comfortably goes on your "good" list. Okay. You've got the picture. Now, get ready, check your starting time, and start writing down items on your expanded "good" list.

Strong Test Subjects: An Expanded List

_____   _____

_____   _____

_____   _____

_____   _____

After you've stopped, check your time. Did you find yourself going beyond the two minutes allotted? Did you write down more things than you thought you knew? Is it possible you know more than you've given yourself credit for? Could that mean you've found a number of areas in which you feel strong?

You just took an active step toward helping yourself. Notice any increased feelings of confidence? Enjoy them.

Here's another way to think about your writing exercise. Every area of strength and confidence you can identify is much like having a reserve of solid gold at Fort Knox. You'll be able to draw on your reserves as you need them. You can use your reserves to solve difficult questions, maintain confidence, and keep test stress and anxiety at a distance. The encouraging thing is that every time you recognize another area of strength, succeed at coming up with a solution, or get a good score on a test, you increase your reserves. And, there is absolutely no limit to how much self-confidence you can have or how good you can feel about yourself.

Imagine Yourself Succeeding

This next little group of exercises is both physical and mental. It's a natural follow-up to what you've just accomplished with your lists.

First, get yourself into a comfortable sitting position in a quiet setting. Wear loose clothes. If you wear glasses, take them off. Then, close your eyes and breathe in a deep, satisfying breath of air. Really fill your lungs until your rib cage is fully expanded and you can't take in any more. Then, exhale the air completely. Imagine you're blowing out a candle with your last little puff of air. Do this two or three more times, filling your lungs to their maximum and emptying them totally. Keep your eyes closed, comfortably but not tightly. Let your body sink deeper into the chair as you become even more comfortable.

With your eyes shut you can notice something very interesting. You're no longer dealing with the worrisome stuff going on in the world outside of you. Now you can concentrate on what happens *inside* you. The more you recognize your own physical reactions to stress and anxiety, the more you can do about them. You might not realize it, but you've begun to regain a sense of being in control.

Let images begin to form on the "viewing screens" on the back of your eyelids. You're experiencing visualizations from the place in your mind that makes pictures. Allow the images to come easily and naturally; don't force them. Imagine yourself in a relaxing situation. It might be in a special place you've visited before or one you've read about. It can be a fictional location that you create in your imagination, but a real-life memory of a place or situation you know is usually better. Make it as detailed as possible, and notice as much as you can.

Stay focused on the images as you sink farther back into your chair. Breathe easily and naturally. You might have the sensations of any stress or tension draining from your muscles and flowing downward, out your feet and away from you.

Take a moment to check how you're feeling. Notice how comfortable you've become. Imagine how much easier it would be if you could take the test feeling this relaxed and in this state of ease. You've coupled the images of your special place with sensations of comfort and relaxation. You've also found a way to become relaxed simply by visualizing your own safe, special place.

Now, close your eyes and start remembering a real-life situation in which you did well on a test. If you can't come up with one, remember a situation in which you did something (academic or otherwise) that you were really proud of—a genuine accomplishment. Make the memory as detailed as possible. Think about the sights, the sounds, the smells, even the tastes associated with this remembered experience. Remember how confident you felt as you accomplished your goal. Now start thinking about the upcoming test. Keep your thoughts and feelings in line with that successful experience. Don't make comparisons between them. Just imagine taking the upcoming test with the same feelings of confidence and relaxed control.

This exercise is a great way to bring the test down to Earth. You should practice this exercise often, especially when the prospect of taking the exam starts to bum you out. The more you practice it, the more effective the exercise will be for you.

Ocean Dumping

Visualize a beautiful beach, with white sand, blue skies, sparkling water, a warm sun, and seagulls. See yourself walking on the beach, carrying a small plastic pail. Stop at a good spot and put your worries and whatever may be bugging you into the pail. Drop it at the water's edge and watch it drift out to sea. When the pail is out of sight, walk on.

Counseling

Don't forget that your school probably has counseling available. If you can't conquer test stress on your own, make an appointment at the counseling center. That's what counselors are there for.

Take a Hike, Pal

When you're in the middle of studying and hit a wall, take a short, brisk walk. Breathe deeply and swing your arms as you walk. Clear your mind. (And, don't forget to look for flowers that grow in the cracks of the sidewalk.)

Play the Music

If you want to play music, keep it low and in the background. Music with a regular, mathematical rhythm—reggae, for example—aids the learning process. A recording of ocean waves is also soothing.

Exercise Your Frustrations Away

Whether it is jogging, walking, biking, mild aerobics, pushups, or a pickup basketball game, physical exercise is a very effective way to stimulate both your mind and body and to improve your ability to think and concentrate. A surprising number of students get out of the habit of regular exercise, ironically because they're spending so much time prepping for exams. Also, sedentary people—this is a medical fact—get less oxygen to the blood and hence to the head than active people. You can live fine with a little less oxygen; you just can't think as well.

Any big test is a bit like a race. Thinking clearly at the end is just as important as having a quick mind early on. If you can't sustain your energy level in the last sections of the exam, there's too good a chance you could blow it. You need a fit body that can weather the demands any big exam puts on you. Along with a good diet and adequate sleep, exercise is an important part of keeping yourself in fighting shape and thinking clearly for the long haul.

There's another thing that happens when students don't make exercise an integral part of their test preparation. Like any organism in nature, you operate best if all your "energy systems" are in balance. Studying uses a lot of energy, but it's all mental. When you take a study break, do something active instead of raiding the fridge or vegging out in front of the TV. Take a 5- to 10-minute activity break for every 50 or 60 minutes that you study. The physical exertion gets your body into the act, which helps to keep your mind and body in sync. Then, when you finish studying for the night and hit the sack, you won't lie there, tense and unable to sleep because your head is overtired and your body wants to pump iron or run a marathon.

One warning about exercise, however: It's not a good idea to exercise vigorously right before you go to bed. This could easily cause sleep onset problems. For the same reason, it's also not a good idea to study right up to bedtime. Make time for a "buffer period" before you go to bed: For 30 to 60 minutes, just take a hot shower, meditate, simply veg out.

The Dangers of Drugs

Using drugs (prescription or recreational) specifically to prepare for and take a big test is definitely self-defeating. (And if they're illegal drugs, you can end up with a bigger problem than the SAT II Chemistry test on your hands.) Except for the drugs that occur naturally in your brain, every drug has major drawbacks—and a false sense of security is only one of them.

You may have heard that popping uppers helps you study by keeping you alert. If they're illegal, definitely forget about it. They wouldn't really work anyway, since amphetamines make it hard to retain information. Mild stimulants, such as coffee, cola, or over-the-counter caffeine pills can sometimes help as you study, since they keep you alert. On the down side, they can also lead to agitation, restlessness, and insomnia. Some people can drink a pot of high-octane coffee and sleep like a baby. Others have one cup and start to vibrate. It all depends on your tolerance for caffeine. Remember, a little anxiety is a good thing. The adrenaline that gets pumped into your bloodstream helps you stay alert and think more clearly. But, too much anxiety and you can't think straight at all.

Instead, go for endorphins—the "natural morphine." Endorphins have no side effects and they're free—you've already got them in your brain. It just takes some exercise to release them. Running around on the basketball court, bicycling, swimming, aerobics, power walking—these activities cause endorphins to occupy certain spots in your brain's neural synapses. In addition, exercise develops staying power and increases the oxygen transfer to your brain. Go into the test naturally.

Take a Deep Breath . . .

Here's another natural route to relaxation and invigoration. It's a classic isometric exercise that you can do whenever you get stressed out—just before the test begins, even *during* the test. It's very simple and takes just a few minutes.

Close your eyes. Starting with your eyes and—without holding your breath—gradually tighten every muscle in your body (but not to the point of pain) in the following sequence:

1. Close your eyes tightly.

2. Squeeze your nose and mouth together so that your whole face is scrunched up. (If it makes you self-conscious to do this in the test room, skip the face-scrunching part.)

3. Pull your chin into your chest, and pull your shoulders together.

4. Tighten your arms to your body, then clench your hands into tight fists.

5. Pull in your stomach.

6. Squeeze your thighs and buttocks together, and tighten your calves.

7. Stretch your feet, then curl your toes (watch out for cramping in this part).

At this point, every muscle should be tightened. Now, relax your body, one part at a time, *in reverse order*, starting with your toes. Let the tension drop out of each muscle. The entire process might take five minutes from start to finish (maybe a couple of minutes during the test). This clenching and unclenching exercise should help you to feel very relaxed.

Cyberstress

If you spend a lot of time in cyberspace anyway, do a search for the phrase *stress management*. There's a ton of stress advice on the Net, including material specifically for students.

Nutrition and Stress: The Dos and Don'ts

Do eat:
- Fruits and vegetables (raw is best, or just lightly steamed or nuked)
- Low-fat protein such as fish, skinless poultry, beans, and legumes (like lentils)
- Whole grains such as brown rice, whole wheat bread, and pastas (no bleached flour)

Don't eat:
- Refined sugar; sweet, high-fat snacks (simple carbohydrates like sugar make stress worse and fatty foods lower your immunity)
- Salty foods (they can deplete potassium, which you need for nerve functions)

The Relaxation Paradox

Forcing relaxation is like asking yourself to flap your arms and fly. You can't do it, and every push and prod only gets you more frustrated. Relaxation is something you don't work at. You simply let it happen. Think about it. When was the last time you tried to force yourself to go to sleep, and it worked?

Enlightenment

A lamp with a 75-watt bulb is optimal for studying. But don't put it so close to your study material that you create a glare.

And Keep Breathing

Conscious attention to breathing is an excellent way of managing test stress (or any stress, for that matter). The majority of people who get into trouble during tests take shallow breaths. They breathe using only their upper chests and shoulder muscles, and may even hold their breath for long periods of time. Conversely, the test taker who by accident or design keeps breathing normally and rhythmically is likely to be more relaxed and in better control during the entire test experience.

So, now is the time to get into the habit of relaxed breathing. Do the next exercise to learn to breathe in a natural, easy rhythm. By the way, this is another technique you can use during the test to collect your thoughts and ward off excess stress. The entire exercise should take no more than three to five minutes.

With your eyes still closed, breathe in slowly and *deeply* through your nose. Hold the breath for a bit, and then release it through your mouth. The key is to breathe slowly and deeply by using your diaphragm (the big band of muscle that spans your body just above your waist) to draw air in and out naturally and effortlessly. Breathing with your diaphragm encourages relaxation and helps minimize tension. Try it and notice how relaxed and comfortable you feel.

CHAPTER 18

THE FINAL COUNTDOWN

Quick Tips for the Days Just Before the Exam

- The best test takers do less and less as the test approaches. Taper off your study schedule and take it easy on yourself. You want to be relaxed and ready on the day of the test. Give yourself time off, especially the evening before the exam. By then, if you've studied well, everything you need to know is firmly stored in your memory banks.

- Positive self-talk can be extremely liberating and invigorating, especially as the test looms closer. Tell yourself things such as, "I choose to take this test" rather than "I have to"; "I will do well" rather than "I hope things go well"; "I can" rather than "I cannot." Be aware of negative, self-defeating thoughts and images and immediately counter any you become aware of. Replace them with affirming statements that encourage your self-esteem and confidence. Create and practice visualizations that build on your positive statements.

- Get your act together sooner rather than later. Have everything (including choice of clothing) laid out days in advance. Most important, know where the test will be held and the easiest, quickest way to get there. You will gain great peace of mind if you know that all the little details—gas in the car, directions, etcetera—are firmly in your control before the day of the test.

- Experience the test site a few days in advance. This is very helpful if you are especially anxious. If at all possible, find out what room your part of the alphabet is assigned to, and try to sit there (by yourself) for a while. Better yet, bring some practice material and do at least a section or two, if not an entire practice test, in that room. In this situation, familiarity doesn't breed contempt, it generates comfort and confidence.

Dress for Success

On the day of the test, wear loose layers. That way, you'll be prepared no matter what the temperature of the room is. (An uncomfortable temperature will just distract you from the job at hand.) And, if you have an item of clothing that you tend to feel "lucky" or confident in—a shirt, a pair of jeans, whatever—wear it. A little totem couldn't hurt.

What Are "Signs of a Winner," Alex?

Here's some advice from a Kaplan instructor who won big on *Jeopardy!*™ In the green room before the show, he noticed that the contestants who were quiet and "within themselves" were the ones who did great on the show. The contestants who did not perform as well were the ones who were fact-cramming, talking a lot, and generally being manic before the show. Lesson: Spend the final hours leading up to the test getting sleep, meditating, and generally relaxing.

- Forego any practice on the day before the test. It's in your best interest to marshal your physical and psychological resources for 24 hours or so. Even race horses are kept in the paddock and treated like princes the day before a race. Keep the upcoming test out of your consciousness, go to a movie, take a pleasant hike, or just relax. Don't eat junk food or tons of sugar. And—of course—get plenty of rest the night before. Just don't go to bed too early. It's hard to fall asleep earlier than you're used to, and you don't want to lie there thinking about the test.

Handling Stress During the Test

The biggest stress monster will be the test itself. Fear not; there are methods of quelling your stress during the test.

- Keep moving forward instead of getting bogged down in a difficult question. You don't have to get everything right to achieve a fine score. The best test takers skip difficult material temporarily in search of the easier stuff. They mark the ones that require extra time and thought. This strategy buys time and builds confidence so you can handle the tough stuff later.

- Don't be thrown if other test takers seem to be working more furiously than you are. Continue to spend your time patiently thinking through your answers; it's going to lead to better results. Don't mistake the other people's sheer activity as signs of progress and higher scores.

- *Keep breathing!* Weak test takers tend to forget to breathe properly as the test proceeds. They start holding their breath without realizing it, or they breathe erratically or arrhythmically. Improper breathing interferes with clear thinking.

- Some quick isometrics during the test—especially if concentration is wandering or energy is waning—can help. Try this: Put your palms together and press intensely for a few seconds. Concentrate on the tension you feel through your palms, wrists, forearms, and up into your biceps and shoulders. Then, quickly release the pressure. Feel the difference as you let go. Focus on the warm relaxation that floods through the muscles. Now you're ready to return to the task.

- Here's another isometric that will relieve tension in both your neck and eye muscles. Slowly rotate your head from side to side, turning your head and eyes to look as far back over each shoulder as you can. Feel the muscles stretch on one side of your neck as they contract on the other. Repeat five times in each direction.

FULL-LENGTH
PRACTICE TESTS

- The two tests that follow offer realistic practice for the SAT II: Chemistry Subject Test. To get the most out of them, take them under testlike conditions:

  —Take the tests in a quiet room with no distractions. Bring some No. 2 pencils.

  —Time yourself. Spend no more than one hour on the 85 questions on each test.

  —Use the answer sheets provided to mark your answers.

- Answers and explanations follow the test.

- Scoring instructions are in the "Compute Your Practice Test Score" sections immediately following the Answer Keys of each test.

PRACTICE TEST 2

Periodic Table of the Elements

1 H 1.0																	2 He 4.0
3 Li 6.9	4 Be 9.0											5 B 10.8	6 C 12.0	7 N 14.0	8 O 16.0	9 F 19.0	10 Ne 20.2
11 Na 23.0	12 Mg 24.3											13 Al 27.0	14 Si 28.1	15 P 31.0	16 S 32.1	17 Cl 35.5	18 Ar 39.9
19 K 39.1	20 Ca 40.1	21 Sc 45.0	22 Ti 47.9	23 V 50.9	24 Cr 52.0	25 Mn 54.9	26 Fe 55.8	27 Co 58.9	28 Ni 58.7	29 Cu 63.5	30 Zn 65.4	31 Ga 69.7	32 Ge 72.6	33 As 74.9	34 Se 79.0	35 Br 79.9	36 Kr 83.8
37 Rb 85.5	38 Sr 87.6	39 Y 88.9	40 Zr 91.2	41 Nb 92.9	42 Mo 95.9	43 Tc (98)	44 Ru 101.1	45 Rh 102.9	46 Pd 106.4	47 Ag 107.9	48 Cd 112.4	49 In 114.8	50 Sn 118.7	51 Sb 121.8	52 Te 127.6	53 I 126.9	54 Xe 131.3
55 Cs 132.9	56 Ba 137.3	57 La* 138.9	72 Hf 178.5	73 Ta 180.9	74 W 183.9	75 Re 186.2	76 Os 190.2	77 Ir 192.2	78 Pt 195.1	79 Au 197.0	80 Hg 200.6	81 Tl 204.4	82 Pb 207.2	83 Bi 209.0	84 Po (209)	85 At (210)	86 Rn (222)
87 Fr (223)	88 Ra 226.0	89 Ac† 227.0	104 Unq (261)	105 Unp (262)	106 Unh (263)	107 Uns (262)	108 Uno (265)	109 Une (267)									

	58 Ce 140.1	59 Pr 140.9	60 Nd 144.2	61 Pm (145)	62 Sm 150.4	63 Eu 152.0	64 Gd 157.3	65 Tb 158.9	66 Dy 162.5	67 Ho 164.9	68 Er 167.3	69 Tm 168.9	70 Yb 173.0	71 Lu 175.0
*														
†	90 Th 232.0	91 Pa (231)	92 U 238.0	93 Np (237)	94 Pu (244)	95 Am (243)	96 Cm (247)	97 Bk (247)	98 Cf (251)	99 Es (252)	100 Fm (257)	101 Md (258)	102 No (259)	103 Lr (260)

ANSWER SHEET

1 Ⓐ Ⓑ Ⓒ Ⓓ Ⓔ   19 Ⓐ Ⓑ Ⓒ Ⓓ Ⓔ   37 Ⓐ Ⓑ Ⓒ Ⓓ Ⓔ   55 Ⓐ Ⓑ Ⓒ Ⓓ Ⓔ
2 Ⓐ Ⓑ Ⓒ Ⓓ Ⓔ   20 Ⓐ Ⓑ Ⓒ Ⓓ Ⓔ   38 Ⓐ Ⓑ Ⓒ Ⓓ Ⓔ   56 Ⓐ Ⓑ Ⓒ Ⓓ Ⓔ
3 Ⓐ Ⓑ Ⓒ Ⓓ Ⓔ   21 Ⓐ Ⓑ Ⓒ Ⓓ Ⓔ   39 Ⓐ Ⓑ Ⓒ Ⓓ Ⓔ   57 Ⓐ Ⓑ Ⓒ Ⓓ Ⓔ
4 Ⓐ Ⓑ Ⓒ Ⓓ Ⓔ   22 Ⓐ Ⓑ Ⓒ Ⓓ Ⓔ   40 Ⓐ Ⓑ Ⓒ Ⓓ Ⓔ   58 Ⓐ Ⓑ Ⓒ Ⓓ Ⓔ
5 Ⓐ Ⓑ Ⓒ Ⓓ Ⓔ   23 Ⓐ Ⓑ Ⓒ Ⓓ Ⓔ   41 Ⓐ Ⓑ Ⓒ Ⓓ Ⓔ   59 Ⓐ Ⓑ Ⓒ Ⓓ Ⓔ
6 Ⓐ Ⓑ Ⓒ Ⓓ Ⓔ   24 Ⓐ Ⓑ Ⓒ Ⓓ Ⓔ   42 Ⓐ Ⓑ Ⓒ Ⓓ Ⓔ   60 Ⓐ Ⓑ Ⓒ Ⓓ Ⓔ
7 Ⓐ Ⓑ Ⓒ Ⓓ Ⓔ   25 Ⓐ Ⓑ Ⓒ Ⓓ Ⓔ   43 Ⓐ Ⓑ Ⓒ Ⓓ Ⓔ   61 Ⓐ Ⓑ Ⓒ Ⓓ Ⓔ
8 Ⓐ Ⓑ Ⓒ Ⓓ Ⓔ   26 Ⓐ Ⓑ Ⓒ Ⓓ Ⓔ   44 Ⓐ Ⓑ Ⓒ Ⓓ Ⓔ   62 Ⓐ Ⓑ Ⓒ Ⓓ Ⓔ
9 Ⓐ Ⓑ Ⓒ Ⓓ Ⓔ   27 Ⓐ Ⓑ Ⓒ Ⓓ Ⓔ   45 Ⓐ Ⓑ Ⓒ Ⓓ Ⓔ   63 Ⓐ Ⓑ Ⓒ Ⓓ Ⓔ
10 Ⓐ Ⓑ Ⓒ Ⓓ Ⓔ   28 Ⓐ Ⓑ Ⓒ Ⓓ Ⓔ   46 Ⓐ Ⓑ Ⓒ Ⓓ Ⓔ   64 Ⓐ Ⓑ Ⓒ Ⓓ Ⓔ
11 Ⓐ Ⓑ Ⓒ Ⓓ Ⓔ   29 Ⓐ Ⓑ Ⓒ Ⓓ Ⓔ   47 Ⓐ Ⓑ Ⓒ Ⓓ Ⓔ   65 Ⓐ Ⓑ Ⓒ Ⓓ Ⓔ
12 Ⓐ Ⓑ Ⓒ Ⓓ Ⓔ   30 Ⓐ Ⓑ Ⓒ Ⓓ Ⓔ   48 Ⓐ Ⓑ Ⓒ Ⓓ Ⓔ   66 Ⓐ Ⓑ Ⓒ Ⓓ Ⓔ
13 Ⓐ Ⓑ Ⓒ Ⓓ Ⓔ   31 Ⓐ Ⓑ Ⓒ Ⓓ Ⓔ   49 Ⓐ Ⓑ Ⓒ Ⓓ Ⓔ   67 Ⓐ Ⓑ Ⓒ Ⓓ Ⓔ
14 Ⓐ Ⓑ Ⓒ Ⓓ Ⓔ   32 Ⓐ Ⓑ Ⓒ Ⓓ Ⓔ   50 Ⓐ Ⓑ Ⓒ Ⓓ Ⓔ   68 Ⓐ Ⓑ Ⓒ Ⓓ Ⓔ
15 Ⓐ Ⓑ Ⓒ Ⓓ Ⓔ   33 Ⓐ Ⓑ Ⓒ Ⓓ Ⓔ   51 Ⓐ Ⓑ Ⓒ Ⓓ Ⓔ   69 Ⓐ Ⓑ Ⓒ Ⓓ Ⓔ
16 Ⓐ Ⓑ Ⓒ Ⓓ Ⓔ   34 Ⓐ Ⓑ Ⓒ Ⓓ Ⓔ   52 Ⓐ Ⓑ Ⓒ Ⓓ Ⓔ
17 Ⓐ Ⓑ Ⓒ Ⓓ Ⓔ   35 Ⓐ Ⓑ Ⓒ Ⓓ Ⓔ   53 Ⓐ Ⓑ Ⓒ Ⓓ Ⓔ
18 Ⓐ Ⓑ Ⓒ Ⓓ Ⓔ   36 Ⓐ Ⓑ Ⓒ Ⓓ Ⓔ   54 Ⓐ Ⓑ Ⓒ Ⓓ Ⓔ

	I	II	CE*		I	II	CE*
101	Ⓣ Ⓕ	Ⓣ Ⓕ	◯	109	Ⓣ Ⓕ	Ⓣ Ⓕ	◯
102	Ⓣ Ⓕ	Ⓣ Ⓕ	◯	110	Ⓣ Ⓕ	Ⓣ Ⓕ	◯
103	Ⓣ Ⓕ	Ⓣ Ⓕ	◯	111	Ⓣ Ⓕ	Ⓣ Ⓕ	◯
104	Ⓣ Ⓕ	Ⓣ Ⓕ	◯	112	Ⓣ Ⓕ	Ⓣ Ⓕ	◯
105	Ⓣ Ⓕ	Ⓣ Ⓕ	◯	113	Ⓣ Ⓕ	Ⓣ Ⓕ	◯
106	Ⓣ Ⓕ	Ⓣ Ⓕ	◯	114	Ⓣ Ⓕ	Ⓣ Ⓕ	◯
107	Ⓣ Ⓕ	Ⓣ Ⓕ	◯	115	Ⓣ Ⓕ	Ⓣ Ⓕ	◯
108	Ⓣ Ⓕ	Ⓣ Ⓕ	◯	116	Ⓣ Ⓕ	Ⓣ Ⓕ	◯

Use the answer key following the test to count up the number of questions you got right and the number you got wrong. (Remember not to count omitted questions as wrong.) "Compute Your Score" at the back of this section will show you how to find your score.

[ ]
right

[ ]
wrong

PART A

Directions: Each set of lettered choices below refers to the numbered formulas or statements immediately following it. Select the one lettered choice that best fits each formula or statement; then fill in the corresponding oval on the answer sheet. In each set, a choice may be used once, more than once, or not at all.

Note: For all questions involving solutions and/or chemical equations, you can assume that the system is in water unless otherwise stated.

Questions 1–5

(A) Brønsted-Lowry Acid
(B) Arrhenius Acid
(C) Lewis Acid
(D) Brønsted-Lowry Base
(E) Lewis Base

1. Produces H^+ in solution

2. Accepts protons

3. An electron pair donor

4. An electron pair acceptor

5. Donates protons

Questions 6–9

(A) Boyle's Law
(B) Charles's Law
(C) Avogadro's Principle
(D) Ideal Gas Law
(E) Dalton's Law

6. Total pressure of a gaseous mix is equal to the sum of the partial pressures.

7. Volume is inversely proportional to pressure.

8. Volume is directly proportional to temperature.

9. All gases have the same number of moles in the same volume at constant temperature and pressure.

Questions 10–13

(A) Ionic compound
(B) Noble gas
(C) Polar covalent compound
(D) Nonpolar covalent compound
(E) Metallic substance

10. Ethanol

11. CsCl

12. Butane

13. Argon

Questions 14–17

(A) pH
(B) pOH
(C) K_a
(D) conjugate acid
(E) conjugate base

14. Acid dissociation constant

15. Measure of hydrogen ion concentration

16. Formed when an acid loses a proton

17. Measure of hydroxide ion concentration

GO ON TO THE NEXT PAGE

Questions 18–20

(A) An increase in the reactant concentration
(B) An increase in the temperature
(C) A decrease in pressure
(D) Catalysts
(E) pH

18. Increases effective collisions without increasing average energy

19. Decreases activation energy

20. Increases average kinetic energy

Questions 21–24

(A) positive ΔH
(B) negative ΔH
(C) positive ΔG
(D) negative ΔG
(E) positive ΔS

21. Describes a spontaneous reaction

22. Describes an endothermic reaction

23. Describes a nonspontaneous reaction

24. Is multiplied by temperature in the equation which calculates free energy

PART B

Directions: Each question below consists of two statements, statement I in the left-hand column and statement II in the right-hand column. For each question, determine whether statement I is true or false and whether statement II is true or false. Then, fill in the corresponding T or F ovals on your answer sheet. Fill in oval CE only if statement II is a correct explanation of statement I.

Examples:

I		II
EX 1. HCl is a strong acid	**BECAUSE**	HCl contains sulfur.
EX 2. An atom of nitrogen is electrically neutral	**BECAUSE**	a nitrogen atom contains the same number of protons and electrons.

	I	II	CE
SAMPLE ANSWERS			
EX 1	● F	T ●	○
EX 1	● F	● F	●

I		II
101. The ideal gas law does not hold under low temperatures and high pressure	BECAUSE	interactions between particles cannot be neglected under these conditions.
102. An increase in entropy leads to a decrease of randomness	BECAUSE	the lower energy state of ordered crystals has a high entropy.
103. Two electrons in the 2s subshell must have opposite spins	BECAUSE	the Pauli Exclusion Principle states that no two electrons in the same atom can have identical quantum numbers.

GO ON TO THE NEXT PAGE

	I		II
104.	Atom A with 7 valence electrons forms AB_2 with atom B with two valence electrons	BECAUSE	B donates its electrons to fill the outer shell of A.
105.	CO_2 is able to sublimate at atmospheric pressure	BECAUSE	its liquid form is impossible to produce.
106.	A gas is more random than a liquid	BECAUSE	entropy increases from gas to liquid.
107.	A mixture of pure metals is called an alloy. The melting point of an alloy will be lower than either of the component metals	BECAUSE	the new bonds are weaker.
108.	A salt dissolved in an organic solvent will be a good electrical conductor	BECAUSE	salts will not dissolve appreciably in an organic solvent.
109.	A super saturated solution of glucose in boiling water crystallizes as it cools	BECAUSE	the solubility increases as the temperature decreases.
110.	In a second order reaction with respect to A, when you double [A], the rate is quadrupled	BECAUSE	the rate equation is $r = k[A]^2$ for such a reaction.
111.	Ca is a neutral atom	BECAUSE	it has the same number of protons and electrons.
112.	Salt dissolved in water depresses the freezing point	BECAUSE	the solute particles interfere with ice crystal formation.
113.	Ionic bonds are always stronger than covalent bonds	BECAUSE	they only break when bombarded with electrons.
114.	Substances with hydrogen bonding tend to have unusually low boiling points	BECAUSE	extra energy is necessary to break the hydrogen bonds.
115.	A solution with a pH of 12 has a higher concentration of hydroxide ions than one with a pH of 10	BECAUSE	$[H^+][OH^-]$ must equal 10^{-14}.
116.	Catalysts decrease the rate of a chemical reaction	BECAUSE	catalysts decrease the activation energy.

RETURN TO THE SECTION OF YOUR ANSWER SHEET YOU
STARTED FOR CHEMISTRY AND ANSWER QUESTIONS 25–65.

GO ON TO THE NEXT PAGE

PART C

Directions: Each of the incomplete statements or questions below is followed by five suggested completions or answers. Select the one that is best for each case and fill in the corresponding oval on the answer sheet.

25. What is the percentage by mass of sulfur in H_2SO_4?

    (A) 16%

    (B) 33%

    (C) 36%

    (D) 42%

    (E) Cannot be determined

26. $BaCl_2$ dissociates in water to give one Ba^{2+} ion and two Cl^- ions. If concentrated HCl is added to this solution:

    (A) $[Ba^{2+}]$ increases

    (B) $[Ba^{2+}]$ remains constant

    (C) $[OH^-]$ increases

    (D) the number of moles of undissociated $BaCl_2$ increases

    (E) $[H^+]$ decreases

27. What is the percent composition by weight of Al in $Al_2(SO_4)_3$?

    (A) 7.9%

    (B) 31.6%

    (C) 15.8%

    (D) 12.7%

    (E) 22.3%

28. The properties of light can best be explained by assuming that light is composed of:

    I.   Particles
    II. Waves
    III. Atoms

    (A) I only

    (B) II only

    (C) I and II

    (D) II and III

    (E) I, II, and III

29. Which of the following aqueous solutions would have a pH greater than 7.0?

    (A) 1.0M HCl

    (B) 0.5M NH_4Cl

    (C) 0.25M HCN

    (D) 0.1M KCN

    (E) 1.0M H_2SO_4

GO ON TO THE NEXT PAGE

30. Avogadro's number is NOT equal to:

    (A) the number of atoms in 11.2 L of O_2 at STP

    (B) the number of atoms in 1 mole of He at STP

    (C) the number of electrons in 96,5000 coulombs

    (D) the number of SO_4^{2-} ions in 1L of 0.5 M sulfuric acid

    (E) None of the above

31. Which of the following is not a state function?

    (A) Temperature

    (B) Density

    (C) Work

    (D) Volume

    (E) Pressure

32. 10g of liquid at 300K are heated to 350K. The liquid absorbs 6 kcals. What is the specific heat of the liquid (in cal/g°C)?

    (A) 6

    (B) 12

    (C) 60

    (D) 120

    (E) 600

33. What is the range of possible values for the $[OH^-]/[H^+]$ ratio in an aqueous acid solution?

    (A) 0–1

    (B) 0–14

    (C) 1–14

    (D) 1–infinity

    (E) 7–14

34. Boron found in nature has an atomic weight of 10.811 and is made up of the isotopes B^{10} (mass 10.013 amu) and B^{11} (mass 11.0093). What percentage of naturally occurring boron is made up of B^{10} and B^{11} respectively?

    (A) 30:70

    (B) 25:75

    (C) 20:80

    (D) 15:85

    (E) 10:90

35. For the reaction: $A + B \rightarrow C$, determine the order or the reaction with respect to B from the information given below:

Initial [A]	Initial [B]	Initial rate of formation of C
1.00	1.00	2.0
1.00	2.00	8.1
2.00	2.00	15.9

    (A) zero order

    (B) first order

    (C) second order

    (D) third order

    (E) fourth order

GO ON TO THE NEXT PAGE

36. Which of the following is NOT a correct Lewis dot diagram?

(A) H—C≡C—H

(B) :Ö=N—Ö·

(C)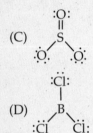

(D)

(E) H $\overset{\cdot\ddot{O}\cdot}{\diagup\diagdown}$ H

37. A flask contains three times as many moles of H_2 gas as it does O_2 gas. If hydrogen and oxygen are the only gases present, what is the total pressure in the flask if the partial pressure due to oxygen is P?

(A) 4P

(B) 3P

(C) 4/3P

(D) 3/4P

(E) 7P

38. How many mL of a 0.25N solution of HCl are needed to completely neutralize 5 grams of crystalline NaOH dissolved in water?

(A) 150

(B) 250

(C) 500

(D) 750

(E) 3,750

39. Using the table below, what is the heat of combustion of one mole of C_2H_4 at 298K and 1 atm?

Compound	ΔH of formation (kcal/mol)
H_2O (g)	−57.8
C_2H_6 (g)	−20/2
C_2H_4 (g)	12.5
C_2H_2 (g)	54.2
CO (g)	−26.4
CO_2 (g)	−94.1

(A)   316.3 kcal

(B)   12.5 kcal

(C)   −291.3 kcal

(D)   −316.3 kcal

(E)   57.8 kcal

40. Which of the following molecules is polar?

(A) BH_3

(B) NF_3

(C) C_2H_6

(D) SF_6

(E) CCl_4

GO ON TO THE NEXT PAGE

KAPLAN

41. Consider the following gas phase reaction:

$$H_2 (g) + Br_2 (g) \rightleftharpoons 2HBr (g)$$

The concentrations of H_2, Br_2, and HBr are 0.05M, 0.03M, and 500.0 M respectively. The concentration equilibrium constant for this reaction at 400°C is 2.5×10^3. Is this system at equilibrium?

(A) Yes, the system is at equilibrium.

(B) No, the reaction must shift to the right in order to reach equilibrium.

(C) No, the reaction must shift to the left in order to reach equilibrium.

(D) It cannot be determined.

(E) This system will never be at equilibrium.

42. The K_a of acetic acid is 2×10^{-5}. What is the pH of a 0.5M solution of acetic acid?

(A) 1.0

(B) 2.0

(C) 2.5

(D) 3.0

(E) 7.0

43. A Faraday is

(A) the magnitude of the charge of 1 mole of electrons

(B) the magnitude of the electric dipole

(C) a fundamental constant of nature equal to 6.63×10^{-34}Js/photon

(D) a constant that accounts for the existence of ions in solution

(E) the assignment of charges to individual atoms

44. Which of the following is NOT a true statement about the entropy of a system?

(A) Entropy is a measure of the randomness in a system.

(B) The entropy of an amorphous solid is greater than that of a crystalline solid.

(C) The entropy of a spontaneous reaction cannot decrease.

(D) The entropy of an isolated system will spontaneously increase or remain constant.

(E) The entropy of a liquid is generally greater than that of a solid.

45. How many grams of O_2 will it take to oxidize 88 grams of C_3H_8 to CO and H_2O?

(A) 32

(B) 64

(C) 112

(D) 166

(E) 224

46. Which of the following combinations represents an element with a net charge of +1 with a mass number of 75?

(A) 35 neutrons, 35 protons, 34 electrons

(B) 40 neutrons, 40 protons, 39 electrons

(C) 40 neutrons, 35 protons, 34 electrons

(D) 37 neutrons, 38 protons, 39 electrons

(E) 40 neutrons, 35 protons, 35 electrons

GO ON TO THE NEXT PAGE

47. A physicist starts out with 320 grams of a radioactive element Z and after 20 minutes he has only 20 grams left. What is the half-life of element Z?

(A) 2 minutes

(B) 3 minutes

(C) 4 minutes

(D) 5 minutes

(E) 10 minutes

48. The modern periodic table is ordered on the basis of

(A) atomic mass

(B) atomic radius

(C) atomic charge

(D) atomic number

(E) number of neutrons

49. A catalyst

(A) changes ΔG for an equation

(B) acts by increasing the rate of the forward reaction more than the reverse reaction

(C) raises the equilibrium constant of a system

(D) may have a molecular weight as low as 1 or higher than 200,000

(E) does not react chemically during the course of a reaction

50. Element $^{108}_{26}Y$ is formed as a result of 3 α and 2 β^- decays. Which of the following is the parent element?

(A) $^{96}_{22}\psi$

(B) $^{120}_{30}\iota$

(C) $^{120}_{24}\Omega$

(D) $^{18}_{14}\Delta + ^{96}_{18}\vartheta$

(E) $^{120}_{28}\Pi$

51. What is the molecular formula of a compound with the empirical formula $C_3H_6O_2$ and a mass of 148 amu?

(A) $C_6H_{12}O_4$

(B) $C_2H_6O_2$

(C) $C_9H_{18}O_6$

(D) C_2H_3O

(E) None of the above

GO ON TO THE NEXT PAGE

Questions 52–54 refer to the following equation.

A chemist interested in the reactivity of iodine concentrates his study on the decomposition of gaseous hydrogen iodide (Reaction 1).

$$2HI\,(g) \leftrightharpoons H_2(g) + I_2\,(g)$$

Reaction 1

52. What is the equilibrium expression for Reaction 1?

(A) $[H_2]^2[I_2]$

(B) $[H_2]$

(C) $[H_2][I_2]/[HI]^2$

(D) $[H_2][I_2]^2$

(E) $[H_2]^2[I_2]^2$

53. The reaction profile shown below is for an uncatalyzed reaction:

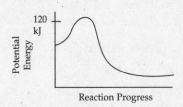

Which of the following is the reaction profile for the same reaction after the addition of a catalyst?

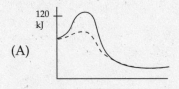

(A)

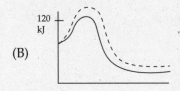

(B)

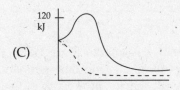

(C)

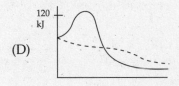

(D)

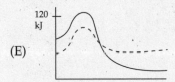

(E)

GO ON TO THE NEXT PAGE

54. An increase in pressure in Reaction 1 would

(A) produce more I^- (aq)

(B) produce more H_2

(C) not affect the system

(D) drive it to the right

(E) drive it to the left

55. For the following reaction, calculate how much heat will be released when 8g of hydrogen reacts. (1 mole H_2 weights approximately 2 grams)

$$2H_2(g) + O_2(g) \rightarrow 2H_2O \ (g);$$
$$\Delta H = -115.60 \text{ kcal}$$

(A) 57.8 kcal

(B) 115.6 kcal

(C) 173.4 kcal

(D) 231.2 kcal

(E) 462.4 kcal

56. The gas in a large cylinder is at a pressure of 3,040 torr. Assuming constant temperature and ideal gas behavior, how much of this gas could you compress into a 100 L box at 8 atm?

(A) 20L

(B) 200L

(C) 5,000L

(D) 50,000L

(E) 500,000L

57. Which of the following atoms does the electron configuration $1s^2 2s^2 2p^6 3s^2 3p^6 3d^2$ represent?

(A) An excited Ca atom

(B) A neutral Ca atom

(C) An excited Sc atom

(D) An excited K atom

(E) None of the above

Question 58–60 refer to the following experimental data.

Half Reaction	Eo
I. $Mg \rightarrow Mg^{2+} + 2e^-$	2.37V
II. $Mn \rightarrow Mn^{2+} + 2e^-$	1.03V
III. $H_2 \rightarrow 2H^+ + 2e^-$	0.00V
IV. $Cu \rightarrow Cu^{2+} + 2e^-$	–.16V

58. Which of the following reactions is spontaneous?

(A) $Mn^{2+} + H_2 \rightarrow$

(B) $Mg + Mn^{2+} \rightarrow$

(C) $Mg^{2+} + Mn \rightarrow$

(D) $Cu + 2H^+ - \rightarrow$

(E) $Mg^{2+} + Cu \rightarrow$

59. What will be the standard reduction potential for the reaction in the answer to question 58?

(A) –.16V

(B) –1.34V

(C) 1.03V

(D) 1.34V

(E) 2.21V

GO ON TO THE NEXT PAGE

60. Which of the following is NOT a resonance structure of the others?

(A) 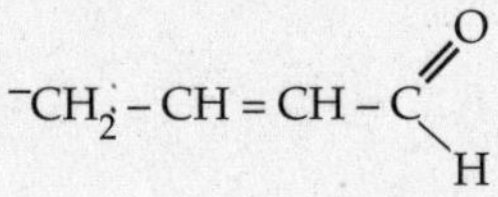 $^-CH_2 - CH = CH - C\begin{smallmatrix} \diagup O \\ \diagdown H \end{smallmatrix}$

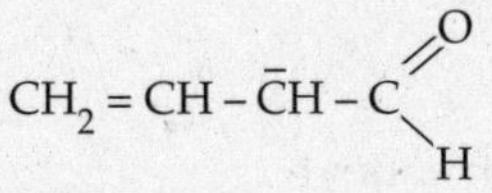

(B) $CH_2 = CH - \bar{C}H - C\begin{smallmatrix} \diagup O \\ \diagdown H \end{smallmatrix}$

(C) $CH_2 = C - C\begin{smallmatrix} \diagup O^- \\ \diagdown H \end{smallmatrix}$
        |
        CH_3

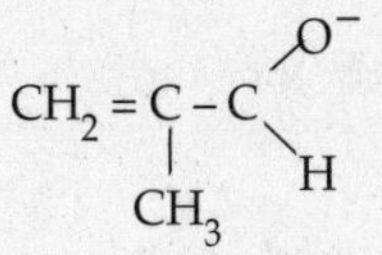

(D) 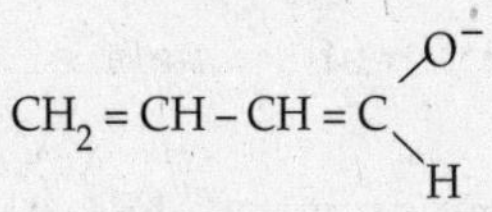 $CH_2 = CH - CH = C\begin{smallmatrix} \diagup O^- \\ \diagdown H \end{smallmatrix}$

(E) They are all resonance structures.

61. If you have 0.15 equivalents of H_3PO_4 (full reaction or full neutralization occurs), how many moles of H_3PO_4 do you have?

(A)   0.05

(B)   0.15

(C)   4.9

(D)   9.8

(E)   98.0

62. Which of the following generalizations CAN-NOT be made about the phase change of a pure substance from solid to liquid?

(A) It involves a change in potential energy.

(B) It involves no change in temperature.

(C) It involves a change in kinetic energy.

(D) It involves a change in entropy.

(E) It may occur at different temperatures for different compounds.

63. What occurs in the following reaction?

$$NH_4^+ + H_2O \leftrightharpoons NH_3 + H_3O^+$$

(A) Electron transfer

(B) Neutralization

(C) Double replacement

(D) Reduction

(E) Proton transfer

64. What is the value of the equilibrium constant for the following reaction if the equilibrium concentrations of nitrogen, hydrogen, and ammonia are 1M, 2M, and 15M, respectively?

$$N_2(g) + 3H_2(g) \leftrightharpoons 2NH_3(g)$$

(A) 0.035

(B) 7.5

(C) 28

(D) 380

(E) None of the above

65. What would be the stoichiometric coefficient of hydrochloric acid in the following equation?

$$....Cl_2 + ...H_2O \rightarrow ...HCl + ...HClO_3$$

(A) 1

(B) 3

(C) 5

(D) 10

(E) 12

GO ON TO THE NEXT PAGE

66. A solid cube of iron whose edge is 4 inches weighs 20.8 lbs. What is the specific gravity of this sample of iron? Water weighs 62.4 pounds per cubic foot.

   (A)   4.5

   (B)   5.2

   (C)   9.0

   (D)   27

   (E)   324

67. Inelastic collisions occur in

   I.   Real gases

   II.   Ideal gases

   III.   Fusion reactions

   (A) I and II

   (B) II and III

   (C) I and III

   (D) I only

   (E) II only

68. Which of the following compounds can act as a Lewis base?

   (A) $HClO_2$

   (B) NH_2NH_2

   (C) NH_4+

   (D) BF_3

   (E) None of the above

69. Two electrons with the same n, l, and ml values

   (A) must be in different atoms

   (B) are in different orbitals of the same subshell

   (C) are in the same orbital of the same subshell with opposite spins

   (D) are indistinguishable from each other

   (E) None of the above

STOP!   END OF TEST.

Turn the page
for answers and explanations
to Practice Test 2.

Answer Key

1. B	15. A	29. D	43. A	57. A
2. D	16. E	30. D	44. C	58. B
3. E	17. B	31. C	45. E	59. D
4. C	18. A	32. B	46. C	60. C
5. A	19. D	33. A	47. D	61. A
6. E	20. B	34. C	48. D	62. C
7. A	21. D	35. C	49. D	63. E
8. B	22. A	36. B	50. B	64. C
9. C	23. C	37. A	51. A	65. C
10. C	24. E	38. C	52. C	66. C
11. A	25. B	39. D	53. A	67. C
12. D	26. E	40. B	54. C	68. B
13. B	27. C	41. C	55. D	69. C
14. C	28. C	42. C	56. B	

101. T, T ,CE	109. T, F
102. F, F	110. T, T, CE
103. T, T, CE	111. T, T, CE
104. F, T	112. T, T, CE
105. T, F	113. F, F
106. T, F	114. F, T
107. T, T, CE	115. T, T, CE
108. F, T	116. F, T

Compute Your Practice Test Score

Step 1: Figure out your raw score. Refer to your answer sheet for the number right and the number wrong on the practice test you're scoring. (If you haven't checked your answers, do that now, using the answer key that follows the test.) You can use the chart below to figure out your raw score. Multiply the number wrong by 0.25 and subtract the result from the number right. Round the result to the nearest whole number. This is your raw score.

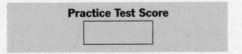

SAT II: CHEMISTRY Practice Test

Number right $-$ (0.25 x Number wrong) $=$ Raw score (rounded)

Step 2: Find your practice test score. Find your raw score in the left column of the table below. The score in the right column is your Practice Test score.

Find Your Practice Test Score

Raw	Scaled	Raw	Scaled	Raw	Scaled	Raw	Scaled	Raw	Scaled	Raw	Scaled
85	800	67	720	49	610	31	510	13	400	−5	290
84	800	66	720	48	610	30	500	12	400	−6	290
83	800	65	720	47	600	29	500	11	390	−7	280
82	800	64	710	46	600	28	490	10	390	−8	270
81	800	63	700	45	590	27	480	9	380	−9	270
80	800	62	700	44	580	26	480	8	370	−10	260
79	790	61	700	43	580	25	470	7	360	−11	250
78	790	60	690	42	570	24	460	6	360	−12	250
77	780	59	680	41	570	23	460	5	350	−13	240
76	780	58	670	40	560	22	460	4	350	−14	240
75	780	57	670	39	560	21	450	3	340	−15	230
74	770	56	660	38	550	20	440	2	330	−16	230
73	760	55	650	37	550	19	440	1	330	−17	220
72	760	54	640	36	540	18	430	0	320	−18	220
71	750	53	640	35	530	17	430	−1	320	−19	210
70	740	52	630	34	520	16	420	−2	310	−20	200
69	740	51	630	33	520	15	420	−3	300	−21	200
68	730	50	620	32	520	14	410	−4	300		

Practice Test Score

A note on your practice test scores: Don't take these scores too literally. Practice test conditions cannot precisely mirror real test conditions. Your actual SAT II: Chemistry Subject Test score will almost certainly vary from your practice test scores. Your scores on the practice tests will give you a rough idea of your range on the actual exam.

Answers and Explanations to Practice Test 2

1. B

An Arrhenius acid produces H^+ in solution.

2. D

A Brønsted-Lowry base accepts protons. For example, NH_3 and Cl^- are both Brønsted-Lowry bases because they will accept protons in solution. They are not Arrhenius bases because they don't make OH^- in solution.

3. E

A Lewis base is an electron pair donor.

4. C

A Lewis acid is an electron pair acceptor; for example, BCl_3 and $AlCl_3$.

5. A

A Brønsted-Lowry acid is a species that can donate protons (H^+) (for example HCl). Brønsted-Lowry acids and bases always occur in pairs, called conjugates; for example, H_3O^+ is an conjugate acid of the base H_2O.

6. E

Dalton's Law states that the sum of the partial pressures of the components of a gaseous mixture must equal the total pressure of the sample.

7. A

Boyle's Law states that at constant temperature, the volume of a gaseous sample is inversely proportional to its pressure.

8. B

Charles's Law states that the volume of a gaseous sample at constant pressure is directly proportional to absolute temperature.

9. C

Avogadro's Principle states that under the same conditions of temperature and pressure, equal volumes of different gases will have the same number of molecules.

10. C

Ethanol, with its –OH group, has a covalent bond between atoms with different electronegativities in which electron density is unevenly distributed, giving the bond both + and – ends.

11. A

CsCl is an ionic compound with Cs donating an electron to Cl and the subsequent chemical bond formed through the electrostatic interaction between the positive and negative ions.

12. D

Butane, C_4H_{10}, is a molecule that exhibits no net separation of charges and therefore has no net dipole moment.

13. B

Argon is an element located in Group 0 (Group VIII) of the Periodic Table. These elements are known as the noble or inert gases. They contain a full octet of valence electrons in their outermost shell, and this electron configuration makes them the least reactive of the elements.

14. C

The acid dissociation constant (K_a) is the equilibrium constant that measures the degree of dissociation for an acid under specific conditions.

15. A

pH is a measure of the hydrogen ion content of an aqueous solution, and is defined as the negative log of the hydrogen ion concentration.

16. E

When an acid loses a proton it becomes its conjugate base. A conjugate acid is defined as the acid formed when the base gains a proton.

17. B

pOH is a measure of hydrogen (OH^-) ion content of an aqueous solution. It is defined as the negative log of the hydroxide ion concentration.

18. A

Within the same volume, an increase in the reactant concentration will increase the number of times molecules bump into each other (elastic collisions) and increases the rate of effective collisions.

19. D

Catalysts increase the rate of a reaction by decreasing the activation energy.

20. B

An increase in temperature will add energy to the system and will therefore increase the average kinetic energy.

21. D

ΔG is the energy of a system available to do work. A negative ΔG denotes a spontaneous reaction, while a positive ΔG denotes a nonspontaneous reaction.

22. A

An endothermic reaction absorbs heat from its surroundings as the reaction proceeds and has a $+\Delta H$.

23. C

See the explanation to question 21.

24. E

The change in Gibbs Free Energy (ΔG) can be determined for a given reaction from the equation $\Delta G = \Delta H - T\Delta S$.

101. T, T, CE

Gases deviate from the Ideal Gas Law when atoms and molecules are forced into close proximity under high pressure and at low temperature because molecular volume and intermolecular attractions become significant.

102. F, F

An increase in entropy means an increase in randomness; ordered crystals have low entropy.

103. T, T, CE

Two electrons in the same orbital must have opposite spins because of the Pauli Exclusion Principle, which states that two electrons within a given atom can never have the same four quantum numbers.

104. F, T

Atom A wants to fill its outer shell with 8 electrons, while B wants to lose its 2 valence electrons. Therefore, B donates one electron to two molecules of A to form A_2B.

105. T, F

CO_2 is able to sublimate because its phase diagram has the region that allows solids to sublimate and gases to deposit at atmospheric temperature and pressure. In the lab at high pressures, liquid CO_2 can be formed.

106. T, F

In order of decreasing entropy: gas > liquid > solid.

107. T, T, CE

Bonds between different metal atoms in an alloy are much weaker than those between atoms in pure metals. Therefore, breaking these bonds requires less energy than breaking the bonds in pure metals. Since melting and freezing points are lowered as the stability of the bonds decreases, they tend to be lower in alloys than in pure metals. Alternatively, you can think of an alloy as a solid solution with impurities that decrease the melting point.

108. F, T

Salts will not dissolve appreciably in organic solvents because they are nonpolar, and salts needs a polar solvent to dissociate into ions. If a salt cannot dissociate, then there will be no electrical conduction.

109. T, F

Glucose will crystallize as the supersaturated solution cools because as heat (energy) is lost, a solution cannot have as much of a solute dissolved in it, so glucose will precipitate out.

110. T, T, CE

In a second order reaction with respect to A, the rate $= k[A]^2$. If you double A, the rate $= k[2A]^2$ which equals $k4[A]^2$. Therefore the rate is quadrupled. If you triple A, rate $= k[3A]^2 = 9k[A]^2$, i.e., rate is increased by a factor of nine.

111. T, T, CE

Ca is a neutral atom. All neutral atoms have the same number of positive protons and negative electrons.

112. T, T, CE

Salt dissolved in H_2O will result in a freezing point depression because solute particles interfere with the process of crystal formation that occurs during freezing.

113. F, F

Covalent bonds are usually stronger than ionic bonds. Ionic bonds are an association of ions after an element has completely transferred its electron to another element. These can dissociates easily; for example, NaCl in water becomes Na^+ and Cl^-.

114. F, T

Hydrogen bonding occurs when hydrogen atoms with a partial positive charge interact with the partial negative charge located on electronegative atoms of nearby molecules. These substances have unusually high boiling points compared to similar compounds that do not hydrogen bonding.

115. T, T, CE

A solution with pH of 12 has 10^{-12} M H^+ ions. Since $[H^+][OH^-]$ must equal 10^{-14}, then the $[OH^-]$ must be 10^{-2}M. Therefore, the OH^- concentration is higher than a solution with a pH of 10 in which the $[H^+]$ will be 10^{-10}M and the $[OH^-]$ concentration will be 10^{-4}M.

116. F, T

Catalysts increase the rate of a reaction by decreasing the activation energy.

25. B

$$2H = 2 \text{ amu}$$
$$S = 32 \text{ amu}$$
$$4O = 64 \text{ amu}$$
$$\text{total} = 98 \text{ total amu}$$

Percent of sulfur: $(32/98) \times 100\% = 33\%$

26. D

To answer this question, you should know that when hydrochloric acid, a very strong acid, is added to a solution, it will dissociate completely into hydrogen ions and chloride ions. This increases the concentration of chloride ions already in the solution from the barium chloride. According to Le

Châtelier's Principle, if the concentration of one reaction species is increased, the reaction will be driven in the opposite direction. For this example and the dissociation reaction of barium chloride, if you increase the concentration of chloride ion, you will drive the reaction in the direction of reassociation of barium chloride. Therefore, barium chloride will precipitate out of the solution.

27. C

$$2Al = 54 \text{ amu}$$
$$3S = 96 \text{ amu}$$
$$12O = 192 \text{ amu}$$
$$\text{total} = 342 \text{ amu}$$

Percent of aluminum: $(54/342) \times 100\% = 15.8\%$

28. C

Light is considered in a class by itself but the different behaviors of light can be explained by thinking of it as either a particle or a wave. The ability of light to show interference is a wave property, while the photoelectric effect is a property best described through a particle model.

29. D

The trick is understanding what should happen to each when it is dissolved in water. Choice D, potassium cyanide, in addition to being highly poisonous, is the salt of a strong base, KOH, and a weak acid, HCN. When the salt dissolves in water, the potassium ion does not react with water, but the cyanide ion picks up hydrogen ions from the water to establish an equilibrium with hydrocyanic acid. The result is that excess hydroxide ions are created, and the solution is basic.

30. D

Avogadro's number, 6.02×10^{23}, is the number of molecules in one gram molecular weight of a substance, or one mole of the substance. So the correct answer choice is the one that does not describe one mole of a species. You probably remember that a

mole of gas at STP occupies a volume of 22.4 liters. Choice A gives the volume of a half a mole of oxygen molecules, but since each molecule of oxygen gas has two oxygen atoms, choice A represents Avogadro's number of atoms. Choice B, one mole of helium, is a mole. Choice C is the definition of a Faraday, which is one mole of electrons. Choice D, the number of sulfate ions in a 0.5 M sulfuric acid solution, is not a mole. The reason is that there are two equivalents of hydrogen in each mole of sulfuric acid. A 0.5 M solution contains one mole of hydrogen ions, but only a half mole of sulfate ions. One mole of sulfate ions from sulfuric acid would have to be 2N.

31. C

A state function is one whose value depends only on the initial and final states of the system; it is independent of the path taken in going from the initial to the final states. Examples of state functions are pressure, volume, temperature, density, viscosity, enthalpy, internal energy, entropy, and free energy. Two import functions that are not state functions are work and q, the amount of heat transferred.

32. B

$$q = mc\Delta T$$
Solve for c: $c = q/m\Delta T$
$$6,000 \text{cal}/(10g \times 50°C) = 12 \text{cal}/g°C$$

33. A

The question asks you to give the range of possible values for the ratio of hydroxide ion concentration to hydrogen ion concentration in an aqueous acid solution. For a solution to be acidic, it must have a pH between about 1 and 6.9999. This means that the pOH of these solutions will be between 13 and 7.0001. We know that pH is equal to the negative log of the hydrogen ion concentration in solution and the pOH is equal to the negative log of the hydroxide ion concentration in solution. From this we can determine that the range of possible hydrogen ion concentrations in an acid solution will be about 10^{-1} to $10^{-6.999}$. The corresponding range of concentrations of hydroxide ions in the same solutions will be 10^{-13} to $10^{-7.0001}$. Now calculate the range of the ratio of hydroxide ions to hydrogen ions. For the most acidic solution, the ratio will be $10^{-13}/10^{-1}$, which is equal to 10^{-12}, or approximately 0. For a slightly acidic solution, one very close to neutrality, the ratio will be $10^{-7.001}/10^{-6.999}$, which is nearly equal to one. Thus, the range of the hydroxide ions to the hydrogen ions in an acidic solution is 0 to 1.

34. C

$20\%(10.013\text{amu}) + 80\%(11.0093) = 10.811\text{amu}$

35. C

With respect to the concentration of B, every time the concentration doubles, the rate is four times faster. $[2]x = 4$, $x = 2$.

36. B

The first thing you should do to verify a Lewis structure of a molecule is to make sure that all the valence electrons are accounted for. For choice A, acetylene, there are two carbons, each having four valence electrons, and 2 hydrogens, each having one valence electron. So, choice A needs to have 10 valence electrons and indeed it has. For choice B, nitrogen dioxide, there is 1 nitrogen, which has 5 valence electrons and 2 oxygens, each having 6 valence electrons—there should be a total of 17 electrons accounted for. Counting the electrons in choice B you can see that it has 18 electrons. Choice C, sulfur trioxide, should have a total of 24 valence electrons, 6 from each element. Counting the valence electrons, you'll see that it has the required 24. Choice D, boron trichloride, should have, and does have, 24 electrons.

37. A

You are asked to determine the total pressure in the flask in terms of the partial pressure of the oxygen gas. To do this, you need to use Dalton's Law of partial pressure. This law says that the sum of the partial pressures of all the gases in a given vessel is equal to the total pressure. Since the partial pressure of oxygen is P and you know that there are three times as many moles of hydrogen as oxygen in the flask, then the partial pressure of hydrogen must be 3P. Thus, the total pressure in the flask is P + 3P, or 4P.

38. C

5g NaOH/40g/mol = 0.125 mol NaOH and therefore 0.125 mole of OH^-. You must have 0.125 mole of H^+ ions to neutralize your OH^- ions.

0.25N HCl; in this case N = M so you have 0.25M HCl.

0.25mol/L; you need 0.125 mol. How many mL will have 0.125 mol in it?

0.125mol/0.25mol/L = 0.5L or 500mL.

Recall that N (normality) is equal to the molarity of the substance of interest for the reaction. In this case, the molarity of H^+ is what we're concerned with.

39. D

$C_2H_4 + 3O_2 \rightarrow 2CO_2 + 2H_2O$

$\Delta H_c = (2\text{mole } CO_2)([H_f (CO_2)] + (2\text{mol } H_2O)([H_f (H_2O)] - [(1\text{mole } C_2H_4)([H_f(C_2H_4)]$

$$\Delta H_c = 2(-94.1) + 2(-57.8) - 1(12.5)$$

$$\Delta H_c = -316.3\text{kcal}$$

40. B

To answer questions about a molecule's polarity, you must think about the geometry of the molecule as well as the polarity of the individual bonds. Of the molecules given in the choices, all but choice B are symmetrical and so, even if their bonds are polar, the molecules themselves will not have a dipole moment. According to the Valence Shell Electron Pair Repulsion Theory, the three N-F bonds in NF_3 point toward the vertices of a tetrahedron and the nitrogen's lone pair of electrons points toward the fourth vertex, so the molecule has a trigonal pyramidal conformation. The NF_3 molecule is thus asymmetrical and will have a net dipole moment. The electronegativity difference between nitrogen and fluorine shows that fluorine exerts a greater pull on the electrons in the NF bonds, so the dipole moment will put a partial negative charge on the fluorine end of the molecule and a partial positive charge on the nitrogen, despite its lone pair of electrons.

41. C

The first thing that you need to do to answer this question correctly is to construct the equilibrium constant expression. The equilibrium constant expression for this reaction is the concentration of hydrogen bromide squared over the product of the hydrogen concentration and the bromine concentration. Plugging in the appropriate concentration values, you get 500^2 over the product of 0.05 times 0.03. You should be able to tell that this figures out to be in the 10^5 region, much larger than the equilibrium constant. In order for this reaction to reach equilibrium, it must use up all that excess hydrogen bromide. Consequently, the equilibrium will shift to the left.

42. C

.5M solution of acetic acid (CH_3COOH)

$$K_a = [H^+][C_2H_3O_2^-]/[CH_3COOH]$$
$$[H^+] = [CH_3COO^-]$$
$$2 \times 10^{-5} = [H^+][H^+]/.5M$$
$$1 \times 10^{-5} = [H^+]^2$$
$$1 \times 10^{-2.5} = [H^+]$$

Take the negative log of both sides to determine pH.

$$2.5 = pH$$

43. A

A Faraday is the magnitude of the charge of one mole of electrons. The Faraday was named after scientist Michael Faraday who was responsible for the discovery of electromagnetic induction, the laws of electrolysis, the discovery of benzene, and many other important discoveries.

44. C

This question gives you statements concerning the entropy of a system and asks you to determine which one is incorrect. Choice C says that a spontaneous reaction will never have a decrease in entropy. This is a false statement. Depending on the temperature and the enthalpy change of a reaction, the entropy change can be negative. For instance, in phase changes from gas to liquid or liquid to solid, a decreasing temperature spontaneously drives the system into the more ordered state. That would be a decrease in entropy.

45. E

$$2C_3H_8 + 7O_2 \rightarrow 6CO + 8H_2O$$

88g C_3H_8/44g/mol = 2 mol; therefore, 7 mol of oxygen are needed.

7 mol $O_2 \times$ 32g/mol = 224 g of O_2

46. C

A net charge of +1 has one more proton than electrons. Therefore, only choices B or C may be correct. Mass number of 75 means that the protons plus neutrons must equal 75. This eliminates answer choice B.

47. D

half lives	0	1	2	3	4
grams remaining	320	160	80	40	20

four half lives in 20 minutes

20 minutes divided by 4 = 5 minutes

48. D

When the Periodic Table was first being designed, it was thought that the periodicity of the elements could be explained on the basis of atomic mass. Mendeleev discovered that when the elements were arranged in order or increasing atomic mass, certain chemical properties were repeated at regular intervals. However, certain elements could not be fit into any group of a table based on increasing atomic mass. It was the discovery of the nucleus and its components that led scientists to order the elements by increasing atomic number, the number of protons.

49. D

A catalyst is a substance that in relatively small amounts will accelerate both the forward and reverse reactions by lowering the activation energy, without being changed itself. It can be as simple as a proton (acid catalysis) or a big biomolecule (in living organisms).

50. B

An alpha particles is defined as ^4_2He and a beta particle is defined as $^0_{-1}e$ (an electron).

3 α and 2 β emissions will be ($^4_2He + ^4_2He + ^4_2He + ^0_{-1}e + ^0_{-1}e$). This describes what is lost from the parent, to result in $^{108}_{26}Y$. 108 + 12 = 120 and 26 + 4 = 30. This describes $^{120}_{30}l$.

51. A

The empirical formula of a compound is the formula that shows the smallest whole number relationship between the elements of a molecule. The empirical formula for this compound, $C_3H_6O_2$, tells us that for every 3 carbon atoms in the molecule there are 6 hydrogen atoms and 2 oxygen atoms. This does not mean that a molecule actually contains 3 carbons, 6 hydrogens, and 2 oxygens (though it might); instead, it tells us the ratio among

them. Therefore, the molecular formula could be choice C, $C_9H_{18}O_6$, where the carbon, hydrogen, and oxygen ratio is 3:6:2 as in the empirical formula, but might also be choice A, where the ratio is correct. Knowing that the mass of the molecule is 148 amu can help you distinguish between choices A and C. Amu stands for atomic mass units, and means that the compound has a molecular weight of 148 grams/mole. Choice A has MW = 148 grams/mole while choice C has a MW = 222 grams/mole.

52. C

The equilibrium constant expression shows the product of the product concentrations raised to their stoichiometric coefficients over the product of the reactant concentrations raised to their stoichiometric coefficients. For heterogeneous equilibria, pure solids and liquids are not included in this equation; gases and aqueous components are included. You need to construct the correct equilibrium constant expression and calculate the equilibrium constant: $[H_2][I_2]/[HI]^2$.

53. A

The addition of a catalyst lowers the activation energy for a reaction.

54. C

Le Châtelier's principal says that a system will try to relieve stress put on it. An increase in pressure will drive the reaction to the side of an equation with fewer moles of gas. Since both sides have 2 moles of gas, an increase in pressure would not shift the equilibrium.

55. D

$$8g\ H_2/2g/mol = 4mol$$

$$2\ mole = 115.6\ kcal$$

$$4\ mole = 231.1\ kcal$$

56. B

$$760 \text{ torr} = 1 \text{ atm}$$
$$3{,}040 \text{ torr} = 4 \text{ atm}$$
$$P_1V_1 = P_2V_2$$
$$4(V_1) = 8(100)$$
$$V_1 = 200L$$

57. A

Twenty total electrons describes Ca. But it is an excited calcium molecule, because the $3d$ subshell is filled before the $4s$ subshell (usually the $4s$ subshell is filled first then the $3d$ subshell). Those two electrons must have jumped from the $4s$ to the $3d$ subshell.

58. B

$2.37V - 1.03V = +1.34V$; a spontaneous reaction. All of the other equations have negative values.

59. D

See the explanation to question 58.

60. C

Remember that in resonance forms, only electrons (bonds and nonbonding electron pairs) can move. If you examine the answer choices, you can see that the atomic linkages are all the same except for answer choice C; the carbon second from the left in choice C now has a methyl group bonded to it, whereas the other choices have hydrogens in this position.

61. A

One equivalent of an acid is the amount of the acid that is able to furnish one mole of hydrogen ion, so 1 mole of H_3PO_4 will furnish 3 mol of H^+ ions.

1 mol $H_3PO_4/3eq = x/.15eq$

0.05 mol H_3PO_4

62. C

This questions asks you which generalization CANNOT be made about the phase change of a pure substance from solid to liquid. When a phase change occurs, the internal energy of the system—that is, the total energy contained in the system—will change. The potential energy of the system during the phase change is the same as the internal energy. Therefore, when a solid melts into a liquid, the potential energy of the substance will change. Solids have a defined temperature, known as the melting point, at which they change to liquid. At this temperature, any energy added to the solid will go toward changing the phase, not changing the temperature, until all the solid has changed to liquid. We have already shown that the phase change occurs at a constant temperature, and because a change in kinetic energy is associated with a temperature change, the kinetic energy in the solid to liquid transition will remain the same.

63. E

Reading from left to right, the NH_4^+ loses a proton to become NH_3 while the H_2O picks up a proton to become H_3O^+. Thus, one proton is transferred from NH_4^+ to H_2O. Choices A and D are wrong because the oxidation states of the atoms are not changed. Double displacement would require that two species be exchanged; however, only one species, the proton, is being exchanged. Neutralization is a special case of a double displacement reaction.

64. C

The equilibrium constant expression shows the product of the product concentrations raised to their stoichiometric coefficients over the product of the reactant concentrations raised to their stoichiometric coefficients. For heterogeneous equilibria, pure solids and liquids are not included in this equation, but gases and aqueous components are included. So for this question you need to construct the correct equilibrium constant expression and calculate the equilibrium constant: The concentration of ammonia squared over the product of the nitrogen concentration and the cube of the hydrogen concentration. Plugging in the appropriate values, you get 15^2 over the product of 1 times 2^3. This gives an equilibrium constant of 28.

65. C

To answer this question, you must balance the equation given to find the stoichiometric coefficient for hydrochloric acid. Start with oxygen since it's the only element that is present in only one compound on each side of the equation. Since there are three oxygens on the right side of the equation, you have to place a "3" before the water molecule on the left. There are now six hydrogen atoms on the left side of the equation, so a "5" must be placed before the HCl on the right side for a total on six hydrogens on that side also. Before jumping right to the answer, we should balance the entire equation, since the answer might be five or some multiple of five, depending on the other compounds. There are six chlorines on the right, so you must place a "3" before the Cl_2 on the left. The equation is now fully balanced and it's clear that the answer is indeed five, choice C.

66. C

Specific gravity = density of substance/density of water

$$4 \text{ inches} = 1/3 \text{ foot}$$

$$\text{volume of cube } (1/3)^3 = 1/27 \text{ cubic inches}$$

$$\text{Specific gravity} = (20.8 \text{ lbs}/\tfrac{1}{27} \text{ cubic feet}) \div$$
$$62.4 \text{ lbs/cubic feet}$$
$$= 9$$

67. C

Ideal gases can be described by the Kinetic Molecular Theory of Gas. One part of this model is that collisions are elastic; there is no overall net gain or loss of energy. Inelastic collisions do occur in real gases. Also, a nuclear fusion reaction is a reaction in which two nuclei collide to form a new, heavier nucleus.

68. B

A Lewis base is a compound that can donate an electron pair. Choice B, hydrazine, consists of two nitrogens bonded to each other and to two hydrogen atoms each. A nitrogen atoms has a valence of five, and when it's bonded to another nitrogen and two hydrogens, it's only using three of its five valence electrons. Therefore, each nitrogen will have a pair of unbonded electrons. This makes hydrazine able to act as an electron pair donor, or Lewis base.

69. C

You know from the Pauli Exclusion Principle that no two electrons in the same atom can have identical quantum number. So, with n, l, and ml the same, you may be tempted to say the two electrons must be in different atoms. However, there is a fourth quantum number that could be different. The n value indicates that these electrons are in the same shell. The value of n also limits the value of l, defining the possible subshells the electrons can occupy. If they have the same l value, they must be in the same subshell if they are in the same atom. Well, the l value limits the ml value, which defines the specific spatial orientation of the orbital the electrons are in. If the ml value is the same for two electrons, they are in the same orbital. So, the identical ml value puts these two electrons in the same orbital. You know that the last quantum number, ms, must be different for these two electrons. So one will have an ms value equal to +1/2 and the other will be –1/2.

PRACTICE TEST 3

Periodic Table of the Elements

1 H 1.0																	2 He 4.0
3 Li 6.9	4 Be 9.0											5 B 10.8	6 C 12.0	7 N 14.0	8 O 16.0	9 F 19.0	10 Ne 20.2
11 Na 23.0	12 Mg 24.3											13 Al 27.0	14 Si 28.1	15 P 31.0	16 S 32.1	17 Cl 35.5	18 Ar 39.9
19 K 39.1	20 Ca 40.1	21 Sc 45.0	22 Ti 47.9	23 V 50.9	24 Cr 52.0	25 Mn 54.9	26 Fe 55.8	27 Co 58.9	28 Ni 58.7	29 Cu 63.5	30 Zn 65.4	31 Ga 69.7	32 Ge 72.6	33 As 74.9	34 Se 79.0	35 Br 79.9	36 Kr 83.8
37 Rb 85.5	38 Sr 87.6	39 Y 88.9	40 Zr 91.2	41 Nb 92.9	42 Mo 95.9	43 Tc (98)	44 Ru 101.1	45 Rh 102.9	46 Pd 106.4	47 Ag 107.9	48 Cd 112.4	49 In 114.8	50 Sn 118.7	51 Sb 121.8	52 Te 127.6	53 I 126.9	54 Xe 131.3
55 Cs 132.9	56 Ba 137.3	57 La* 138.9	72 Hf 178.5	73 Ta 180.9	74 W 183.9	75 Re 186.2	76 Os 190.2	77 Ir 192.2	78 Pt 195.1	79 Au 197.0	80 Hg 200.6	81 Tl 204.4	82 Pb 207.2	83 Bi 209.0	84 Po (209)	85 At (210)	86 Rn (222)
87 Fr (223)	88 Ra 226.0	89 Act 227.0	104 Unq (261)	105 Unp (262)	106 Unh (263)	107 Uns (262)	108 Uno (265)	109 Une (267)									

*	58 Ce 140.1	59 Pr 140.9	60 Nd 144.2	61 Pm (145)	62 Sm 150.4	63 Eu 152.0	64 Gd 157.3	65 Tb 158.9	66 Dy 162.5	67 Ho 164.9	68 Er 167.3	69 Tm 168.9	70 Yb 173.0	71 Lu 175.0
†	90 Th 232.0	91 Pa (231)	92 U 238.0	93 Np (237)	94 Pu (244)	95 Am (243)	96 Cm (247)	97 Bk (247)	98 Cf (251)	99 Es (252)	100 Fm (257)	101 Md (258)	102 No (259)	103 Lr (260)

ANSWER SHEET

1 (A) (B) (C) (D) (E)   19 (A) (B) (C) (D) (E)   37 (A) (B) (C) (D) (E)   55 (A) (B) (C) (D) (E)
2 (A) (B) (C) (D) (E)   20 (A) (B) (C) (D) (E)   38 (A) (B) (C) (D) (E)   56 (A) (B) (C) (D) (E)
3 (A) (B) (C) (D) (E)   21 (A) (B) (C) (D) (E)   39 (A) (B) (C) (D) (E)   57 (A) (B) (C) (D) (E)
4 (A) (B) (C) (D) (E)   22 (A) (B) (C) (D) (E)   40 (A) (B) (C) (D) (E)   58 (A) (B) (C) (D) (E)
5 (A) (B) (C) (D) (E)   23 (A) (B) (C) (D) (E)   41 (A) (B) (C) (D) (E)   59 (A) (B) (C) (D) (E)
6 (A) (B) (C) (D) (E)   24 (A) (B) (C) (D) (E)   42 (A) (B) (C) (D) (E)   60 (A) (B) (C) (D) (E)
7 (A) (B) (C) (D) (E)   25 (A) (B) (C) (D) (E)   43 (A) (B) (C) (D) (E)   61 (A) (B) (C) (D) (E)
8 (A) (B) (C) (D) (E)   26 (A) (B) (C) (D) (E)   44 (A) (B) (C) (D) (E)   62 (A) (B) (C) (D) (E)
9 (A) (B) (C) (D) (E)   27 (A) (B) (C) (D) (E)   45 (A) (B) (C) (D) (E)   63 (A) (B) (C) (D) (E)
10 (A) (B) (C) (D) (E)   28 (A) (B) (C) (D) (E)   46 (A) (B) (C) (D) (E)   64 (A) (B) (C) (D) (E)
11 (A) (B) (C) (D) (E)   29 (A) (B) (C) (D) (E)   47 (A) (B) (C) (D) (E)   65 (A) (B) (C) (D) (E)
12 (A) (B) (C) (D) (E)   30 (A) (B) (C) (D) (E)   48 (A) (B) (C) (D) (E)   66 (A) (B) (C) (D) (E)
13 (A) (B) (C) (D) (E)   31 (A) (B) (C) (D) (E)   49 (A) (B) (C) (D) (E)   67 (A) (B) (C) (D) (E)
14 (A) (B) (C) (D) (E)   32 (A) (B) (C) (D) (E)   50 (A) (B) (C) (D) (E)   68 (A) (B) (C) (D) (E)
15 (A) (B) (C) (D) (E)   33 (A) (B) (C) (D) (E)   51 (A) (B) (C) (D) (E)   69 (A) (B) (C) (D) (E)
16 (A) (B) (C) (D) (E)   34 (A) (B) (C) (D) (E)   52 (A) (B) (C) (D) (E)
17 (A) (B) (C) (D) (E)   35 (A) (B) (C) (D) (E)   53 (A) (B) (C) (D) (E)
18 (A) (B) (C) (D) (E)   36 (A) (B) (C) (D) (E)   54 (A) (B) (C) (D) (E)

right

wrong

	I	II	CE*		I	II	CE*
101	(T) (F)	(T) (F)	◯	109	(T) (F)	(T) (F)	◯
102	(T) (F)	(T) (F)	◯	110	(T) (F)	(T) (F)	◯
103	(T) (F)	(T) (F)	◯	111	(T) (F)	(T) (F)	◯
104	(T) (F)	(T) (F)	◯	112	(T) (F)	(T) (F)	◯
105	(T) (F)	(T) (F)	◯	113	(T) (F)	(T) (F)	◯
106	(T) (F)	(T) (F)	◯	114	(T) (F)	(T) (F)	◯
107	(T) (F)	(T) (F)	◯	115	(T) (F)	(T) (F)	◯
108	(T) (F)	(T) (F)	◯	116	(T) (F)	(T) (F)	◯

Use the answer key following the test to count up the number of questions you got right and the number you got wrong. (Remember not to count omitted questions as wrong.) "Compute Your Score" at the back of this section will show you how to find your score.

PART A

Directions: Each set of lettered choices below refers to the numbered formulas or statements immediately following it. Select the one lettered choice that best fits each formula or statement; then fill in the corresponding oval on the answer sheet. In each set, a choice may be used once, more than once, or not at all.

Note: For all questions involving solutions and/or chemical equations, you can assume that the system is in water unless otherwise stated.

Questions 1–5

    (A) Ca^+ and K
    (B) H^+ and He
    (C) Cl^- and F
    (D) O^- and S^+
    (E) Na^+ and O^-

1. Difference of 6 electrons

2. Same number of electrons

3. Difference of 9 electrons

4. Difference of 2 electrons

5. Difference of 1 electron

Questions 6–9

    (A) Solute
    (B) Solvent
    (C) Solubility
    (D) Aqueous solution
    (E) Solvation

6. Is present in a lesser amount in a solution

7. Solvent is water

8. Is present is greater quantity in a solution

9. Interaction between the solute and solvent molecules

GO ON TO THE NEXT PAGE

Questions 10–12

    (A) Sublimation
    (B) Condensation
    (C) Evaporation
    (D) Deposition
    (E) Melting

10.   Gas → Solid

11.   Gas → Liquid

12.   Solid → Gas

Questions 13–16

    (A) Proton
    (B) Neutron
    (C) Electron
    (D) Isotope
    (E) Ion

13.   Neutral charge; 1 amu

14.   Positive charge

15.   Negligible weight

16.   Negative charge; pairs with opposite spin

Questions 17–19

    (A) K
    (B) As
    (C) Be
    (D) Sc
    (E) Ir

17.   Metalloid

18.   Nonmetal

19.   Class IA Metal

Questions 20–23

    (A) O_2
    (B) CO_2
    (C) N_2
    (D) H_2SO_4
    (E) He

20.   Inert gas, not very soluble in water

21.   Very soluble in water, forming a very acidic solution

22.   When ignited, burns with a blue flame and is not very soluble in water

23.   Will not burn, is not very soluble in water, and makes a weakly acidic solution

GO ON TO THE NEXT PAGE

PART B

Directions: Each question below consists of two statements, statement I in the left-hand column and statement II in the right-hand column. For each question, determine whether statement I is true or false and whether statement II is true or false. Then, fill in the corresponding T or F ovals on your answer sheet. Fill in oval CE only if statement II is a correct explanation of statement I.

Examples:		
I		**II**
EX 1. HCl is a strong acid	**BECAUSE**	HCl contains sulfur.
EX 2. An atom of nitrogen is electrically neutral	**BECAUSE**	a nitrogen atom contains the same number of protons and electrons.

	I	II	CE
SAMPLE ANSWERS			
EX 1	● (F)	(T) ●	○
EX 1	● (F)	● (F)	●

	I		**II**
101.	The most important factor in determining the chemical properties of an element is the number of electrons in the outermost shell	BECAUSE	the number of electrons in the outer shell determines the bonding characteristics of the element.
102.	NH_3 is a Lewis base	BECAUSE	it can accept a proton.
103.	An element (X) with an atomic number of 16 has 14 electrons in X^{2-}	BECAUSE	two protons bind the two outermost electrons.
104.	Bromine is a stronger oxidizing agent than chlorine	BECAUSE	it has a large atomic radius.
105.	An exothermic reaction has a positive ΔH	BECAUSE	heat must be added to the system for the reaction to occur.
106.	Iron is an element	BECAUSE	it cannot be broken into smaller units and retain its physical and chemical properties.

GO ON TO THE NEXT PAGE

<table>
<tr><td colspan="3" align="center">I</td><td align="center">II</td></tr>
</table>

	I		II
107.	CCl_4 is a nonpolar compound	BECAUSE	the dipole moments are canceled out.
108.	A basic solution has more hydrogen ions than an acidic solution	BECAUSE	pH is defined as the $-\log [H^+]$.
109.	Atomic radii decrease down a group	BECAUSE	the higher the atomic number within a group, the smaller the atom.
110.	When an ideal gas is cooled its volume will increase	BECAUSE	temperature and volume are proportional.
111.	A 0.2M solution of carbonic acid is a weaker conductor of electricity than a 0.2M solution of HBr	BECAUSE	in solutions with the same concentration of solute molecules, H_2CO_3 is less dissociated than HBr.
112.	An equation where two gas molecules combine to form one gas molecule in equilibrium will increase the yield of the product when the pressure is increased	BECAUSE	increased pressure always favors products.
113.	Water makes a good buffer	BECAUSE	a good buffer will resist changes in pH.
114.	In the kinetic theory of gases, collisions between gas particles and the walls of the container are considered elastic	BECAUSE	gas molecules are considered pointlike, volumeless particles with no intermolecular forces and in constant, random motion.
115.	The oxidation state of Cr in $Al_2(Cr_2O_7)_3$ is +3	BECAUSE	as a neutral compound, the sum of oxidation numbers of all the atoms must equal zero.
116.	The entropy of a solid increases when it is dissolved in solvent	BECAUSE	it becomes less ordered.

RETURN TO THE SECTION OF YOUR ANSWER SHEET YOU
STARTED FOR CHEMISTRY AND ANSWER QUESTIONS 24–69.

GO ON TO THE NEXT PAGE

PART C

Directions: Each of the incomplete statements or questions below is followed by five suggested completions or answers. Select the one that is best for each case and fill in the corresponding oval on the answer sheet.

24. Which of the following is a Lewis base?

    (A) NH_4^+

    (B) CH_4

    (C) PH_3

    (D) CH_3CH_3

    (E) HCl

25. When one mole of sulfur burns to form SO_2, 1,300 calories are released. When one mole of sulfur burns to form SO_3, 3,600 calories are released. What is the ΔH when one mole of SO_2 is burned to form SO_3?

    (A) 3,900 cal

    (B) –1,950 cal

    (C) 1,000 cal

    (D) –500 cal

    (E) –2,300 cal

26. How many atoms are in a 36.5 g sample of SF_6 gas?

    (A) 1.51×10^{22}

    (B) 1.06×10^{22}

    (C) 1.51×10^{23}

    (D) 1.06×10^{24}

    (E) 1.51×10^{-24}

27. Which of the following statements about molecular and empirical formulas is (are) false?

    I.   A given compound can have the same molecular and empirical formula.

    II. The molecular formula is a whole number multiple of the empirical formula.

    III. H_2O_2 represents the empirical formula of hydrogen peroxide.

    (A) III only

    (B) I and II only

    (C) II and III only

    (D) I, II, an III

    (E) I only

28. The product formed when oxygen and hydrogen are mixed in a test tube at room temperature is

    (A) hydrogen peroxide

    (B) water

    (C) a base

    (D) a zwitterion

    (E) no reaction takes place

GO ON TO THE NEXT PAGE

29. All of the following statement are consistent with Bohr's model of the atom EXCEPT

   (A) an electron may assume an infinite number of velocities

   (D) an atom is most stable when its electronic configuration is that of the ground state

   (C) the electron shell numbers represent the principal energy levels

   (D) electrons in orbitals closest to the nucleus have the lowest energy

   (E) they are all consistent

30. What is the amount of heat given off by 100 g of O_2 when it is used to burn an excess of sulfur according to the following reaction?

   $S(s) + O_2 \rightarrow SO_2 (g) \quad \Delta H = -296kJ/mole$

   (A) 925,000J

   (B) 29,000J

   (C) 1,850J

   (D) 296J

   (E) 100J

31. If electricity costs 10 cents/coulomb, which of the following would be have the highest cost/mole?

   (A) copper from $CuSO_4$

   (B) sodium from NaCl

   (C) chlorine from KCl

   (D) hydrogen from H_2O

   (E) iron from $FeCl_3$

32. What volume of water would be needed to dilute 50mL of a 3M HCl solution to 1M?

   (A) 25mL

   (B) 50mL

   (C) 75mL

   (D) 100mL

   (E) 150mL

33. If the pressure of a gas sample is doubled at constant temperature, the volume will be

   (A) 4 times the original

   (B) 2 times the original

   (C) 1/2 of the original

   (D) 1/4 of the original

   (E) 1/8 of the original

34. Which of the following molecules has a trigonal pyramidal geometry?

   (A) BH_3

   (B) H_2O

   (C) CH_4

   (D) NH_3

   (E) $AlCl_3$

35. The field of organic chemistry focuses mainly on

   (A) oxygen-containing compounds

   (B) carbon-containing compounds

   (C) proteins and enzymes

   (D) carbohydrates

   (E) nitrogen-containing compounds

GO ON TO THE NEXT PAGE

36. Which of the following is the most electronegative element?

    (A) He
    (B) I
    (C) N
    (D) O
    (E) C

37. Which of the following has the greatest affinity for electrons?

    (A) Na
    (B) Cl
    (C) Br
    (D) K
    (E) C

38. What would be the approximate weight of 1.204×10^{24} bromine atoms?

    (A) 80 grams
    (B) 120 grams
    (C) 160 grams
    (D) 180 grams
    (E) 200 grams

39. What is the oxidation state of Mn in $KMnO_4$?

    (A) –7
    (B) –3
    (C) 0
    (D) +3
    (E) +7

40. What is the daughter element produced by technetium-99m (atomic number-43; mass number-99) after γ-decay?

    (A) $^{98}_{43}Tc$

    (B) $^{99}_{42}Mo$

    (C) $^{99}_{44}Ru$

    (D) $^{95}_{41}Nb$

    (E) $^{99}_{43}Tc$

41. What volume of water would be needed to dilute 50 ml of 3M H_2SO_4 to 0.75M?

    (A) 50 mL
    (B) 100 mL
    (C) 150 mL
    (D) 200 mL
    (E) 250 mL

42. All members of Group IA have similar reactivity because

    (A) they have the same number of protons
    (B) they have the same number of electrons
    (C) they have similar outer shell electron configurations
    (D) they have valence electrons with the same quantum numbers
    (E) they have the same number of neutrons

GO ON TO THE NEXT PAGE

43. If one mole of a gas originally at STP is placed in a container where the pressure is doubled and the temperature in K is tripled, what is the new volume in L?

(A) 2.2

(B) 5.6

(C) 7.5

(D) 11.2

(E) 33.6

44. Chlorophyll, a green pigment involved in the light reactions of photosynthesis, consists of 2.4312 percent Mg. If you are given a 100 g sample of chlorophyll, how many atoms of Mg will it contain?

(A) 6.02×10^{22} atoms

(B) 6.02×10^{23} atoms

(C) 6.02×10^{24} atoms

(D) 6.02×10^{25} atoms

(E) None of the above

45. $H_2O + H_2O \rightleftharpoons H_3O^+ + OH^-$

The reverse reaction for the one shown above is exothermic. If the temperature is lowered, what will occur?

(A) the pH will decrease

(B) the equilibrium will shift to the right

(C) the concentration of H_3O^+ ions will increase

(D) the equilibrium will shift to the left

(E) temperature does not affect equilibrium

46. Which of the following reactions shows a decrease in entropy?

(A) $C(s) + 2H_2 (g) \rightarrow CH_4 (g)$

(B) $H_2O(g) \rightarrow H_2(g) + 1/2O_2(g)$

(C) $2NI_3(s) \rightarrow N_2(g) + 3I_2(g)$

(D) $2O_3(g) \rightarrow 3O_2 (g)$

(E) None of the above

47. Which of the following is an incorrect association?

(A) Mendelev–periodic table

(B) Faraday–electrolytic cells

(C) Millikan–charge of electrons

(D) Rutherford–photoelectric effect

(E) They are all correct.

48. Which of the following is true of an electrolytic cell?

(A) An electric current causes an otherwise nonspontanteous chemical reaction to occur.

(B) Reduction occurs at the anode.

(C) A spontaneous electrochemical reaction produces an electric current.

(D) The electrode to which electrons flow is where oxidation occurs.

(E) None of the above

GO ON TO THE NEXT PAGE

49.

	1st Ionization Energy (eV)	2nd Ionization Energy (eV)
U	5.6	5.6
V	1.7	2.9
X	1.1	13.6
Y	12.4	2.8
Z	2.9	1.7

From the information given in the table above, which of the following is most probably a Group IA metal?

(A) U

(B) V

(C) X

(D) Y

(E) Z

50. Three canisters, A, B, and C, are all at the same temperature, with volumes of 2.0, 4.0, and 6.0 liters, respectively. Canister A contains 0.976 g of argon gas at a pressure of 120 torr, Canister B contains 1.37 g of nitrogen gas at a pressure of 210 torr, and canister C is completely empty at the start. Assuming ideality, what would the pressure become in canister C if the contents of A and B are completely transferred to C?

(A) 180 torr

(B) 330 torr

(C) 675 torr

(D) 0.25 atm

(E) None of the above

51. H_2O has a higher boiling point than HF because

(A) H_2O is more polar than HF

(B) H_2O can form more hydrogen bonds

(C) H_2O has a higher molecular weight

(D) H_2O has more atoms

(E) H_2O does not have a higher boiling point that HF

Questions 52–54 refer to the following graph.

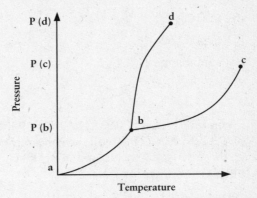

52. In what pressure range will the compound sublime?

(A) pressures less than $P_{(b)}$

(B) pressures between $P_{(b)}$ and $P_{(c)}$

(C) pressures between $P_{(d)}$ and $P_{(c)}$

(D) pressures between $P_{(b)}$ and $P_{(d)}$

(E) The compound will not sublime under any pressure.

GO ON TO THE NEXT PAGE

53. The phase change represented by crossing curve bd is:

(A) evaporation

(B) sublimation

(C) condensation

(D) melting

(E) boiling

54. The phase change represented by crossing curve bc is

(A) freezing

(B) melting

(C) deposition

(D) sublimation

(E) evaporation

55. If you mix 3 liters of 0.5M NaCl with 9 liters of 0.2777M NaCl what will the concentration of the final solution be, assuming that volumes are additive?

(A) 0.33M

(B) 0.39M

(C) 0.5777M

(D) 0.5777m

(E) None of the above

56. Which of the following statements about boiling point elevation is NOT true?

I.   Addition of a nonvolatile solute raises the vapor pressure of a solution. Therefore, a higher temperature will be needed to produce boiling.

II. The molal concentration of solute particles dissolved in the solvent is an important factor in determining the molal boiling point elevation constant.

III. The identity of solute particles dissolved in the solvent is an important factor in determining the molar boiling point elevation constant.

(A) I only

(B) II only

(C) I and III only

(D) II and III only

(E) I, II, and III

57. What is the sum of the coefficients of the products for the following reaction, after balancing?

$...K_2Cr_2O_7 + ...HCl \rightarrow ...KCl + ...CrCl_3 + ...H_2O + ...Cl_2$

(A) 10

(B) 12

(C) 13

(D) 14

(E) 15

GO ON TO THE NEXT PAGE

Question 58–60 refer to the following equation:

$$^{14}_{6}C \rightarrow ^{14}_{7}N + X$$

58. What is X?

(A) $^{4}_{2}He$

(B) $^{0}_{+1}e$

(C) $^{0}_{-1}e$

(D) $^{1}_{1}H$

(E) $^{12}_{6}C$

59. This nuclear reaction is an example of:

(A) α decay

(B) β^- decay

(C) β^+ decay

(D) fusion

(E) γ decay

60. Water is formed by the addition of 4.0g of $H_2(g)$ to an excess of $O_2(g)$. If 27 g of H_2O is recovered, what is the percent yield for the reaction?

(A) 25%

(B) 50%

(C) 75%

(D) 100%

(E) Cannot be determined

61. For which of the following is there an increase in entropy?

(A) $Na\ (s) + H_2O\ (l) \rightarrow NaOH\ (aq) + H_2(g)$

(B) $I_2\ (g) \rightarrow I_2\ (s)$

(C) $H_2SO_4\ (aq) + Ba(OH)_2\ (aq) \rightarrow BaSO_4\ (s) + H_2O\ (l)$

(D) $H_2\ (g) + 1/2\ O_2\ (g) \rightarrow H_2O(l)$

(E) None of the above

62. Which of the following compounds contains the greatest percentage of oxygen by weight?

(A) $C_3H_6O_5Cl$

(B) $C_3H_6O_2$

(C) $C_5H_{10}O_5$

(D) $C_4H_8O_3$

(E) They are all equal

63. In the reaction $2SO_2 + O_2 \rightarrow 2SO_3$, 0.25 mole of sulfur dioxide is mixed with 0.25 mole of oxygen and allowed to react. What is the maximum number of moles of SO_3 that can be produced?

(A) 0.0625 moles

(B) 0.125 moles

(C) 0.25 moles

(D) 0.5 moles

(E) 1.0 mole

GO ON TO THE NEXT PAGE

64. What happens to the pH of a buffer system if one halves the concentration of both the acid and the salt?

(A) Nothing

(B) pH goes up because there is less total acid in the solution.

(C) pH goes down because there is less conjugate base to mask the presence of the acid.

(D) It depends upon the original concentration of acid and salt.

(E) It is impossible to predict.

65. What is the mass of nitrogen in a 50.0 g sample of sodium nitrite ($NaNO_2$)?

(A) 20.2g

(B) 16.4g

(C) 10.1g

(D) 8.23g

(E) 23.4g

66. Which of the following conditions guarantee a spontaneous reaction?

(A) Positive ΔH, positive ΔS

(B) Positive ΔH, negative ΔS

(C) Negative ΔH, negative ΔS

(D) Negative ΔH, positive ΔS

(E) None of the above

67. Arrange the following elements in order of decreasing nonmetallic character:

Ge, Sn, Pb, Si

(A) Pb, Sn, Ge, Si

(B) Ge, Sn, Pb, Si

(C) Si, Ge, Sn, Pb

(D) They all have equal nonmetallic character since they are all in the same column of the periodic table

(E) None of the above

68. Which of the following is not a property of Group IA elements?

(A) low ionization energies

(B) low electronegativities

(C) high melting points

(D) metallic bonding

(E) electrical conductivity

69. Electron affinity is defined as

(A) the change in energy when a gaseous atom in its ground state gains an electron

(B) the pull an atom has on the electrons in a chemical bond

(C) the energy required to remove a valence electron from a neutral gaseous atom in its ground state

(D) the energy difference between an electron in its ground and excited states

(E) None of the above

STOP!   **END OF TEST.**

Turn the page
for answers and explanations
to Practice Test 3.

Answer Key

1.	D	15.	C	29.	A	43.	E	57.	D
2.	A	16.	C	30.	A	44.	A	58.	C
3.	C	17.	B	31.	E	45.	D	59.	B
4.	B	18.	D	32.	D	46.	A	60.	C
5.	E	19.	A	33.	C	47.	D	61.	A
6.	A	20.	E	34.	D	48.	A	62.	C
7.	D	21.	D	35.	B	49.	C	63.	C
8.	B	22.	C	36.	D	50.	A	64.	A
9.	E	23.	B	37.	A	51.	B	65.	C
10.	D	24.	C	38.	C	52.	A	66.	D
11.	B	25.	E	39.	E	53.	D	67.	C
12.	A	26.	D	40.	E	54.	E	68.	C
13.	B	27.	A	41.	C	55.	A	69.	A
14.	A	28.	E	42.	C	56.	C		

101.	T, T, CE	109.	F, F
102.	T, T	110.	F, T
103.	F, F	111.	T, T, CE
104.	F, T	112.	T, F
105.	F, F	113.	F, T
106.	T, T, CE	114.	T, T
107.	T, T, CE	115.	F, T
108.	F, T	116.	T, T, CE

Compute Your Practice Test Score

Step 1: Figure out your raw score. Refer to your answer sheet for the number right and the number wrong on the practice test you're scoring. (If you haven't checked your answers, do that now, using the answer key that follows the test.) You can use the chart below to figure out your raw score. Multiply the number wrong by 0.25 and subtract the result from the number right. Round the result to the nearest whole number. This is your raw score.

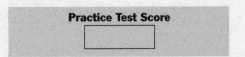

SAT II: CHEMISTRY Practice Test

Number right $-$ (0.25 x Number wrong) $=$ Raw score (rounded)

Step 2: Find your practice test score. Find your raw score in the left column of the table below. The score in the right column is your Practice Test score.

Find Your Practice Test Score

Raw	Scaled	Raw	Scaled	Raw	Scaled	Raw	Scaled	Raw	Scaled	Raw	Scaled
85	800	67	720	49	610	31	510	13	400	−5	290
84	800	66	720	48	610	30	500	12	400	−6	290
83	800	65	720	47	600	29	500	11	390	−7	280
82	800	64	710	46	600	28	490	10	390	−8	270
81	800	63	700	45	590	27	480	9	380	−9	270
80	800	62	700	44	580	26	480	8	370	−10	260
79	790	61	700	43	580	25	470	7	360	−11	250
78	790	60	690	42	570	24	460	6	360	−12	250
77	780	59	680	41	570	23	460	5	350	−13	240
76	780	58	670	40	560	22	460	4	350	−14	240
75	780	57	670	39	560	21	450	3	340	−15	230
74	770	56	660	38	550	20	440	2	330	−16	230
73	760	55	650	37	550	19	440	1	330	−17	220
72	760	54	640	36	540	18	430	0	320	−18	220
71	750	53	640	35	530	17	430	−1	320	−19	210
70	740	52	630	34	520	16	420	−2	310	−20	200
69	740	51	630	33	520	15	420	−3	300	−21	200
68	730	50	620	32	520	14	410	−4	300		

Practice Test Score

A note on your practice test scores: Don't take these scores too literally. Practice test conditions cannot precisely mirror real test conditions. Your actual SAT II: Chemistry Subject Test score will almost certainly vary from your practice test scores. Your scores on the practice tests will give you a rough idea of your range on the actual exam.

Answers and Explanations to Practice Test 3

1. D

O^- has 9 electrons, while S^+ has 15 electrons.

2. A

Ca^+ has 19 electrons, as does K.

3. C

Cl^- has 18 electrons, while F has 9 electrons.

4. B

H^+ has no electrons, while He has 2 electrons.

5. E

Na^+ has 10 electrons, while O^- has 9 electrons.

6. A

Solute is defined as something that is dissolved in a solvent to make a solution. It is the component that is present in a lesser amount than solvent.

7. D

In aqueous solutions, the solvent is H_2O.

8. B

Solvent is the component of a solution present in the greatest amount. It is the substance in which the solute is dissolved.

9. E

Solvation is the interaction between the solute and the solvent molecules; it is often the development of a cagelike network of a solution's solvent molecules about a molecule of ion of the solute.

10. D

Deposition is the conversion of a gas directly into a solid without passing through the liquid state.

11. B

Condensation is the conversion of a gas into a liquid state, an instance of which you can see on a pot lid during the boiling of water.

12. A

Sublimation is the conversion of a solid into a gas without going through a liquid phase.

13. B

A neutron is a subatomic particle with a charge of zero and a mass of 1 amu. Neutrons can be found in all nuclei, except for one of the hydrogen isotopes.

14. A

A proton is a subatomic particle with a charge of +1 and mass of 1 amu. It is found in all atomic nuclei.

15. C

An electron is a subatomic particle with a charge of −1 and negligible weight. It is found around the nucleus.

16. C

See the explanation to question 15.

17. B

Metalloids' properties lie in between those of metals and nonmetals and are found between them on the Periodic Table. They are semiconductors in that they do conduct electricity, but not very well. They lie in a diagonal from B to At.

18. D

Nonmetals have no metallic character. Three examples are N_2, O_2, and carbon. They are typically poor conductors of heat and electricity.

19. A

Class IA metals are found on the first column of the Periodic Table. Metals tend to have shine and luster, which makes them easily recognizable. They conduct heat and electricity very well.

20. E

Helium is an inert gas that is not very soluble in H_2O.

21. D

H_2SO_4 is a very strong acid (sulfuric acid) that when bubbled through H_2O will make a very acidic solution.

22. C

N_2 is not very soluble, but will burn with a blue flame.

23. B

CO_2 will extinguish a glow and will slightly associate with H_2O to form carbonic acid ($H_2CO_3 \rightarrow H^+ + HCO_3^-$).

101. T, T, CE

In the Periodic Table, groups represent elements that have the same electronic configuration in their outermost shell and share similar chemical properties. These valence electrons are involved in chemical bonding and determine the chemical reactivity of an element.

102. T, T

NH_3 is a Lewis base because it can donate an electron pair; it is also a Brønsted-Lowry base because it can accept a H^+ from a solution, but this is not what makes it a Lewis base.

103. F, F

An element with the atomic number of 16 will have 16 protons and electrons in a neutral atom. The anion with –2 charge is formed upon the addition of 2 more electrons.

104. F, T

Chlorine is a better oxidizing agent because it has a greater electronegativity, which is the ability to attract electrons. Remember that as you go up and from left to right on the Periodic Table, electronegativity increases. Bromine has a larger atomic radius than chlorine. Also remember that as you go from right to left and down on the Periodic Table, atomic radius increases.

105. F, F

Exothermic reactions have a negative ΔH and give off heat to their surroundings as the reaction progresses.

106. T, T, CE

An element such as Fe is a substance in which all atoms have the same atomic number. It cannot be broken down by chemical reactions into anything that is more stable or simpler.

107. T, T, CE

CCl_4 is a nonpolar covalent bond. CCl_4 has four polar C–Cl bonds so that the shape is tetrahedral. The four bond dipoles point to the vertices of the tetrahedron and cancel each other out, resulting in a nonpolar molecule.

108. F, T

A basic solution has more OH^- ions than H^+ ions, but pH is defined as $-\log[H^+]$. pOH is $-\log[OH^-]$.

109. F, F

Atomic radii increase down a group because the valence electrons are farther form the nucleus and are able to resist the positive charge of the protons, which results in a larger atom.

110. F, T

$V_1/T_1 = V_2/T_2$. In an ideal gas, temperature and volume are proportional, so when a gas is cooled, the volume decreases.

111. T, T, CE

Carbonic acid will not dissociate completely in H_2O. $H_2CO_3 \leftrightharpoons H^+ + HCO_3^-$. Therefore, in a 0.2 M solution, there will be fewer ions than in a 0.2 M solution of HBr (which dissociates completely). Electrical conductivity depends on the number of ions in solution, so HBr would be the better electrical conductor.

112. T, F

In the equation $A + B \leftrightharpoons C$, when you increase pressure, the system will try to relieve the stress and skew the equilibrium to the side of the equation with fewer moles of gas formed, and more C will be produced.

113. F, T

Water does not make a good buffer. A good buffer is defined as a pair of solutes (salts) that can keep the pH of a solution almost constant if either acid or base is added. Water is unable to do this.

114. T, T

The kinetic molecular theory of gases has five assumptions to it:

1. Gases are made up of particles whose volumes are negligible compared to the container volume.

2. Gas atoms or molecules exhibit no intermolecular attractions or repulsions.

3. Gas particles are in continuous, random motion, undergoing collisions with other particles and the container walls.

4. Collisions between any two gas particles are elastic, meaning that there is no overall gain or loss of energy.

5. The average kinetic energy of gas particles is proportional to the absolute temperature of the gas, and is the same for all gases at a given temperature. That

collisions between the particles and the container walls are elastic is not necessarily a consequence of the other assumptions.

115. F, T

$Al_2(Cr_2O_7)_3$

The oxidation number of Al is 3+ while Cr_2O_7 must be –2.

Oxygen   7(–2)   = –14

Therefore, 2Cr must be +12 to result in a net charge of –2.

2Cr = +12; Cr = +6

116. T, T, CE

An increase in entropy describes an increase in randomness; gas > liquid > solid. When a solid is dissolved in water, it dissociates and entropy increases as it becomes less ordered.

24. C

A Lewis base is a substance that can donate a pair of electrons. Among the answer choices, only PH_3 has a pair of nonbonding electrons, making choice C the correct response. The easiest way to determine if a substance is a Lewis base is to draw its Lewis dot structure and see if a lone pair of electrons exists.

25. E

The difference between SO_2 and SO_3 is 2,300 cal. This quantity must be a negative, since this is an exothermic reaction.

26. D

While this is a relatively simple calculation, the question tests your ability to read carefully. The molecular weight of sulfur hexafluoride is 146 grams/mole. If you divide the mass given, 36.5 grams, by the molecular weight, you find that you have 0.25 mole of sulfur hexafluoride molecules. A mole is Avogadro's number of particles, in other words, there are 6.022×10^{23} particles per mole. If you multiply the 0.25 mole of sulfur hexafluoride by Avogadro's Number, you'll find the number of molecules of sulfur hexafluoride, 1.511×10^{23}. 1.51×10^{23} is the number of MOLECULES of sulfur hexafluoride in the sample, not the number of atoms. There are seven atoms in each molecule, so you must multiply the number of molecules by the number of atoms per molecule, seven. This trick is in making sure all your units cancel. The answer is 1.06×10^{24} atoms.

27. A
Hydrogen peroxide has the empirical formula, HO.

28. E
No reaction takes place. In order for a reaction to take place to produce H_2O, energy must be added to the system (typically in the form of heat).

29. A
Choice A, which says that an electron may assume an infinite number of different velocities, is true in classical mechanics but not in Bohr's model. Bohr used quantum theory in developing his atomic model and placed specific conditions on the possible values of the electron velocity. Since we're looking for the incorrect statement, choice A is the correct answer.

30. A
100g O_2 divided by 32g/mol = 3.125 moles of oxygen. You're given 296kJ/mol in the equation (a negative value means heat is released) so multiply 296kJ/mole x 3.125 mole to get 925kJ. 925kJ is equal to 925,000 Joules of heat.

31. E
At 10 cents/coulomb, the question is really asking you which reaction involves the transfer of the most electrons.

Cu from Cu^{2+}	2 electrons
Na from Na^+	1 electron
Cl from Cl^-	1 electron
H from H^+	1 electron
Fe from Fe^{3+}	3 electrons

Therefore, the production of Fe from Fe^{3+} involves the transfer of the most electrons and would cost the most.

32. D
$$M_1V_1 = M_2V_2$$
$$3M(50mL) = 1M(V_2 mL)$$
$$V_2 = 150mL$$

Therefore 100 mL need to be added.

33. C
This question is an application of Boyle's Law. This states that at constant temperature, the pressure and volume of a gas are inversely proportional to each other. Therefore, since the pressure of the gas in the question is increased, the volume must decrease.

34. D
The Lewis Structure of NH_3 is:

The central atom has three bonding electron pairs and one nonbonding pair for a total of four pairs. The four electron pairs will be farthest apart when they occupy the four corners of a tetrahedron. Since one of the four pairs is a lone pair, the observed geometry is trigonal pyramidal.

35. B
Organic chemistry focuses only on carbon containing compounds such as carbohydrates, alcohols, and ethers, and on their reactions.

36. D
Electronegativity increases in going from the lower left to the upper right of the Periodic Table. So all you need to do here is locate the answer choices in the periodic table and find the one that is the closest to the upper right hand corner. When you do this, you'll see that choice D is the correct answer.

37. B

The atoms with the greatest affinity for electrons are found in the upper right corner of the Periodic Table (excluding the noble gases). Fluorine is an exception to the trend because of electron-electron repulsions in small atoms.

38. C

There are 6.023×10^{23} atoms in a mole of a compound, so 1.204×10^{24} atoms would be approximately 2 moles of Br. 2 moles of Br $\times$ 80g/mole = 160 g.

39. E

$$K = +1$$
$$4O = 4(-2) = -8$$

Neutral atom must add up to 0: $+1 + -8 + Mn = 0$
$$Mn = +7$$

40. E

γ decay is a high energy emission that has no mass or charge. Therefore $^{99}_{43}$Tc after γ decay would still be $^{99}_{43}$Tc.

41. C

The number of moles of solute in the solution will be the same after dilution as before, and the number of moles in each case is equal to the molar concentration multiplied by the volume of solution. This means that the initial concentration times the initial volume will be equal to the final concentration times the final volume ($M_1V_1 = M_2V_2$). So, the final volume will equal $M_1(V_1)/M_2$. Plugging into this equation, we find that $V_2 = 200$ml (50ml+150ml).

42. C

All members of Group IA have similar reactivities because they have a similar valence shell configuration (one loosely bound electron). They lose it easily to form univalent cations and react readily with nonmetals, especially halogens.

43. E

One mole of a gas at STP will have a volume of 22.4L. If the pressure is doubled, the volume will halve since they are inversely proportional. If the temperature is tripled, the volume will triple since they are proportional. 3/2(22.4L) = 33.6L

44. A

$$100g(2.4312\%Mg) = 2.4312g\ Mg$$
$$2.4312g/24.3g/mol = 0.1mol$$
$$6.022 \times 10^{23}atom/mol \times 0.1mole$$
$$= 0.6 \times 10^{23}\ atoms$$
$$= 6 \times 10^{22}\ atoms$$

45. D

If the reverse reaction is exothermic, the equation can be written as such:
$$heat + H_2O + H_2O \Leftrightarrow H_3O^+ + OH^-$$
If you remove heat, the equilibrium will shift to the left in an attempt to produce more heat.

46. A

An increase or decrease of entropy is easy to predict when the chemical reactions involve gases. If there are more moles of gas on the products side than on the reactants side, there is an increase in entropy. If there are fewer moles of gas on the products side than on the reactants side, there is a decrease in entropy. For choice A, there are two mol of gas on the reactants side and one mole of gas on the products side—there is, therefore, a decrease in entropy for this reaction.

47. D

Mendeleev discovered the properties of the elements and how they had regular intervals, and formulated the first Periodic Table. Faraday found that chemical changes could occur when an electrical current was sent through certain chemical solutions. Milikan did the famous oil drop experiment which determined the charge on an electron. Rutherford discovered the nucleus by striking thin metal foils with alpha particles.

KAPLAN

48. A

Electrochemical reactions that are nonspontaneous; those having a positive ΔG can be driven to completion by passing an electric current through the solution. This process is known as electrolysis and the cell is called an electrolytic cell. In an electrolytic cell, the anode is positively charged and the cathode is negatively charged—the opposite of a galvanic cell. Just like a galvanic cell, oxidation occurs at the anode and reduction occurs at the cathode.

49. C

Ionization energy is defined as the amount of energy required to remove an electron from a given species. Ionization energy is usually expressed in energy per particle (as it is here) or in energy per mole, and energy, in turn, is usually expressed in electron volts (eV), joules, or kilojoules. The first ionization energy of an element is the energy required to remove an electron from a neutral atom of that element, while the second ionization energy is the energy required to remove a second electron, i.e., to remove an electron from the +1 cation. A particularly useful piece of logic/knowledge here is that it will be more difficult to remove an electron from a positively charged species than it will be to remove an electron from a neutral version of the same species. In other words, the second ionization energy of an element is always greater than its first ionization energy. Based on this fact alone, choices A, D, and E can be eliminated. To choose between the remaining two choices, we must reason that a small value for ionization energy corresponds to relative ease of removal of the electron. A Group IA metal atom lose its first electron with relative ease, but after that it will possess an electronic configuration similar to that of a noble gas. It will therefore be difficult to remove another electron, implying a high second ionization energy. Looking at the numbers in the table, we see that both remaining choices have relatively small values for first ionization energy, but that X has a much higher second ionization energy. It follows that element V is most likely a Group IIA element (e.g., Mg, Ca) while element X is most likely a member of Group IA (e.g., Na, K). Choice C, then, is the best choice.

50. A

$P_1V_1 = P_2V_2$

Argon gas	$120(2.0) = P_2(6)$
Canister C	$40 \text{ torr} = P_2$
Nitrogen gas	$210(4) = P_2(6)$
Canister C	$140 \text{ torr} = P_2$

Pressures are additive as described by Dalton's Law of partial pressures:

$40 \text{torr} + 140 \text{torr} = 180 \text{ torr}$

51. B

This question asks you to determine why water has a higher boiling point than hydrogen fluoride. Choice B says that water has a higher boiling point because it can form more hydrogen bonds. Hydrogen bonding does affect boiling points by increasing the attraction between the molecules of a compound. Water is capable of forming as many as four hydrogen bonds per molecule, while hydrogen fluoride can only form two. This intermolecular attraction leads to a complexation of water molecules and contributes to the high boiling point of water.

52. A

curve abd = solid
curve dbc = liquid
curve cba = gas

At pressures less than (b), solids are able to convert directly to gas (sublimation) and gases are able to convert directly to solid (deposition).

53. D

Line bd describes the conversion between liquid and solid and therefore between melting and freezing.

54. E

Line bc describes the conversion between liquid and gas and therefore between evaporation (boiling) and condensation.

55. A

In a total volume of 12L, you must calculate the number of moles you have.

$$3L(.5mol/L) = 1.5 \text{ mol}$$
$$9L(.277mol/L) = \sim2.5 \text{ mol}$$
$$\text{Total} = \sim4.0 \text{ mole}$$
$$4.0mol/12L = \sim.3M \ (.3333M)$$

56. C

Boiling point elevation is a colligative property, one due solely to the number of particles and not the nature of the particles. Therefore, III is false and II is true. I is false because vapor pressure is lowered by addition of solute which increases boiling point. An increase in vapor pressure would lead to boiling point depression.

57. D

$$1K_2Cr_2O_7 + 14HCl \rightarrow 2KCl + 2CrCl_3 + 7H_2O + 3Cl_2$$
$$2 + 2 + 7 + 3 = 14$$

58. C

In the reaction, X is an example of β^- decay which is an electron ($_{-1}^{0}e$) emitted by radioactive decay.

59. B

See the explanation to question 58.

60. C

In the equation $2H_2 + O_2 \rightarrow 2H_2O$, if you begin with 4g of H_2 at 2g/mole, you have 2 mol of H_2. According to the equation, a 100 percent yield would result in 2 moles of H_2O. Two moles of H_2O at 18g/mol would weight 36 grams. Percent yield is defined as actual/theoretical $\times$ 100%. $(27/36) \times 100\% = 75\%$

61. A

An increase in entropy is an increase in randomness and loss of order. Gas going to a solid is an increase in order and two molecules combining to form one molecule is also an increase in order. A solid (Na) becoming an ionic compound (NaOH) and a gas (H_2) is a decrease in order and an increase in entropy.

62. C

This question requires you to know how to calculate the percentage by weight of an element in a compound. You need to first calculate the weight of oxygen in each compound and then divide that value by the compound's molecular weight. The weight of oxygen in choice A is 80 grams/mole and the MW = 157 grams/mole. This is 51% oxygen. Doing the same with the other answer choices, you get 43% oxygen (choice B), 53% oxygen (choice C), and 46% oxygen (choice D). Obviously, choice C contains the most oxygen by weight and is the correct answer.

63. C

If you have 0.25 mole of SO_2 and 0.25 mole of O_2, SO_2 is the limiting reagent since you need two moles per reaction compared to one mole of O_2. Since 2 moles of SO_2 yields 2 moles of SO_3, then 0.25 moles of SO_2 would yield 0.25 moles of SO_3.

64. A

A buffer solution is prepared from a weak acid and its conjugate base in near equal quantities. As long as these conditions are met, the pH should remain the same. A buffer solution with the concentrations of each of these components halved may have less ability to buffer, but the initial solution will have the same pH.

65. C

The molecular weight of sodium nitrite is 69 grams/mole. Since there is one nitrogen atom per formula unit, we can find the weight fraction of sodium nitrite that is nitrogen by dividing the atomic weight of nitrogen, 14 grams/mole, by the molecular weight of sodium nitrite, 69 grams/mole. If we multiply this fraction, 14/69, by 50 grams, we get 10.1 grams.

66. D

This question asks you to predict which combination of ΔH and $T\Delta S$ values will always give a spontaneous reaction. Recall that spontaneous reactions have negative values of Gibbs free energy, ΔG, and that $\Delta G = \Delta H - T\Delta S$, where ΔH is the change in enthalpy, ΔS is the change in entropy, and T is the absolute temperature. This is the key equation, which you probably have memorized. From this equation, it's clear that the best way to guarantee a negative ΔG is to have a negative ΔH value and a positive ΔS value since T, in Kelvin, is always positive.

67. C

The most nonmetallic compound is Si; these elements are found in the right-hand corner of the Periodic Table. The most metallic (it actually is a metal) is Pb; these elements are found on the left-hand side of the Periodic Table and through all the transition metals. Also, in Groups III through VII, nonmetallic character increases as you go up the Periodic Table. Therefore, the order of these elements in decreasing nonmetallic character would be Si, Ge, Sn, Pb.

68. C

Group IA elements are alkali metals; they have low densities, large atomic radii, low ionization energies, and low electronegativities. They are metals and have metallic bonding, and are good conductors of electricity. However, they have low melting points.

69. A

This question boils down to definitions. The electron affinity of an atom is defined as the change in energy that occurs when an electron is added to gaseous neutral atom in its ground state. So electron affinities can be positive or negative, depending on whether energy is released when an atom spontaneously accepts an electron or energy is gained when an electron is forced onto an atom. So choice A is the correct answer. Electronegativity is a derived quantity, usually scaled for all atoms between 0 and 4, that characterizes the pull an atom has for the electrons in a bond. The electronegativity has nothing to do with how likely an atom is to gain an electron, just how strong its pull is on an electron in a bond. The "in a bond" part is very important. The concept of electronegativity can only be applied to atoms that are already bonded. It characterizes the polarity of the bond, not the likelihood of bond formation. So choice B is the definition of electronegativity. Choice C is the definition of the first ionization energy. This is sort of the opposite of electron affinity, since in electron affinity, electrons are gained; in ionization, electrons are lost. Choice D would give the energy of a photon released when an electron relaxed from an excited state to a lower-lying state. Since there are many quantities for this energy, depending on which excited state the electron is in, it hardly makes a good answer to this question.

APPENDIX

GLOSSARY

Absolute zero
The temperature at which all substances have no thermal energy; 0K or − 273.15° C.

Absorption spectrum
The series of discrete lines at characteristic frequencies representing the energy required to make an atom undergo a transition to a higher energy state.

Acid
A species that donates hydrogen ions and/or accepts electrons. See Acidic solution; Arrhenius acid; Brønsted-Lowry acid; Lewis acid.

Acid dissociation constant (Ka)
The equilibrium constant that measures the degree of dissociation for an acid under specific conditions. For an acid HA:

$$K_a = \frac{[H^+][A^-]}{[HA]}$$

Acidic anhydride
An oxide that dissolves in water to form an acidic solution.

Acidic solution
An aqueous solution that contains more H^+ ions than OH^- ions. The pH of an acidic solution is less than 7 at 25 °C.

Activated complex
The transition state of a reaction in which old bonds are partially broken and new bonds are partially formed. The activated complex has a higher energy than the reactants or products of the reaction.

Activation energy (Ea)
The minimum amount of energy required for a reaction to occur.

Adiabatic process
A process that occurs without the transfer of heat to or from the system.

Alcohols
Organic compounds of the general formula ROH.

Aldehydes
Organic compounds of the general formula RCHO.

Alkali metals
Elements found in Group IA of the periodic table. They are highly reactive, readily losing their one valence electron to form ionic compounds with nonmetals.

Alkaline earth metals
Elements found in Group IIA of the periodic table. Their chemistry is similar to that of the alkali metals, except that they have two valence electrons, and thus form 2+ cations.

Alkanes
Hydrocarbons with only single bonds. The general formula for alkanes is CnH_2n_{+2}.

Alkenes
Hydrocarbons with at least one carbon-carbon double bond. Their general formula is CnH_2n.

Alkynes

Hydrocarbons with at least one carbon-carbon triple bond. Their general formula is C_nH_{2n-2}.

Alpha (α) particle

A particle ejected from the nucleus in one form of radioactive decay, identical to the helium-4 nucleus.

Amines

Compounds of the general formula RNH_2, R_2NH, or R_3N.

Amino acids

Building blocks of proteins with the general formula $NH_2CRHCOOH$.

Amorphous solids

Solids that do not possess long-range order. Compare to crystals or crystalline solids.

Amphoteric species

A species capable of reacting either as an acid or as a base.

Anhydride

A compound obtained by the removal of water from another compound.

Anion

An ionic species with a negative charge.

Anode

The electrode at which oxidation occurs. Compare to cathode.

Aqueous solution

A solution in which water is the solvent.

Aromatic compounds

Planar, cyclic organic compounds that are unusually stable because of the delocalization of π electrons.

Arrhenius acid

A species that donates protons (H^+) in an aqueous solution; e.g., HCl.

Arrhenius base

A species that gives off hydroxide ions (OH^-) in an aqueous solution; e.g., NaOH.

Atom

The most elementary form of an element; it cannot be further broken down by chemical means.

Atomic mass

The averaged mass of the atoms of an element, taking into account the relative abundance of the various isotopes in a naturally occurring substance. Also called the atomic weight.

Atomic mass units (amu)

A unit of mass defined as $\frac{1}{12}$ the mass of a carbon-12 atom; approximately equal to the mass of one proton or one neutron.

Atomic number

The number of protons in a given element.

Atomic orbital

The region of space around the nucleus in an atom in which there is a high probability of finding the electron.

Atomic radius

The radius of an atom. The average distance between a nucleus and the outermost electron. Usually measured as one-half the distance between two nuclei of an element in its elemental form.

Aufbau Principle

The principle that electrons fill energy levels in a given atom in order of increasing energy, completely filling one sublevel before beginning to fill the next.

Avogadro's Number

The number corresponding to a mole. It is the number of carbon-12 atoms in exactly 12 g of carbon-12, approximately 6.022×10^{23}.

Avogadro's Principle

The law stating that under the same conditions of temperature and pressure, equal volumes of different gases will have the same number of molecules.

Azimuthal quantum number (l)

The second quantum number, denoting the sublevel or subshell in which an electron can be found. Reveals the shape of the orbital. This quantum number represents the orbital angular momentum of the motion of the electron about a point in space.

Balanced equation

An equation for a chemical reaction in which the number of atoms for each element in the reaction and the total charge are the same for the reactants and the products.

Barometer

An instrument for measuring atmospheric pressure.

Base

A species that donates hydroxide ions or electrons, or that accepts protons. See Arrhenius base; Basic solution; Brønsted-Lowry base; Lewis base.

Base dissociation constant (Kb)

The equilibrium constant that measures the degree of dissociation for a base under specific conditions. For a base BOH:

$$K_b = \frac{[B^+][OH^-]}{[BOH]}$$

Basic anhydride

An oxide that dissolves in water to form a basic solution.

Basic solution

An aqueous solution that contains more OH^- ions than H^+ ions. The pH of a basic solution is greater than 7 at 25 °C.

Beta (β) particle

An electron produced and ejected from the nucleus during radioactive beta decay.

Binding energy

The energy required to break a nucleus apart into its constituent neutrons and protons.

Bohr model

The model of the hydrogen atom postulating that atoms are composed of electrons that assume certain circular orbits about a positive nucleus.

Boiling point

The temperature at which the vapor pressure of a liquid is equal to the surrounding pressure. The normal boiling point of any liquid is defined as temperature at which its vapor pressure is 1 atmosphere.

Boiling-point elevation

The amount by which a given quantity of solute raises the boiling point of a liquid; a colligative property.

Bond energy

The energy (enthalpy change) required to break a particular bond under given conditions.

Boyle's Law

The law stating that at constant temperature, the volume of a gaseous sample is inversely proportional to its pressure.

Brønsted-Lowry acid

Proton donor, e.g., H_3PO_4.

Brønsted-Lowry base

Proton acceptor, e.g., OH^-.

Buffer

A solution containing a weak acid and its salt (or a weak base and its salt) which tends to resist changes in pH.

Buffer region

The region of a titration curve in which the concentration of a conjugate acid is approximately equal to that of the corresponding base. The pH remains relatively constant when small amounts of H^+ or OH^- are added because of the combination of these ions with the buffer species already in solution.

Calorie (cal)

A unit of thermal energy (1 cal = 4.184 J).

Calorimeter

An apparatus used to measure the heat absorbed or released by a reaction.

Carbohydrates

A compound with the general formula $Cn(H_2O)_m$.

Carbon dating

A technique for estimating the age of (ancient) objects by measuring the amount of radioactive carbon-14 remaining.

Carbonyl group

C=O group found in aldehydes, ketones, etcetera. The C=O bond is known as the carbonyl bond, and organic compounds containing this group are known as carbonyl compounds.

Carboxylic acids

Compounds of the general formula RCOOH.

Catalysis

Increasing a reaction rate by adding a substance (the catalyst) not permanently changed by the reaction. The catalyst lowers the activation energy.

Catalyst

A substance that increases the rates of the forward and reverse directions of a specific reaction but is itself left unchanged.

Cathode

The electrode at which reduction takes place.

Cation

An ionic species with a positive charge.

Celsius (°C)

A temperature scale defined by having 0°C equal to the freezing point of water and 100°C equal to the boiling point of water; also the units of that scale. Otherwise known as the centigrade temperature scale. 0 °C = 273.15K.

Charles's Law

The law stating that the volume of a gaseous sample at constant pressure is directly proportional to its absolute (Kelvin) temperature.

Chemical bond

The interaction between two atoms resulting from the overlap of electron orbitals, holding the two atoms together at a specific average distance from each other.

Chemical properties

Those properties of a substance describing its reactivity.

Closed system

A system that can exchange energy but not matter with its surroundings.

Colligative properties

Those properties of solutions that depend only on the number of solute particles present but not on the nature of those particles. See Boiling-point elevation; Freezing-point depression; Vapor-pressure lowering.

Common ion effect

A shift in the equilibrium of a solution due to the addition of ions of a species already present in the reaction mixture.

Compound

A pure substance that can be decomposed to produce elements, other compounds, or both.

Concentration

The amount of solute per unit of solvent (denoted by square brackets), or the relative amount of one component in a mixture.

Conjugate acid-base pair

Brønsted-Lowry acid and base related by the transfer of a proton, e.g., H_2CO_3 and HCO_3^-.

Coordination complex

A compound in which a central metal atom or ion is bonded by coordinate covalent bonds to other atoms or groups.

Coordinate covalent bond

A covalent bond in which both electrons of the bonding pair are donated by only one of the bonded atoms.

Covalent bond

A chemical bond formed by the sharing of an electron pair between two atoms. See Coordinate covalent bond; Nonpolar covalent bond; Polar covalent bond.

Critical pressure

The vapor pressure at the critical temperature of a given substance.

Critical temperature

The highest temperature at which the liquid and vapor phases of a substance can coexist; above this temperature the substance does not liquefy at any pressure.

Crystal

A solid whose atoms, ions, or molecules are arranged in a regular three-dimensional lattice structure.

Cycloalkanes

Saturated cyclic compounds of the formula CnH_2n.

d subshell

The subshells corresponding to the angular momentum quantum number $l = 2$, found in the third and higher principal energy levels; each containing five orbitals.

Dalton's Law

The law stating that the sum of the partial pressures of the components of a gaseous mixture must equal the total pressure of the sample.

Daniell cell

An electrochemical cell in which the anode is the site of Zn metal oxidation, and the cathode is the site of Cu^{2+} ion reduction.

Degenerate orbitals

Orbitals that possess equal energy.

Density (ρ)

A physical property of a substance, defined as the mass contained in a unit of volume.

Diamagnetic

A condition that arises when a substance has no unpaired electrons and is slightly repelled by a magnetic field.

Diffusion

The random motion of gas or solute particles across a concentration gradient, leading to uniform distribution of the gas or solute throughout the container.

Dipole

In chemistry, a species containing bonds between elements of different electronegativities, resulting in an unequal distribution of charge in the species.

Dipole-dipole interaction

The attractive force between two dipoles whose magnitude is dependent on both the dipole moments and the distance between the two species.

Dipole moment

A vector quantity whose magnitude is dependent on the product of the charges and the distance between them. The direction of the moment is from the positive to the negative pole.

Dispersion force

A weak intermolecular force that arises from interactions between temporary and/or induced dipoles. Also called London force.

Dissociation

The separation of a single species into two separate species; this term is usually used in reference to salts or weak acids or bases.

Dissolution

The process of dissolving a substance. The opposite of precipitation.

Dynamic equilibrium

A state of balance (no macroscopic change observable) that arises when opposing processes occur at equal rates.

Electrochemical cell

A cell within which a redox reaction takes place, containing two electrodes between which there is an electrical potential difference. See Electrolytic cell; Voltaic cell.

Electrode

An electrical conductor through which an electric current enters or leaves a medium.

Electrolysis

The process in which an electric current is passed though a solution, resulting in chemical changes that would not otherwise occur spontaneously.

Electrolyte

A compound that ionizes in water.

Electrolytic cell

An electrochemical cell that uses an external voltage source to drive a nonspontaneous redox reaction.

Electromagnetic radiation

A wave composed of electric and magnetic fields oscillating perpendicular to each other and to the direction of propagation.

Electromagnetic spectrum

The range of all possible frequencies or wavelengths of electromagnetic radiation.

Electromotive force (EMF)

The potential difference developed between the cathode and the anode of an electrochemical cell.

Electron (e–)

A subatomic particle that remains outside the nucleus and carries a single negative charge. In most cases its mass is considered to be negligible ($\frac{1}{1837}$ that of the proton).

Electron affinity

The amount of energy that is released when an electron is added to an atom.

Electron configuration

The symbolic representation used to describe the electron occupancy of the various energy sublevels in a given atom.

Electronegativity

A measure of the ability of an atom to attract the electrons in a bond.

Electron spin

The intrinsic angular momentum of an electron, having arbitrary values of $+\frac{1}{2}$ and $-\frac{1}{2}$. See Spin quantum number.

Element

A substance that cannot be further broken down by chemical means. All atoms of a given element have the same number of protons.

Emission spectrum

The spectrum produced by a species emitting energy as it relaxes from an excited to a lower energy state.

Empirical formula

The simplest whole number ratio of the different elements in a compound.

Endothermic reaction

A reaction which absorbs heat from the surroundings as the reaction proceeds (positive ΔH).

End point

The point in a titration at which the indicator changes color, showing that enough reactant has been added to the solution to complete the reaction.

Enthalpy (H)

The heat content of a system at constant pressure. The change in enthalpy (ΔH) in the course of a reaction is the difference between the enthalpies of the products and the reactants.

Entropy (S)

A property related to the degree of disorder in a system. Highly ordered systems have low entropies. The change in entropy (ΔS) in the course of a reaction is the difference between the entropies of the products and the reactants.

Equilibrium

The state of balance in which the forward and reverse reaction rates are equal. In a system at equilibrium, the concentrations of all species will remain constant over time unless there is a change in the reaction conditions. See Le Châtelier's principle.

Equilibrium constant

The ratio of the concentration of the products to the concentration of the reactants for a certain reaction at equilibrium, all raised to their stoichiometric coefficients.

Equivalence point

The point in a titration at which the number of equivalents of the species being added to the solution is equal to the number of equivalents of the species being titrated.

Esters

Compounds of the general formula RCOOR'.

Ethers

Compounds of the general formula ROR'.

Excess reagent

In a chemical reaction, any reagent whose amount does not limit the amount of product that can be formed. Compare to Limiting reagent.

Excited state

An electronic state having a higher energy than the ground state.

Exothermic reaction

A reaction that gives off heat (negative ΔH) to the surroundings as the reaction proceeds.

f subshell

The subshells corresponding to the angular momentum quantum number $l = 3$, found in the fourth and higher principal energy levels, each containing seven orbitals.

Faraday (F)

The total charge on 1 mole of electrons (1 F = 96,487 coulombs).

Fatty acids

Carboxylic acids with long hydrocarbon chains, derived from the hydrolysis of fats.

First law of thermodynamics

The law stating that the total energy of a system and its surroundings remains constant. Also expressed as $\Delta E = Q - W$: the change in energy of a system is equal to the heat added to it minus the work done by it.

Formal charge

The conventional assignment of charges to individual atoms of a Lewis formula for a molecule, used to keep track of valence electrons. Defined as the total number of valence electrons in the free atom minus the total number of nonbonding electrons minus one-half the total number of bonding electrons.

Freezing point

At a given pressure, the temperature at which the solid and liquid phases of a substance coexist in equilibrium.

Freezing-point depression

Amount by which a given quantity of solute lowers the freezing point of a liquid. A colligative property.

Galvanic cell

An electrochemical cell that uses a spontaneous redox reaction to do work, i.e., produce an electrical current. Also called a Voltaic cell.

Gamma (γ) radiation

High energy photons often emitted in radioactive decay.

Gas

The physical state of matter possessing a high degree of disorder, in which molecules interact only slightly; found at relatively low pressure and high temperatures. Also called vapor. See Ideal gas.

Gas constant (R)

A proportionality constant that appears in the ideal gas law, $PV = nRT$. Its value depends upon the units of pressure, temperature, and volume used in a given situation.

Geiger counter

An instrument used to measure radioactivity.

Gibbs free energy (G)

The energy of a system available to do work. The change in Gibbs free energy, ΔG, is determined for a given reaction from the equation $\Delta G = \Delta H - T\Delta S$. ΔG is used to predict the spontaneity of a reaction: A negative ΔG denotes a spontaneous reaction, while positive ΔG denotes a nonspontaneous reaction.

Graham's law

The law stating that the rate of effusion or diffusion of a gas is inversely proportional to the square root of the gas's molecular weight.

Gram-equivalent weight

The amount of a compound that contains 1 mole of reacting capacity when fully dissociated. One GEW equals the molecular weight divided by the reactive capacity per formula unit.

Group

A vertical column of the periodic table, containing elements that are similar in their chemical properties.

Half-life

The time required for the amount of a reactant to decrease to one-half of its former value.

Half-reaction

Either the reduction half or oxidation half of a redox reaction. Each half-reaction occurs at one electrode of an electrochemical cell.

Halogens

The active nonmetals in Group VIIA of the periodic table, which have high electronegativities and highly negative electron affinities.

Heat

The energy representing the kinetic energy of molecules that is transferred spontaneously from a warmer sample to a cooler sample. See Temperature.

Heat of formation (ΔHf)

The heat absorbed or released during the formation of a pure substance from the elements in their standard states.

Heat of fusion (ΔHfus)

The ΔH for the conversion of a solid to a liquid.

Heat of sublimation (ΔHsub)
The ΔH for the conversion of a solid directly to a gas.

Heat of vaporization (ΔHvap)
The ΔH for the conversion of a liquid to a vapor.

Heisenberg uncertainty principle
The principle that states that it is impossible to simultaneously determine with perfect accuracy both the momentum and position of a particle.

Henry's Law
The law stating that the mass of a gas that dissolves in a solution is directly proportional to the partial pressure of the gas above the solution.

Hess' Law
The law stating that the energy change in an overall reaction is equal to the sum of the energy changes in the individual reactions which comprise it.

Heterogeneous
Nonuniform in composition.

Homogeneous
Uniform in composition.

Hund's Rule
The rule that electrons will occupy all degenerate orbitals in a subshell with single electrons having parallel spins before entering half-filled orbitals.

Hybridization
The combination of two or more atomic orbitals to form new orbitals for bonding purposes.

Hydrate
A compound with associated water molecules.

Hydrocarbons
Organic compounds containing only carbon and hydrogen.

Hydrogen bonding
The strong attraction between a hydrogen atom bonded to a highly electronegative atom, such as fluorine or oxygen, in one molecule, and a highly electronegative atom in another molecule.

Hydrolysis
A reaction between water and a species in solution.

Hydronium ion
The H_3O^+ ion in aqueous solution.

Hydroxide ion
The OH^- ion.

Ideal gas
A hypothetical gas whose behavior is described by the ideal gas law under all conditions. An ideal gas would have particles of zero volume that do not exhibit interactive forces.

Ideal Gas Law
The law stating that $PV = nRT$, where R is the gas constant. It can be used to describe the behavior of many real gases at moderate pressures and temperatures significantly above absolute zero. See Kinetic molecular theory.

Indicator, acid-base
A substance used in low concentration during a titration that changes color over a certain pH range. The color change, which occurs as the indicator undergoes a dissociation reaction, is used to identify the end point of the titration reaction.

Inert gases
The elements located in Group 0 (or Group VIII) of the Periodic Table. They contain a full octet of valence electrons in their outermost shell; this electron configuration makes them the least reactive of the elements. Also called Noble gases.

Intermolecular forces
The attractive and repulsive forces between molecules. See van der Waals forces.

Intramolecular forces

The attractive forces between atoms within a single molecule.

Ion

A charged atom or molecule that results from the loss or gain of electrons.

Ionic bonding

A chemical bond formed through electrostatic interaction between positive and negative ions.

Ionic solid

A solid consisting of positive and negative ions arranged into crystals that are made up of regularly repeated units and held together by ionic bonds.

Ionization product

The general term for the dissociation of salts or of weak acids or bases; the ratio of the concentration of the ionic products to the concentration of the reactant for a reaction, all raised to their stoichiometric coefficients.

Ionization energy

The energy required to remove an electron from the valence shell of a gaseous atom.

Isobaric process

A process that occurs at constant pressure.

Isolated system

A system that can exchange neither matter nor energy with its surroundings.

Isomers

Compounds with the same molecular formula but different structures.

Isothermal process

Process that occurs at constant temperature.

Isotopes

Atoms containing the same number of protons but different numbers of neutrons; e.g., nitrogen-14 and nitrogen-15.

Joule (J)

A unit of energy; $1 \, J = 1 \, kg \, m^2/s^2$

Kelvin (K)

A temperature scale with units equal in magnitude to the units of the Celsius scale and absolute zero defined as 0 K; also the units of that temperature scale. Otherwise known as the absolute temperature scale. $0 \, K = -273.15° \, C$.

Ketones

Compounds of the general formula RCOR'.

Kinetic energy

The energy a body has as a result of its motion, equal to $\frac{1}{2}mv^2$.

Kinetic molecular theory

The theory proposed to account for the observed behavior of gases. The theory considers gas molecules to be pointlike, volumeless particles, exhibiting no intermolecular forces and in constant random motion, undergoing only completely elastic collisions with the container or other molecules. See Ideal Gas Law.

Law of conservation of mass

The law stating that in a given reaction, the mass of the products is equal to the mass of the reactants.

Law of constant composition

The law stating that the elements in a pure compound are found in specific weight ratios.

Le Châtelier's Principle

The observation that when a system at equilibrium is disturbed or stressed, the system will react in such a way as to relieve the stress and restore equilibrium. See Equilibrium.

Lewis acid

A species capable of accepting an electron pair; e.g., BF_3.

Lewis base
A species capable of donating an electron pair; e.g., NH_3.

Lewis structure
A method of representing the shared and unshared electrons of an atom, molecule, or ion.

Limiting reagent
In a chemical reaction, the reactant present in such quantity as to limit the amount of product that can be formed.

Liquid
The state of matter in which intermolecular attractions are intermediate between those in gases and in solids, distinguished from the gas phase by having a definite volume and from the solid phase in that the molecules may mix freely.

Litmus
An organic substance that is used as an acid-base indicator , most often in paper form. It turns red in acidic solution and blue in basic solution.

London force
See Dispersion force.

Magnetic quantum number (*ml*)
The third quantum number, defining the particular orbital of a subshell in which an electron resides. It conveys information about the orientation of the orbital in space.(e.g., p_x versus p_y)

Manometer
An instrument used to measure the pressure of a gas.

Mass
A physical property representing the amount of matter in a given sample.

Mass defect
The difference between the sum of the masses of neutrons and protons forming a nucleus and the mass of that nucleus; the mass equivalence of bind-

ing energy, with the two related via the equation $E = mc^2$.

Mass number
The total number of protons and neutrons in a nucleus.

Maxwell-Boltzmann distribution
The distribution of the molecular speeds of gas particles at a given temperature.

Melting point
The temperature at which the solid and liquid phases of a substance coexist in equilibrium.

Metal
One of a class of elements located on the left side of the periodic table, possessing low ionization energies and electronegativities. Metals readily give up electrons to form cations; they possess relatively high electrical conductivity and are lustrous and malleable.

Metallic bonding
The type of bonding in which the valence electrons of metal atoms are delocalized throughout the metallic lattice.

Metalloid
An element possessing properties intermediate between those of a metal and those of a nonmetal. Also called a semimetal.

Micelles
Clusters of molecules possessing hydrophilic ionic heads facing the surface of a sphere where they can interact with water, and possessing hydrophobic hydrocarbon tails in the interior. Soaps form micelles, facilitating the dissolution of oils and fats.

Miscible
Able to mix in any proportion.

Molality (m)
A concentration unit equal to the number of moles of solute per kilogram of solvent.

Molarity (M)

A concentration unit equal to the number of moles of solute per liter of solution.

Molar mass

The mass in grams of 1 mole of an element or compound.

Mole (mol)

One mole of a substance contains Avogadro's number of molecules or atoms. The mass of 1 mole of substance in grams is the same as the mass of one molecule or atom in atomic mass units.

Mole fraction (X)

A unit of concentration equal to the ratio of the number of moles of a particular component to the total number of moles for all species in the system.

Molecular formula

A formula showing the actual number and identity of all atoms in each molecule of a compound.

Molecular weight

The sum of the atomic weights of all the atoms in a molecule.

Molecule

The smallest polyatomic unit of an element or compound that exists with distinct chemical and physical properties.

Monoprotic acid

An acid that can donate only one proton, e.g., HNO_3. The molarity of a monoprotic acid solution is equal to its normality.

Monosaccharides

Simple sugars that cannot be hydrolyzed to simpler compounds.

Net ionic equation

A reaction equation showing only the species actually participating in the reaction.

Nucleon

A particle found in the nucleus of an atom; can be either a neutron or a proton.

Neutralization reaction

A reaction between an acid and base in which H^+ ions and OH^- ions combine to produce water and a salt solution.

Neutral solution

An aqueous solution in which the concentration of H^+ and OH^- ions are equal (pH = 7).

Neutron

A subatomic particle contained within the nucleus of an atom. It carries no charge and has a mass very slightly larger than that of a proton.

Noble gases

See Inert gases.

Nonelectrolyte

A compound that does not ionize in water.

Nonmetal

One of a class of elements with high ionization potentials and very negative electron affinities that generally gains electrons to form anions. Nonmetals are located on the upper right side of the Periodic Table.

Nonpolar bond

A covalent bond between elements of the same electronegativity. There is no charge separation and the atoms do not carry any partially positive or partially negative charge. Compare to Polar bond.

Nonpolar molecule

A molecule which exhibits no net separation of charge, and therefore no net dipole moment.

Normality (N)

A concentration unit equal to the number of gram equivalent weights of solute per liter of solution.

Nucleus

The small central region of an atom; a dense, positively charged area containing protons and neutrons.

Octet

Eight valence electrons in a subshell around a nucleus.

Octet rule

A rule stating that bonded atoms tend to undergo reactions that will produce a complete octet of valence electrons. Applies without exception only to C, N, O, and F with zero or negative formal charges.

Open system

A system that can exchange both energy and matter with its surroundings.

Orbital

A region of electron density around an atom or molecule, containing no more than two electrons of opposite spin. See Atomic orbital.

Order of reaction

In a calculation of the rate law for a reaction, the sum of the exponents to which the concentrations of reactants must be raised.

Osmosis

The movement of a solvent or solute through a semipermeable membrane across its concentration gradient, i.e., from a container in which the concentration is high to a container in which the concentration is low.

Osmotic pressure

The pressure that must be applied to a solution to prevent the passage of a pure solvent through a semipermeable membrane across its concentration gradient.

Oxidation

A reaction involving the net loss of electrons or, equivalently, an increase in oxidation number.

Oxidation number

The number assigned to an atom in an ion or molecule that denotes its real or hypothetical charge. Atoms, alone or in molecules, of standard state elements have oxidation numbers of zero. Also called the oxidation state.

Oxidizing agent

In a redox reaction, a species that gains electrons and is thereby reduced.

p subshell

The subshells corresponding to the angular momentum quantum number $l = 1$, found in the second and higher principal energy levels. Each subshell contains three dumbbell-shaped p orbitals oriented perpendicular to each other, and referred to as the px, py, and pz orbitals.

Paired electrons

Two electrons in the same orbital with assigned spins of $+\frac{1}{2}$ and $-\frac{1}{2}$. See Orbital; Hund's rule.

Paramagnetism

A property of a substance that contains unpaired electrons, whereby the substance is attracted by a magnetic field.

Partial pressure

The pressure that one component of a gaseous mixture would exert if it were alone in the container.

Pauli exclusion principle

The principle stating that no two electrons within an atom may have an identical set of all four quantum numbers.

Peptides

Molecules that consist of two or more amino acids linked to each other by peptide bonds.

Percent composition

The percentage of the total formula weight of a compound attributed to a given element.

Percent yield

The percentage of the theoretical product yield that is actually recovered when a chemical reaction occurs.

Period

A horizontal row of the periodic table, containing elements with the same number of electron shells.

Periodic law

The law stating that the chemical properties of an element depend on the atomic number of the element, and change in a periodic fashion.

Periodic table

The table displaying all known chemical elements arranged in rows (periods) and columns (groups) according to their electronic structure.

pH

A measure of the hydrogen ion content of an aqueous solution, defined to be equal to the negative log of the H^+ concentration.

Phase

One of the three states of matter: solid, liquid, or gas. (Plasma is often considered a fourth phase of matter.)

Phase diagram

A plot, usually of pressure versus temperature, showing which phases a compound will exhibit under any set of conditions.

Phase equilibrium

For a particular substance, any temperature and pressure at which two or three phases coexist in equilibrium. See Triple point.

Photon

A quantum of energy in the form of light with a value of Planck's constant multiplied by the frequency of the light.

Physical property

A property of a substance related to its physical, not chemical, characteristics; e.g., density, smell, color.

Pi (π) bond

A covalent bond formed by parallel overlap of two unhybridized atomic p orbitals.

pOH

A measure of the hydroxide (OH^-) ion content of an aqueous solution, defined to be equal to the negative log of the OH^- concentration.

Polar covalent bond

A covalent bond between atoms with different electronegativities in which electron density is unevenly distributed, giving the bond positive and negative ends.

Polar molecule

A molecule possessing one or more polar covalent bond(s) and a geometry that allows the bond dipole moments to add up to a net dipole moment, e.g., H_2O.

Polyprotic acid

An acid capable of donating more than one proton, e.g., H_2CO_3.

Positron

An "antielectron": It has the same mass but opposite charge as an electron, and is emitted during a particular form of radioactive decay.

Potential energy diagram

An energy diagram that relates the potential energy of the reactants and products of a reaction to details of the reaction pathway. By convention, the x axis shows the progression of the reaction, and the y axis shows potential energy.

Precipitate

An insoluble solid that separates from a solution, generally the result of mixing two or more solutions or of a temperature change.

Pressure

Average force per unit area measured in atmospheres, torr (mm Hg), or pascals (Pa). 1 atm = 760 torr = 760 mm Hg = 1.01×10^2 kPa.

Primary structure

The amino acid sequence of a protein.

Principal quantum number (n)

The first quantum number, defining the energy level or shell occupied by an electron.

Proteins

Long-chain polypeptides with high molecular weights.

Proton (H+)

A subatomic particle that carries a single positive charge and has a mass defined as one or as the hydrogen ion, H+, which is simply a hydrogen nucleus, consisting of one proton. These species are considered to be equivalent.

Quantum

A discrete bundle of energy, such as a photon.

Quantum number

A number used to describe the energy levels available to electrons. The state of any electron is described by four quantum numbers. See Principal quantum number; Azimuthal quantum number; Magnetic quantum number; Spin quantum number.

Radioactivity

A phenomenon exhibited by certain unstable isotopes in which they undergo spontaneous nuclear transformations via emission of one or more particle(s).

Raoult's law

A law stating that the partial pressure of a component in a solution is proportional to the mole fraction of that component in the solution.

Rate constant

The proportionality constant in the rate law of a reaction; specific to a particular reaction under particular conditions.

Rate-determining step

The slowest step of a reaction mechanism. The rate of this step limits the overall rate of the reaction.

Rate law

A mathematical expression giving the rate of a reaction as a function of the concentrations of the reactants. The rate law of a given reaction must be determined experimentally.

Reaction intermediate

A species that does not appear among the final products of a reaction but is present temporarily during the course of the reaction.

Reaction mechanism

The series of steps that occur in the course of a chemical reaction, often including the formation and destruction of reaction intermediates.

Reaction rate

The speed at which a substance is produced or consumed by a reaction.

Real gas

A gas that exhibits deviations from the Ideal Gas Law.

Redox reaction

A reaction combining reduction and oxidation processes. Also called oxidation-reduction reaction.

Reducing agent

In a redox reaction, a species that loses electrons and is thereby oxidized.

Reduction

A reaction involving the net gain of electrons or, equivalently, a decrease in oxidation number.

Resonance

Delocalization of electrons within a compound that cannot be adequately represented by Lewis structures.

s subshell

Subshell corresponding to the angular momentum quantum number $l = 0$, and containing one spherical orbital; found in all energy levels.

Salt

An ionic substance, i.e., one consisting of anions and cations, but not hydrogen or hydroxide ions. Any salt can be formed by the reaction of the appropriate acid and base, e.g., KBr from HBr and KOH.

Saturated hydrocarbon

A hydrocarbon with only single bonds.

Saturated solution

A solution containing the maximum amount of solute that can be dissolved in a particular solvent at a particular temperature.

Scintillation counter

An instrument used to measure radioactivity by the amount of fluorescence produced.

Second law of thermodynamics

The law stating that all spontaneous processes lead to an increase in the entropy of the universe.

Semimetal

See Metalloid.

Semipermeable

A quality of a membrane allowing only some components of a solution, usually including the solvent, to pass through, while limiting the passage of other species.

Sigma (σ) bond

Bond formed by head-to-head overlap of orbitals from separate atoms.

Solid

The phase of matter possessing the greatest order, in which molecules are fixed in a rigid structure.

Solubility

A measure of the amount of solute that can be dissolved in a solvent at a certain temperature.

Solubility product (K_{sp})

The equilibrium constant for the ionization reaction of a slightly soluble electrolyte.

Solute

The component of a solution that is present in lesser amount than the solvent.

Solution

A homogeneous mixture of two or more substances.

Solvation

The aggregation of solvent molecules around a solute particle in the process of dissolution.

Solvent

The component of a solution present in the greatest amount; the substance in which the solute is dissolved.

Specific heat

The amount of heat required to raise the temperature of a unit mass of a substance by 1°C.

Spectrum

The characteristic wavelengths of electromagnetic radiation emitted or absorbed by an object, atom, or molecule.

Spin quantum number (m_s)

The fourth quantum number, indicating the orientation of the intrinsic angular momentum of an electron in an atom. The spin quantum number can only assume values of $+\frac{1}{2}$ or $-\frac{1}{2}$.

Spontaneous process

A process that will occur on its own without energy input from the surroundings.

Standard conditions

Conditions defined as 25°C and 1 M concentration for each reactant in solution, and a partial pressure of 1 atm for each gaseous reactant. Used for measuring the standard Gibbs free energy, enthalpy, entropy, and cell EMF.

Standard free energy (G°)

The Gibbs free energy for a reaction under standard conditions. See Gibbs free energy.

Standard hydrogen electrode (SHE)

The electrode defined as having a potential of zero under standard conditions. All redox potentials are measured relative to the standard hydrogen electrode. The potentials are measured relative to the standard hydrogen electrode at 25°C and with 1.0 M of each ion in solution.

Standard potential

The voltage associated with a half-reaction of a specific redox reaction. Generally tabulated as a reduction potential, compared to the SHE.

Standard temperature and pressure (STP)

0°C (273 K) and 1 atm. Used for measuring gas volume and density.

State

The set of defined macroscopic properties of a system that must be specified in order to reproduce the system exactly. Sometimes also used as a synonym for Phase.

State function

A function that depends on the state of a system but not on the path used to arrive at that state.

Strong acid

An acid that undergoes complete dissociation in an aqueous solution; e.g., HCl.

Strong base

A base that undergoes complete dissociation in an aqueous solution; e.g., KOH.

Sublimation

A change of phase from solid to gas without passing through the liquid phase.

Subshell

The division of electron shells or energy levels defined by a particular value of the azimuthal quantum number; e.g., s, p, d, and f subshells. Composed of orbitals. Also called Sublevels. See Orbitals.

Surroundings

All matter and energy in the universe not included in the particular system under consideration.

System

The matter and energy under consideration.

Temperature

A measure of the average energy of motion of the particles in a system.

Third law of thermodynamics

The law stating that the entropy of a perfect crystal at absolute zero is zero.

Titrant

A solution of known concentration that is slowly added to a solution containing an unknown amount of a second species to determine its concentration.

Titration, acid-base

A method used to determine the concentration of an unknown solution.

Titration curve

A plot of the pH of a solution versus the volume of acid or base added in an acid-base titration.

Torr

A pressure unit equal to 1 mm Hg. 760 torr = 1 atm.

Transition metal

Any of the elements in the B Groups of the Periodic Table, all of which have partially filled d sublevels.

Triple point

The pressure and temperature at which the solid, liquid, and vapor phases of a particular substance coexist in equilibrium. See Phase equilibrium.

Unit cell

A three-dimensional representation of the repeating units in a crystalline solid.

Unsaturated compound

A compound with double or triple bonds.

Unsaturated solution

A solution into which more solute may be dissolved.

Valence electron

An electron in the highest occupied energy level of an atom, whose tendency to be held or lost determines the chemical properties of the atom.

Van der Waals forces

The weak forces that contribute to intermolecular bonding, including hydrogen bonding, dipole-dipole interactions, and dispersion forces.

Vapor pressure

The pressure exerted by a vapor when it is in equilibrium with the liquid or solid phase of the same substance; the partial pressure of the substance in the atmosphere above the liquid or solid.

Vapor-pressure lowering

The decrease in the vapor pressure of a liquid caused by the presence of dissolved solute; a colligative property. See Raoult's Law.

Voltaic cell

See Galvanic cell.

VSEPR theory

Stands for Valence Shell Electron-Pair Repulsion theory. It predicts/explains the geometry of molecules in terms of the repulsion that electron pairs have that cause them to be as far apart as possible from one another.

Water dissociation constant (Kw)

The equilibrium constant of the water dissociation reaction at a given temperature; 1.00×10^{-14} at $25.00°C$.

Weak acid

An acid that undergoes partial dissociation in an aqueous solution; e.g., CH_3COOH.

Weak base

A base that undergoes partial dissociation in an aqueous solution; e.g., NH_4OH.

Yield

The amount of product obtained from a reaction.

Z

Nuclear charge. Equivalent to atomic number.

Zeff

Effective nuclear charge; the charge perceived by an electron form its orbital. Applies most often to valence electrons, and influences periodic properties such as atomic radius and ionization energy.

NOTES

How Did We Do? Grade Us.

Thank you for choosing a Kaplan book. Your comments and suggestions are very useful to us. Please answer the following questions to assist us in our continued development of high-quality resources to meet your needs.

The title of the Kaplan book I read was: _____

My name is: _____

My address is: _____

_____

My e-mail address is: _____

What overall grade would you give this book?   Ⓐ   Ⓑ   Ⓒ   Ⓓ   Ⓕ

How relevant was the information to your goals?   Ⓐ   Ⓑ   Ⓒ   Ⓓ   Ⓕ

How comprehensive was the information in this book?   Ⓐ   Ⓑ   Ⓒ   Ⓓ   Ⓕ

How accurate was the information in this book?   Ⓐ   Ⓑ   Ⓒ   Ⓓ   Ⓕ

How easy was the book to use?   Ⓐ   Ⓑ   Ⓒ   Ⓓ   Ⓕ

How appealing was the book's design?   Ⓐ   Ⓑ   Ⓒ   Ⓓ   Ⓕ

What were the book's strong points? _____

How could this book be improved? _____

Is there anything that we left out that you wanted to know more about?

Would you recommend this book to others?   ☐ YES   ☐ NO

Other comments: _____

_____

_____

Do we have permission to quote you?   ☐ YES   ☐ NO

Thank you for your help.
Please tear out this page and mail it to:

Managing Editor
Kaplan, Inc.
888 Seventh Avenue
New York, NY 10106

KAPLAN®

Thanks!

About Kaplan

KAPLAN TEST PREPARATION & ADMISSIONS

With 3,000 classroom locations throughout the U.S. and abroad, Kaplan has served more than three million students in its classes over the past 60-plus years. Kaplan's nationally-recognized programs for roughly 35 standardized tests include entrance exams for secondary school, college and graduate school as well as English language and professional licensing exams. Kaplan also offers private tutoring and one-on-one admissions guidance and is a leader in test prep for computerized exams. Kaplan is the first major player to provide online test prep to students across the globe, as well as admissions courses and other resources at **www.kaptest.com.**

SCORE! LEARNING, INC.

SCORE! Learning, Inc. is a national provider of customized learning programs for students. SCORE! Educational Centers help students in K-10 build confidence along with academic skills in a motivating, sports-oriented environment after school and on weekends. SCORE! Prep provides in-home, one-on-one tutoring for high school academic subjects and standardized tests. SCORE! Educational Centers and SCORE! Prep share a highly personalized approach, proven educational techniques, and the goal of cultivating a love of learning in children.

THE KAPLAN COLLEGES

The Kaplan Colleges system (**www.kaplancollege.edu**) is a collection of institutions offering an extensive array of online and traditional educational programs for working professionals who want to advance their careers. Learners will find programs leading to bachelor and associates degrees, certificates and diplomas in fields such as business, IT, paralegal studies, legal nurse consulting, criminal justice and financial planning. The Kaplan Colleges system includes Concord Law School (**www.concordlawschool.com**), the nation's only online law school, offering J.D., Executive J.D. and LL.M. degrees for working professionals, family caregivers, students in rural communities, and others whose circumstances prevent them from attending a fixed facility law school.

QUEST EDUCATION CORPORATION

Kaplan's Quest Education unit (**www.questeducation.com**) is a leading provider of post-secondary education. Quest offers bachelor and associate degrees and diploma programs designed to provide students with the skills necessary to qualify them for entry-level employment. Programs are primarily in the fields of healthcare, business, information technology, fashion and design.

KAPLAN PUBLISHING

Kaplan Publishing, in a joint venture with Simon & Schuster, publishes more than 150 titles on test preparation, admissions, education, career development, and life skills. Kaplan Publishing emerged as a leader in sales of books for statewide assessments with the publication of dozens of new state test titles. Books are offered in traditional paper form, pre-packaged with computer software, and now in e-book form.

KAPLAN INTERNATIONAL

Kaplan International (**www.kaptest.com**) provides students and professionals with intensive English instruction, university preparation, test preparation programs, housing and activities at 12 city and campus centers in the U.S. and Canada. Kaplan also has a strong presence overseas with 41 centers in 18 countries outside of the United States.

KAPLAN COMMUNITY OUTREACH

Kaplan Community Outreach provides educational resources and opportunities to thousands of economically disadvantaged students annually. Kaplan joins forces with numerous nonprofit groups, educational institutions, government agencies, and other grass-roots organizations on a variety of local and national support programs. These programs help students and professionals from a variety of backgrounds achieve their educational and career goals.

KAPLAN PROFESSIONAL

The Kaplan Professional companies (**www.kaplanprofessional.com**) provide licensing and continuing education, training, certification, professional development courses, and compliance tracking for securities, insurance, financial services, legal, IT, and real estate professionals and corporations. Offering an array of educational tools, from on-site training and classroom instruction to nearly 200 online courses and programs, Kaplan Professional serves professionals who must maintain licenses and comply with regulatory mandates despite busy travel schedules and work obligations.

- **Dearborn Financial Services** provides innovative education and compliance solutions to the financial services industry, including registration services, firm element needs analysis and training plan development, securities and insurance prelicensing training, continuing education, and compliance management services, in classes nationwide, online and via books and software.

- **Dearborn Trade Publishing** publishes approximately 250 titles specializing in finance, business management and real estate, plus well-read consumer real estate books to help homebuyers, sellers and real estate investors make informed decisions.

- **Dearborn Real Estate Education** is the leading real estate content provider for real estate schools and associations, offering practical prelicensing and continuing education training materials on appraisal, home inspection, property management, brokerage, ethics, law, sales approaches, and contracts, and an online real estate campus at **RECampus.com**.

- **Perfect Access Speer** is a leader in software education and consulting, bringing both traditional and e-learning solutions to its clients in the legal, financial, and professional services industries.

- **The Schweser Study Program** offers training tools for the Chartered Financial Analyst (CFA®) examination, with a comprehensive product line of study notes, audiotapes, videotapes, flashcards and live seminars that are developed and taught by a top-notch faculty.

- **Kaplan Professional Real Estate Schools** provide real estate licensing and continuing education programs through live classroom instruction, Internet-based learning, and correspondence courses, to help real estate professionals acquire the skills needed to meet state licensing and educational requirements.

- **Self Test Software** is a world leader in exam simulation software and preparation for technical certifications including Microsoft, Oracle, Cisco, Novell, Lotus, CIW and CompTIA, serving businesses and individuals seeking to attain vendor-sponsored certification.

- **Call Center Solutions** provides assessment and training services to the call center industry.

Want more information about our services, products or the nearest Kaplan center?

1 **Call our nationwide toll-free numbers:**

1-800-KAP-TEST for information on our test prep courses, private tutoring and admissions consulting

1-800-KAP-ITEM for information on our books and software

2 **Connect with us online:**

On the web, go to:

www.kaptest.com

3 **Write to:**

Kaplan
888 Seventh Avenue
New York, NY 10106

KAPLAN